CHEMISTRY RESEARCH AND APPLICATIONS

QUANTITATIVE CHEMISTRY, BIOCHEMISTRY AND BIOLOGY

STEPS AHEAD

CHEMISTRY RESEARCH AND APPLICATIONS

Additional books in this series can be found on Nova's website
under the Series tab.

Additional e-books in this series can be found on Nova's website
under the e-book tab.

BIOCHEMISTRY RESEARCH TRENDS

Additional books in this series can be found on Nova's website
under the Series tab.

Additional e-books in this series can be found on Nova's website
under the e-book tab.

CHEMISTRY RESEARCH AND APPLICATIONS

QUANTITATIVE CHEMISTRY, BIOCHEMISTRY AND BIOLOGY

STEPS AHEAD

GENNADY E. ZAIKOV
OLEG V. STOYANOV
WIKTOR TYSZKIEWICZ
AND
ZBIGNIEW WERTEJUK
EDITORS

nova publishers
New York

For permission to use material from this book please contact us:
Telephone 631-231-7269; Fax 631-231-8175
Web Site: http://www.novapublishers.com

Additional color graphics may be available in the e-book version of this book.

Library of Congress Cataloging-in-Publication Data

Quantitative chemistry, biochemistry and biology : steps ahead / [edited by] Gennady E. Zaikov, Oleg V. Stoyanov, Wiktor Tyszkiewicz and Zbigniew Wertejuk (N.M. Emanuel Institute of Biochemical Physics, Russian Academy of Sciences, Moscow, Russia, and others).
pages cm
Includes bibliographical references and index.
ISBN 978-1-62948-332-0 (hardcover)
1. Biochemistry. 2. Biology. 3. Quantitative research. I. Zaikov, G. E. (Gennadii Efremovich), 1935- editor of compilation.
QH345.Q36 2014
571.7'4--dc23

2013038299

Published by Nova Science Publishers, Inc. † New York

CONTENTS

"One kind word can warm
three winter months"
The proverb from
Perry, OH, US

"We live and enjoy ourselves"
Margarita Eskina

"Every person is born a genius"
J. Goethe

'The motivation of any scientific
discovery is craving for a flight.
It is enough to start a flight but for
the flight to be successful the recognition
of scientific authorities is absolutely inevitable"
Ahmed Zevile
Winner Noble Prize

PREFACE

The monk-Jesuit Escober proposed the slogan "The aim justifies the means". This slogan at large is not correct because it permits any action. However in our case this slogan is acceptable because our aim is to target a reader.

We collected the reviews and original papers about quantitative chemistry, biochemistry and biology. Special attention is given to new ideas in the fields of nanoelements formation and reactivity, synthesis of thermoplastic bio-based polyurethanes on the base of vegetable oils, carvacrol and thymol for fresh food packaging, polymer composites structure and electric properties, some properties of small water clusters in water-starch systems, solubility of aluminum polyhydroxochloride in the water-acetone solution, about diversified fenomena of entropy, quantitative calculation of spatial-energy interactions entropy, bacterial plasmids (cloning of multi-copy CpG motifs), using full-length diguanylate cyclase of Thermotoga Maritime for enzymatic synthesis of 3′,5′-cyclic diguanylate, nanocomposites polyethylene/organoclay on suprasegmental level (the reinforcement mechanism) and nanocomposites polypropylene/carbon nanotubes: fractal model and the crystallization kinetics. This volume also includes information about the segmental and translational movements of kinetic parameters, interaction of hydroxyorganocyclotetrasiloxanes with propyl butyrate side groups and polymer electrolyte membranes, UV-protective properties of nanocomposites films based on low density polyethylene, production, morphology, properties. Biodegradable compositions of starch with synthetic polymers and natural polysaccharides, kinetics and mechanism of the modificated starches during gelatinization, investigation of ethylene copolymers silanol modification, determining the surface free energy parameters, investigation of ethylene–vinylacetate copolymers binary mixtures structure and investigation the kinetics and degradation mechanism of the tertiary hydroperoxides activated by the tetraethylammonium bromide.

We also decided to discuss the following topics: investigation of kinetics and mechanism of biologically awake antioxidants in reactions of esterification 2- (n-acetylamid)-3-(3 ', 5 '-*di-tert.butyl*-4 '-hydroxyphenyl)-propionic acid, the turbulent apparatus for oil neutralization, effect melaphen on a mitochondria of sprouts of peas under stressful influences by some methods, preparation of new antioxidants in reaction 2- (n-methylamide)-3- (3 ', 5 '-*di-tert.butyl*-4 '-hydroxyphenyl)-propionic acid and effect of external influences on the structural and dynamic parameters of polyhydroxybutyrate–hydroxyvalerate-based biocomposites.

Some chapters about quantum-chemical calculation of some molecules, the mechanism of protonation of isobutylene by method MNDO, geometrical and electronic structure of molecule 2-(2,2-dimethoxiethylsulfanil)-5-izopropyl-6-methyl pyrimidine-4(3h)-oh method MNDO, geometrical structure of some molecules aromatic olefins, composite materials based on LDPE with the addition of wood flour and rubbers of different nature, fungicidal influence of ions of Cu^{2+} on microorganisms as a way of biocorrosion protection of metal constructions, interfacial rheology of hen egg white lysozyme-5-methylresorcinol mixtures and their foaming properties, structure and thermophysical properties of blends based on isotactic polypropylene and low-density polyethylene, investigating the effect of fungies on thin films of polyethylene, polyethylene – poly-3-hydroxybutyrate and polyethylene – d_2w, determination of the identity of the signs of natural pomegranate juice, influence of carbon nanofibers on the structure and properties of vulcanized EPDM were included in this volume.

We also discuss the problems of determination diffusion coefficient of Brownian particles using velocity or force autocorrelation function in molecular dynamic simulations, in silico simulation of silver and copper ions interacting with fungal cell wall (in vitro antifungal activity of copper ions and silver ions) and the estimation of antioxidants as nuts quality index were included in this monograph as well, structure of soluble unlinked and cross-linked fibrin oligomers, kinetics and mechanism of the ozone reaction with alcohols, ketones, ethers and hydroxybenzenes, structure and antioxidative properties tert.butyl ester 3-(3 ', 5 '-di-tert.butyl-4 '-hydroxyphenyl)-propionic acid, comparative estimation of Kalanchoe juice antioxidant properties, the mechanism of the selective ethylbenzene oxidation, catalyzed with triple binuclear metallocomplex catalysts, including Ni and redox-inactive metal Na (or Li) (nanostructures {NiII(acac)2·NaSt(or LiSt)·(PhOH)}n formation due to intermolecular H-bonds) and quantum-chemical calculation studies the mechanism of protonation of 4-methylpentene-1, 4-methylhexene-1 and 4,4-dimethylpentene-1 by method MNDO.

The editors and contributors will be happy to receive some comments from the readers which we can take in account in our future research.

Prof. Oleg V. Stoyanov
Kazan National Research Technological University
68 Karl Marx str., 420015 Kazan, Russia
Ov_stoyanov@mail.ru

Dr. Wiktor Tyszkiewicz
Military Institute of Chemistry and Radiometry
al.gen.A.Chruściela "Montera" 105,
00-910 Warsaw ,POLAND
w.tyszkiewicz@wichir.waw.pl

Dr. Zbigniew Wertejuk
Military Institute of Chemistry and Radiometry
al.gen.A.Chruściela "Montera" 105,
00-910 Warsaw ,POLAND
zb.wertejuk@wichir.waw.pl

Prof. Gennady E. Zaikov
N.M. Emanuel Institute of Biochemical Physics
Russian Academy of Sciences
4 Kosygin str., Moscow 119334 Russia
chembio@sky.chph.ras.ru

In: Quantitative Chemistry, Biochemistry and Biology
Editors: G.E. Zaikov, O.V. Stoyanov et al.
ISBN: 978-1-62948-332-0

Chapter 1

NEW IDEAS IN THE FIELD OF NANOELEMENTS FORMATION AND REACTIVITY

A.M. Lipanov*[*] *and A.V. Vakhrushev
Institute of Mechanics, Ural Branch of the Russian Academy of Sciences, Izhevsk, Russia

INTRODUCTION

The properties of a nanocomposite are determined by the structure and properties of the nanoelements, which form it. One of the main tasks in making nanocomposites is building the dependence of the structure and shape of the nanoelements forming the basis of the composite on their sizes. This is because with an increase or a decrease in the specific size of nanoelements (nanofibers, nanotubes, nanoparticles, etc.), their physical-mechanical properties such as coefficient of elasticity, strength, deformation parameter, etc. are varying over one order [1 – 5].

The calculations and experiments show that this is primarily due to a significant rearrangement (which is not necessarily monotonous) of the atomic structure and the shape of the nanoelement. The experimental investigation of the above parameters of the nanoelements is technically complicated and laborious because of their small sizes. In addition, the experimental results are often inconsistent. In particular, some authors have pointed to an increase in the distance between the atoms adjacent to the surface in contrast to the atoms inside the nanoelement, while others observe a decrease in the aforementioned distance [6].

Thus, further detailed systematic investigations of the problem with the use of theoretical methods, i.e. mathematical modeling, are required.

The atomic structure and the shape of nanoelements depend both on their sizes and on the methods of obtaining which can be divided into two main groups:

1) Obtaining nanoelements in the atomic coalescence process by "assembling" the atoms and by stopping the process when the nanoparticles grow to a desired size (the

[*] Email: imp@udman.ru.

so-called "bottom-up" processes). The process of the particle growth is stopped by the change of physical or chemical conditions of the particle formation, by cutting off supplies of the substances that are necessary to form particles, or because of the limitations of the space where nanoelements form.

2) Obtaining nanoelements by breaking or destruction of more massive (coarse) formations to the fragments of the desired size (the so-called "up down" processes).

In fact, there are many publications describing the modeling of the "bottom-up" processes [7-8], while the "up down" processes have been studied very little. Therefore, the objective of this work is the investigation of the regularities of the changes in the structure and shape of nanoparticles formed in the destruction ("up down") processes depending on the nanoparticle sizes, and building up theoretical dependences describing the above parameters of nanoparticles.

When the characteristics of powder nanocomposites are calculated it is also very important to take into account the interaction of the nanoelements since the changes in their original shapes and sizes in the interaction process and during the formation (or usage) of the nanocomposite can lead to a significant change in its properties and a cardinal structural rearrangement. In addition, the experimental investigations show the appearance of the processes of ordering and self-assembling leading to a more organized form of a nanosystem [9-15]. In general, three main processes can be distinguished: the first process is due to the regular structure formation at the interaction of the nanostructural elements with the surface where they are situated; the second one arises from the interaction of the nanostructural elements with one another; the third process takes place because of the influence of the ambient medium surrounding the nanostructural elements. The ambient medium influence can have "isotropic distribution" in the space or it can be presented by the action of separate active molecules connecting nanoelements to one another in a certain order. The external action significantly changes the original shape of the structures formed by the nanoelements. For example, the application of the external tensile stress leads to the "stretch" of the nanoelement system in the direction of the maximal tensile stress action; the rise in temperature, vice versa, promotes a decrease in the spatial anisotropy of the nanostructures [10]. Note that in the self-organizing process, parallel with the linear moving, the nanoelements are in rotary movement. The latter can be explained by the action of moment of forces caused by the asymmetry of the interaction force fields of the nanoelements, by the presence of the "attraction" and "repulsion" local regions on the nanoelement surface, and by the "non-isotropic" action of the ambient as well.

The above phenomena play an important role in nanotechnological processes. They allow developing nanotechnologies for the formation of nanostructures by the self-assembling method (which is based on self-organizing processes) and building up complex spatial nanostructures consisting of different nanoelements (nanoparticles, nanotubes, fullerenes, super-molecules, etc.) [15]. However, in a number of cases, the tendency towards self-organization interferes with the formation of a desired nanostructure. Thus, the nanostructure arising from the self-organizing process is, as a rule, "rigid" and stable against external actions. For example, the "adhesion" of nanoparticles interferes with the use of separate nanoparticles in various nanotechnological processes, the uniform mixing of the nanoparticles from different materials and the formation of nanocomposite with desired properties. In connection with this, it is important to model the processes of static and dynamic interaction of the nanostructure elements. In this case, it is essential to take into consideration the interaction force moments of the nanostructure elements, which causes the mutual rotation of the nanoelements.

The investigation of the above dependences based on the mathematical modeling methods requires the solution of the aforementioned problem on the atomic level. This requires large computational aids and computational time, which makes the development of economical calculation methods urgent. The objective of this work was the development of such a technique.

This chapter gives results of the studies of problems of numeric modeling within the framework of molecular mechanics and dynamics for investigating the regularities of the amorphous phase formation and the nucleation and spread of the crystalline or hypocrystalline phases over the entire nanoparticle volume depending on the process parameters, nanoparticles sizes and thermodynamic conditions of the ambient. Also the method for calculating the interactions of nanostructural elements is offered, which is based on the potential built up with the help of the approximation of the numerical calculation results using the method of molecular dynamics of the pairwise static interaction of nanoparticles. Based on the potential of the pairwise interaction of the nanostructure elements, which takes into account forces and moments of forces, the method for calculating the ordering and self-organizing processes has been developed. The investigation results on the self-organization of the system consisting of two or more particles are presented and the analysis of the equilibrium stability of various types of nanostructures has been carried out. These results are a generalization of the authors' research in [16-24]. A more detailed description of the problem you can obtain in these works.

1. PROBLEM STATEMENT AND MODELLING TECHNIQUE

The problem on calculating the internal structure and the equilibrium configuration (shape) of separate non-interacting nanoparticles by the molecular mechanics and dynamics methods has two main stages:

1) The “initiation” of the task, i.e., the determination of the conditions under which the process of the nanoparticle shape and structure formation begins.
2) The process of the nanoparticle formation.

Note that the original coordinates and initial velocities of the nanoparticle atoms should be determined from the calculation of the macroscopic parameters of the destructive processes at static and dynamic loadings taking place both on the nano-scale and on the macro-scale. Therefore, in the general case, the coordinates and velocities are the result of solving the problem of modeling physical-mechanical destruction processes at different structural levels. This problem due to its enormity and complexity is not considered in this paper. The detailed description of its statement and the numerical results of its solution are given in the works of the authors [16-19].

The problem of calculating the interaction of ordering and self-organization of the nanostructure elements includes three main stages: the first stage is building the internal structure and the equilibrium configuration (shape) of each separate non-interacting nanostructure element; the second stage is calculating the pairwise interaction of two nanostructure elements; and the third stage is establishing the regularities of the spatial structure and evolution with time of the nanostructure as a whole.

Let us consider the above problems in sequence.

1.1. The Calculation of the Internal Structure and the Shape of the Non-Interacting Nanoelement

The initialization of the problem is in giving the initial coordinates and velocities of the nanoparticle atoms

$$\vec{\mathbf{x}}_i = \vec{\mathbf{x}}_{i0}, \vec{\mathbf{V}}_i = \vec{\mathbf{V}}_{i0}, t = 0, \vec{\mathbf{x}}_i \subset \Omega_k , \tag{1}$$

where $\vec{\mathbf{x}}_{i0}, \vec{\mathbf{x}}_i$ are original and current coordinates of the i-th atom; $\vec{\mathbf{V}}_{i0}, \vec{\mathbf{V}}_i$ are initial and current velocities of the i-th atom, respectively; Ω_k is an area occupied by the nanoelement.

The problem of calculating the structure and the equilibrium configuration of the nanoelement will be carried out with the use of the molecular dynamics method taking into consideration the interaction of all the atoms forming the nanoelement. Since, at the first stage of the solution, the nanoelement is not exposed to the action of external forces, it is taking the equilibrium configuration with time, which is further used for the next stage of calculations.

At the first stage, the movement of the atoms forming the nanoparticle is determined by the set of Langevin differential equations at the boundary conditions (1) [25]

$$\begin{aligned} & m_i \cdot \frac{d\vec{\mathbf{V}}_i}{dt} = \sum_{j=1}^{N_k} \vec{\mathbf{F}}_{ij} + \vec{\mathbf{F}}_i(t) - \alpha_i m_i \vec{\mathbf{V}}_i, \qquad i = 1,2,..,N_k, \\ & \frac{d\vec{\mathbf{x}}_i}{dt} = \vec{\mathbf{V}}_i, \end{aligned} \tag{2}$$

where N_k is the number of atoms forming each nanoparticle; m_i is the mass of the i-th atom; α_i is the "friction" coefficient in the atomic structure; $\vec{\mathbf{F}}_i(t)$ is a random set of forces at a given temperature which is given by Gaussian distribution.

The interatomic interaction forces usually are potential and determined by the relation

$$\vec{\mathbf{F}}_{ij} = -\sum_{1}^{n} \frac{\partial \Phi(\vec{\boldsymbol{\rho}}_{ij})}{\partial \vec{\boldsymbol{\rho}}_{ij}} , \ i = 1,2,...,N_k, \ j = 1,2,...,N_k, \tag{3}$$

where $\vec{\boldsymbol{\rho}}_{ij}$ is a radius-vector determining the position of the i-th atom relative to the j-th atom; $\Phi(\vec{\boldsymbol{\rho}}_{ij})$ is a potential depending on the mutual positions of all the atoms; n is the number of interatomic interaction types.

In the general case, the potential $\Phi(\vec{\boldsymbol{\rho}}_{ij})$ is given in the form of the sum of several components corresponding to different interaction types:

$$\Phi(\vec{\boldsymbol{\rho}}_{ij}) = \Phi_{cb} + \Phi_{va} + \Phi_{ta} + \Phi_{pg} + \Phi_{vv} + \Phi_{es} + \Phi_{hb}. \tag{4}$$

Here the following potentials are implied: Φ_{cb} - of chemical bonds; Φ_{va} - of valence angles; Φ_{ta} - of torsion angles; Φ_{pg} - of flat groups; Φ_{vv} - of Van der Waals contacts; Φ_{es} - of electrostatics; Φ_{hb} - of hydrogen bonds.

The above addends have different functional forms. The parameter values for the interaction potentials are determined based on the experiments (crystallography, spectral, calorimetric, etc.) and quantum calculations [25].

Giving original coordinates (and forces of atomic interactions) and velocities of all the atoms of each nanoparticle in accordance with equation (2), at the start time, we find the change of the coordinates and the velocities of each nanoparticle atoms with time from the equation of motion (1). Since the nanoparticles are not exposed to the action of external forces, they take some atomic equilibrium configuration with time that we will use for the next calculation stage.

1.2. The Calculation of the Pairwise Interaction of the Two Nanostructure Elements

At this stage of solving the problem, we consider two interacting nanoelements. First, let us consider the problem statement for symmetric nanoelements, and then for arbitrary shaped nanoelements.

First of all, let us consider two symmetric nanoelements situated at the distance S from one another (Figure 1) at the initial conditions

$$\vec{\mathbf{x}}_i = \vec{\mathbf{x}}_{i0}, \vec{\mathbf{V}}_i = 0, t = 0, \vec{\mathbf{x}}_i \subset \Omega_1 \bigcup \Omega_2 , \tag{5}$$

where Ω_1, Ω_2 are the areas occupied by the first and the second nanoparticle, respectively.

We obtain the coordinates $\vec{\mathbf{x}}_{i0}$ from equation (2) solution at initial conditions (1). It allows calculating the combined interaction forces of the nanoelements

$$\vec{\mathbf{F}}_{b1} = -\vec{\mathbf{F}}_{b2} = \sum_{i=1}^{N_1} \sum_{j=1}^{N_2} \vec{\mathbf{F}}_{ij} , \tag{6}$$

where i, j are the atoms and N_1, N_2 are the numbers of atoms in the first and in the second nanoparticle, respectively.

Forces $\vec{\mathbf{F}}_{ij}$ are defined from equation (3).

In the general case, the force magnitude of the nanoparticle interaction $\left|\vec{\mathbf{F}}_{bi}\right|$ can be written as product of functions depending on the sizes of the nanoelements and the distance between them:

$$\left|\vec{\mathbf{F}}_{bi}\right| = \Phi_{11}(S_c) \cdot \Phi_{12}(D) \quad (7)$$

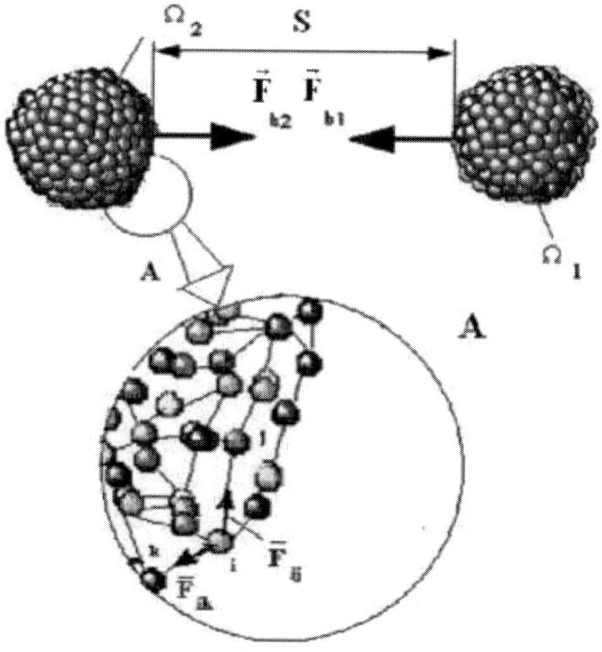

Figure 1. The scheme of the nanoparticle interaction; A – an enlarged view of the nanoparticle fragment.

The $\vec{\mathbf{F}}_{bi}$ vector direction is determined by the direction cosines of a vector connecting the centers of the nanoelements.

Now, let us consider two interacting asymmetric nanoelements situated at the distance S_c between their centers of mass (figure 2) and oriented at certain specified angles relative to each other.

In contrast to the previous problem, the interatomic interaction of the nanoelements leads not only to the relative displacement of the nanoelements but to their rotation as well. Consequently, in the general case, the sum of all the forces of the interatomic interactions of the nanoelements is brought to the principal vector of forces $\vec{\mathbf{F}}_c$ and the principal moment $\vec{\mathbf{M}}_c$

$$\vec{\mathbf{F}}_c = \vec{\mathbf{F}}_{b1} = -\vec{\mathbf{F}}_{b2} = \sum_{i=1}^{N_1} \sum_{j=1}^{N_2} \vec{\mathbf{F}}_{ij} \,, \tag{8}$$

$$\vec{\mathbf{M}}_c = \vec{\mathbf{M}}_{c1} = -\vec{\mathbf{M}}_{c2} = \sum_{i=1}^{N_1} \sum_{j=1}^{N_2} \vec{\boldsymbol{\rho}}_{cj} \times \vec{\mathbf{F}}_{ij} \,, \tag{9}$$

where $\vec{\boldsymbol{\rho}}_{cj}$ is a vector connecting points c and j.

The main objective of this calculation stage is building the dependences of the forces and moments of the nanostructure nanoelement interactions on the distance S_c between the centers of mass of the nanostructure nanoelements, on the angles of mutual orientation of the nanoelements $\Theta_1, \Theta_2, \Theta_3$ (shapes of the nanoelements) and on the characteristic size D of the nanoelement. In the general case, these dependences can be given in the form

$$\vec{\mathbf{F}}_{bi} = \vec{\boldsymbol{\Phi}}_F (S_c, \Theta_1, \Theta_2, \Theta_3, D) \quad , \tag{10}$$

$$\vec{\mathbf{M}}_{bi} = \vec{\boldsymbol{\Phi}}_M (S_c, \Theta_1, \Theta_2, \Theta_3, D) \quad , \tag{11}$$

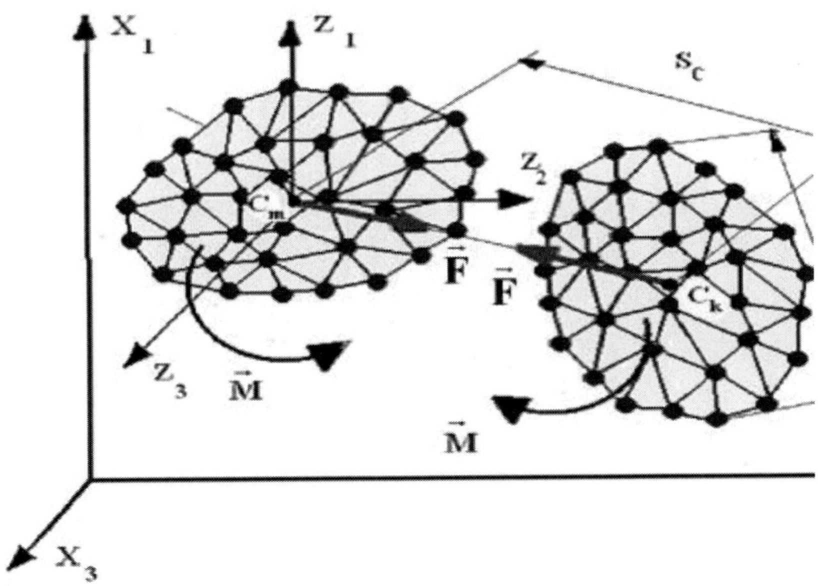

Figure 2. Two interacting nanoelements; $\vec{\mathbf{M}}$, $\vec{\mathbf{F}}$ are the principal moment and the principal vector of the forces, respectively.

For spherical nanoelements, the angles of the mutual orientation do not influence the force of their interaction; therefore, in equation (12), the moment is zero.

In the general case, functions (11) and (12) can be approximated by analogy with (8), as the product of functions $S_0, \Theta_1, \Theta_2, \Theta_3, D$, respectively. For the further numerical solution of the problem of the self-organization of nanoelements, it is sufficient to give the above functions in their tabular form and to use the linear (or non-linear) interpolation of them in space.

2. Problem Formulation for Interaction of Several Nanoelements

When the evolution of the nanosystem as whole (including the processes of ordering and self-organization of the nanostructure nanoelements) is investigated, the movement of each system nanoelement is considered as the movement of a single whole. In this case, the translational motion of the center of mass of each nanoelement is given in the coordinate system X_1, X_2, X_3, and the nanoelement rotation is described in the coordinate system Z_1, Z_2, Z_3, which is related to the center of mass of the nanoelement (Figure 2). The system of equations describing the above processes has the form

$$\begin{cases} M_k \dfrac{d^2 X_1^k}{dt^2} = \sum\limits_{j=1}^{N_e} F_{X_1}^{kj} + F_{X_1}^{ke}, \\ M_k \dfrac{d^2 X_2^k}{dt^2} = \sum\limits_{j=1}^{N_e} F_{X_2}^{kj} + F_{X_2}^{ke}, \\ M_k \dfrac{d^2 X_3^k}{dt^2} = \sum\limits_{j=1}^{N_e} F_{X_3}^{kj} + F_{X_3}^{ke}, \\ J_{Z_1}^k \dfrac{d^2 \Theta_1^k}{dt^2} + \dfrac{d\Theta_2^k}{dt} \cdot \dfrac{d\Theta_3^k}{dt} (J_{Z_3}^k - J_{Z_2}^k) = \sum\limits_{j=1}^{N_e} M_{Z_1}^{kj} + M_{Z_1}^{ke}, \\ J_{Z_2}^k \dfrac{d^2 \Theta_2^k}{dt^2} + \dfrac{d\Theta_1^k}{dt} \cdot \dfrac{d\Theta_3^k}{dt} (J_{Z_1}^k - J_{Z_3}^k) = \sum\limits_{j=1}^{N_e} M_{Z_2}^{kj} + M_{Z_2}^{ke}, \\ J_{Z_3}^k \dfrac{d^2 \Theta_3^k}{dt^2} + \dfrac{d\Theta_2^k}{dt} \cdot \dfrac{d\Theta_1^k}{dt} (J_{Z_2}^k - J_{Z_1}^k) = \sum\limits_{j=1}^{N_e} M_{Z_3}^{kj} + M_{Z_3}^{ke}, \end{cases} \tag{12}$$

where X_i^k, Θ_i^k are coordinates of the centres of mass and angles of the spatial orientation of the principal axes Z_1, Z_2, Z_3 of nanoelements; $F_{X_1}^{kj}, F_{X_2}^{kj}, F_{X_3}^{kj}$ are the interaction forces of nanoelements; $F_{X_1}^{ke}, F_{X_2}^{ke}, F_{X_3}^{ke}$ are external forces acting on nanoelements; N_e is the number of nanoelements; M_k is a mass of a nanoelement; $M_{Z_1}^{kj}, M_{Z_2}^{kj}, M_{Z_3}^{kj}$ is the moment of forces of the nanoelement interaction; $M_{Z_1}^{ke}, M_{Z_2}^{ke}, M_{Z_3}^{ke}$ are external moments acting on nanoelements; $J_{Z_1}^k, J_{Z_2}^k, J_{Z_3}^k$ are moments of inertia of a nanoelement.

The initial conditions for the system of equations (13) and (14) have the form

$$\vec{\mathbf{X}}^k = \vec{\mathbf{X}}_0^k;\ \Theta^k = \Theta_0^k;\ \vec{\mathbf{V}}^k = \vec{\mathbf{V}}_0^k;\ \frac{d\Theta^k}{dt} = \frac{d\Theta_0^k}{dt};\ t = 0, \tag{13}$$

3. Numerical Procedures and Simulation Techniques

In the general case, the problem formulated in the previous sections has no analytical solution at each stage; therefore, numerical methods for solving are used, as a rule. In this work, for the first stages, the numerical integration of the equation of motion of the nanoparticle atoms in the relaxation process are used in accordance with Verlet scheme [26]:

$$\vec{\mathbf{x}}_i^{n+1} = \vec{\mathbf{x}}_i^n + \Delta t\ \vec{\mathbf{V}}_i^n + \left((\Delta t)^2 / 2m_i\right)\left(\sum_{j=1}^{N_k} \vec{\mathbf{F}}_{ij} + \vec{\mathbf{F}}_i - \alpha_i m_i \vec{\mathbf{V}}_i^n\right)^n, \tag{14}$$

$$\vec{\mathbf{V}}_i^{n+1} = (1 - \Delta t \alpha_i)\vec{\mathbf{V}}_i^n + (\Delta t / 2m_i)((\sum_{j=1}^{N_k} \vec{\mathbf{F}}_{ij} + \vec{\mathbf{F}}_i)^n + (\sum_{j=1}^{N_k} \vec{\mathbf{F}}_{ij} + \vec{\mathbf{F}}_i)^{n+1}), \tag{15}$$

where $\vec{\mathbf{x}}_i^n$, $\vec{\mathbf{V}}_i^n$ are a coordinate and a velocity of the i-th atom at the n-th step with respect to the time; Δt is a step with respect to the time.

The solution of the equations (13) also requires the application of numerical methods of integration. In the present work, Runge - Kutta method [27] is used for solving equation (13).

$$(X_i^k)_{n+1} = (X_i^k)_n + (V_i^k)_n \Delta t + \frac{1}{6}(\mu_{1i}^k + \mu_{2i}^k + \mu_{3i}^k)\Delta t, \tag{16}$$

$$(V_i^k)_{n+1} = (V_i^k)_n + \frac{1}{6}(\mu_{1i}^k + 2\mu_{2i}^k + 2\mu_{3i}^k + \mu_{4i}^k). \tag{17}$$

$$\begin{aligned}
\mu_{1i}^k &= \Phi_i^k(t_n; (X_i^k)_n, \ldots; (V_i^k)_n \ldots)\Delta t, \\
\mu_{2i}^k &= \Phi_i^k(t_n + \frac{\Delta t}{2}; (X_i^k + V_i^k \frac{\Delta t}{2})_n, \ldots; (V_i^k)_n + \frac{\mu_{1i}^k}{2}, \ldots)\Delta t, \\
\mu_{3i}^k &= \Phi_i^k(t_n + \frac{\Delta t}{2}; (X_i^k + V_i^k \frac{\Delta t}{2} + \mu_{1i}^k \frac{\Delta t}{4})_n, \ldots; (V_i^k)_n + \frac{\mu_{2i}^k}{2}, \ldots)\Delta t, \qquad (18) \\
\mu_{4i}^k &= \Phi_i^k(t_n + \Delta t; (X_i^k + V_i^k \Delta t + \mu_{2i}^k \frac{\Delta t}{2})_n, \ldots; (V_i^k)_n + \mu_{2i}^k, \ldots)\Delta t.
\end{aligned}$$

$$\Phi_i^k = \frac{1}{M_k}(\sum_{j=1}^{N_e} F_{X_3}^{kj} + F_{X_3}^{ke}) \tag{19}$$

$$(\Theta_i^k)_{n+1} = (\Theta_i^k)_n + (\frac{d\Theta_i^k}{dt})_n \Delta t + \frac{1}{6}(\lambda_{1i}^k + \lambda_{2i}^k + \lambda_{3i}^k)\Delta t \tag{20}$$

$$(\frac{d\Theta_i^k}{dt})_{n+1} = (\frac{d\Theta_i^k}{dt})_n + \frac{1}{6}(\lambda_{1i}^k + 2\lambda_{2i}^k + 2\lambda_{3i}^k + \lambda_{4i}^k) \tag{21}$$

$$\lambda_{1i}^k = \Psi_i^k(t_n;(\Theta_i^k)_n,...;(\frac{d\Theta_i^k}{dt})_n...)\Delta t,$$

$$\lambda_{2i}^k = \Psi_i^k(t_n + \frac{\Delta t}{2};(\Theta_i^k + \frac{d\Theta_i^k}{dt}\frac{\Delta t}{2})_n,...;(\frac{d\Theta_i^k}{dt})_n + \frac{\lambda_{1i}^k}{2},...)\Delta t,$$

$$\lambda_{3i}^k = \Psi_i^k(t_n + \frac{\Delta t}{2};(\Theta_i^k + \frac{d\Theta_i^k}{dt}\frac{\Delta t}{2} + \lambda_{1i}^k\frac{\Delta t}{4})_n,...;(\frac{d\Theta_i^k}{dt})_n + \frac{\lambda_{2i}^k}{2},...)\Delta t, \tag{22}$$

$$\lambda_{4i}^k = \Psi_i^k(t_n + \Delta t;(\Theta_i^k + \frac{d\Theta_i^k}{dt}\Delta t + \lambda_{2i}^k\frac{\Delta t}{2})_n,...;(\frac{d\Theta_i^k}{dt})_n + \lambda_{2i}^k,...)\Delta t.$$

$$\Psi_1^k = \frac{1}{J_{Z_1}^k}(-\frac{d\Theta_2^k}{dt}\cdot\frac{d\Theta_3^k}{dt}(J_{Z_3}^k - J_{Z_2}^k) + \sum_{j=1}^{N_e} M_{Z_1}^{kj} + M_{Z_1}^{ke}),$$

$$\Psi_2^k = \frac{1}{J_{Z_2}^k}(-\frac{d\Theta_1^k}{dt}\cdot\frac{d\Theta_3^k}{dt}(J_{Z_1}^k - J_{Z_3}^k) + \sum_{j=1}^{N_e} M_{Z_2}^{kj} + M_{Z_2}^{ke}), \tag{23}$$

$$\Psi_3^k = \frac{1}{J_{Z_3}^k}(-\frac{d\Theta_1^k}{dt}\cdot\frac{d\Theta_2^k}{dt}(J_{Z_1}^k - J_{Z_2}^k) + \sum_{j=1}^{N_e} M_{Z_3}^{kj} + M_{Z_3}^{ke}),$$

where $i = 1,2,3$; $k = 1,2,\dots N_e$

4. Results and Discussions

Let us consider the realization of the above procedure taking as an example the calculation of the metal nanoparticle.

The potentials of the atomic interaction of Morse (23) and Lennard-Johns (24) were used in the following calculations

$$\Phi(\vec{\rho}_{ij})_m = D_m\,(\exp(-2\lambda_m(|\vec{\rho}_{ij}|-\rho_0)) - 2\exp(-\lambda_m(|\vec{\rho}_{ij}|-\rho_0))), \tag{23}$$

$$\Phi(\vec{\rho}_{ij})_{LD} = 4\varepsilon\left[\left(\frac{\sigma}{|\vec{\rho}_{ij}|}\right)^{12} - \left(\frac{\sigma}{|\vec{\rho}_{ij}|}\right)^{6}\right], \tag{24}$$

where $D_m, \lambda_m, \rho_0, \varepsilon, \sigma$ are the constants of the materials studied.

For sequential and parallel solving the molecular dynamics equations, the program package developed at Applied Mechanics Institute, the Ural Branch of the Russian Academy of Sciences, and the advanced program package NAMD developed at the University of Illinois and Beckman Institute (USA) by the Theoretical Biophysics Group were used. The graphic imaging of the nanoparticle calculation results was carried out with the use of the program package VMD.

4.1. Structure and Forms of Nanoparticles

At the first stage of the problem, the coordinates of the atoms positioned at the ordinary material lattice points [figure 3, (1)] were taken as the original coordinates. During the relaxation process, the initial atomic system is rearranged into a new "equilibrium" configuration [figure 3, (2)] in accordance with the calculations based on equations (6) – (9), which satisfies the condition when the system potential energy is approaching the minimum (Figure 3, the plot).

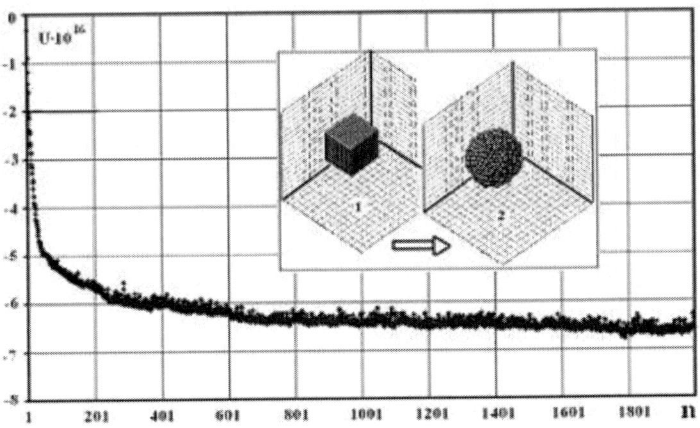

Figure 3. The initial crystalline (1) and cluster (2) structures of the nanoparticle consisting of 1331 atoms after relaxation; the plot of the potential energy U [J] variations for this atomic system in the relaxation process (n - number of iterations with respect to the time).

After the relaxation, the nanoparticles can have quite diverse shapes: globe-like, spherical centered, spherical eccentric, spherical icosahedral nanoparticles and asymmetric nanoparticles (Figure 4).

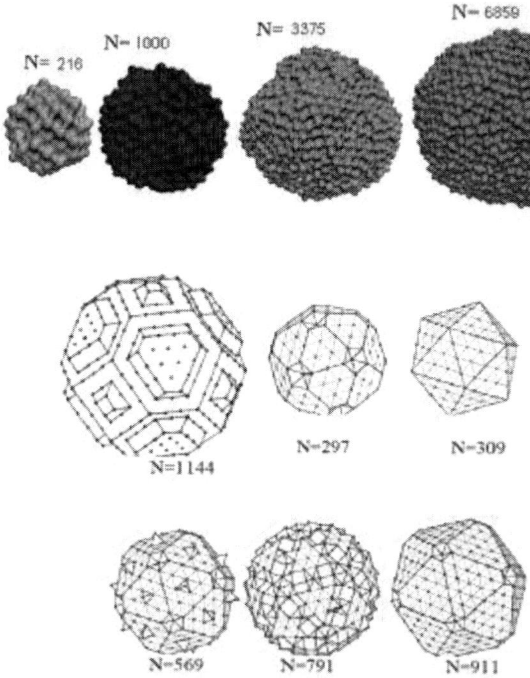

Figure 4. Nanoparticles of diverse shapes, depending on the number of atoms they consist of.

In this case, the number of atoms N significantly determines the shape of a nanoparticle. Note, that symmetric nanoparticles are formed only at a certain number of atoms. As a rule, in the general case, the nanoparticle deviates from the symmetric shape in the form of irregular raised portions on the surface. Besides, there are several different equilibrium shapes for the same number of atoms. The plot of the nanoparticle potential energy change in the relaxation process (Figure 5) illustrates it.

As it follows from this figure, the curve has two areas: the area of the decrease of the potential energy and the area of its stabilization promoting the formation of the first nanoparticle equilibrium shape (1). Then, a repeated decrease in the nanoparticle potential energy and the stabilization area corresponding to the formation of the second nanoparticle equilibrium shape are observed (2). Between them, there is a region of the transition from the first shape to the second one (P). The second equilibrium shape is more stable due to the lesser nanoparticle potential energy. However, the first equilibrium shape also "exists" rather long in the calculation process. The change of the equilibrium shapes is especially characteristic of the nanoparticles with an "irregular" shape. The internal structure of the nanoparticles is of importance since their atomic structure significantly differs from the crystalline structure of the bulk materials: the distance between the atoms and the angles change, and the surface formations of different types appear. In figure 6, the change of the structure of a two-dimensional nanoparticle in the relaxation process is shown.

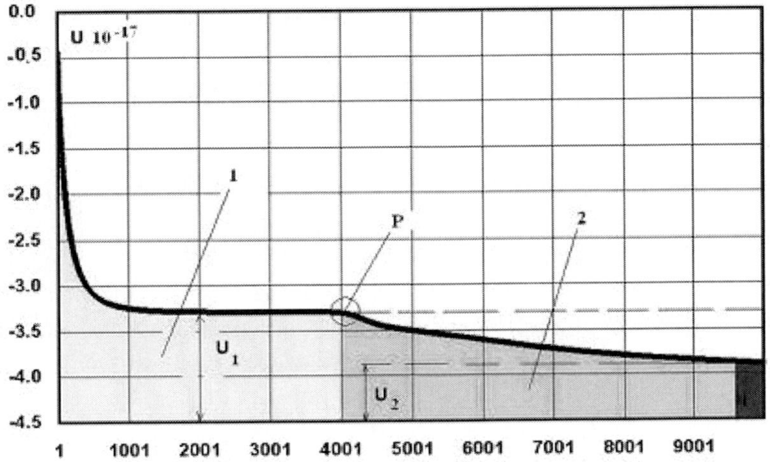

Figure 5. The plot of the potential energy change of the nanoparticle in the relaxation process.

1 – a region of the stabilization of the first nanoparticle equilibrium shape; 2 – a region of the stabilization of the second nanoparticle equilibrium shape; P – a region of the transition of the first nanoparticle equilibrium shape into the second one.

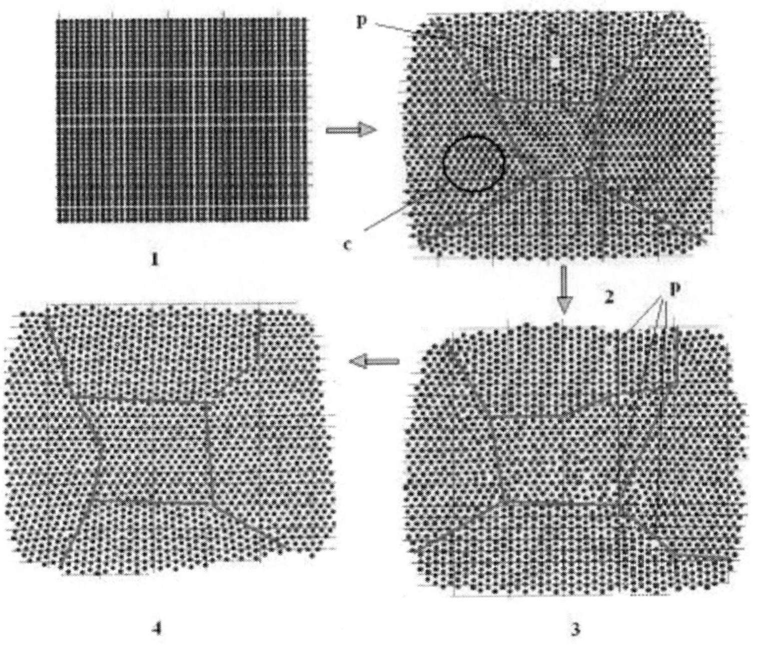

Figure 6. The change of the structure of a two-dimensional nanoparticle in the relaxation process: 1 – the initial crystalline structure; 2, 3, 4 – the nanoparticles structures which change in the relaxation process; p – pores; c – the region of compression.

The figure shows how the initial nanoparticle crystalline structure (1) is successively rearranging with time in the relaxation process (positions 2, 3, 4). Note that the resultant shape of the nanoparticle is not round, i.e., it has "remembered" the initial atomic structure. It is also of interest that in the relaxation process, in the nanoparticle, the defects in the form of pores (designation "p" in the figure) and the density fluctuation regions (designation "c" in the figure) have been formed, which are absent in the final structure.

4.2. Nanoparticles Interaction

Let us consider some examples of nanoparticles interaction. Figure 7 shows the calculation results demonstrating the influence of the sizes of the nanoparticles on their interaction force. One can see from the plot that the larger nanoparticles are attracted stronger, i.e., the maximal interaction force increases with the size growth of the particle. Let us divide the interaction force of the nanoparticles by its maximal value for each nanoparticle size, respectively. The obtained plot of the "relative" (dimensionless) force (Figure 8) shows that the value does not practically depend on the nanoparticle size since all the curves come close and can be approximated to one line.

Figure 9 displays the dependence of the maximal attraction force between the nanoparticles on their diameter that is characterized by non-linearity and a general tendency towards the growth of the maximal force with the nanoparticle size growth.

The total force of the interaction between the nanoparticles is determined by multiplying of the two plots (Figure 8 and Figure 9).

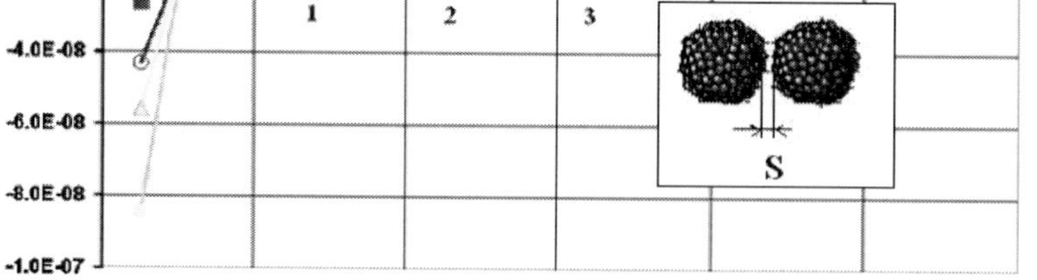

Figure 7. The dependence of the interaction force F [N] of the nanoparticles on the distance S [nm] between them and on the nanoparticle size: 1 – d = 2.04; 2 – d = 2.40; 3 – d = 3.05; 4 – d = 3.69; 5 – d = 4.09 [nm].

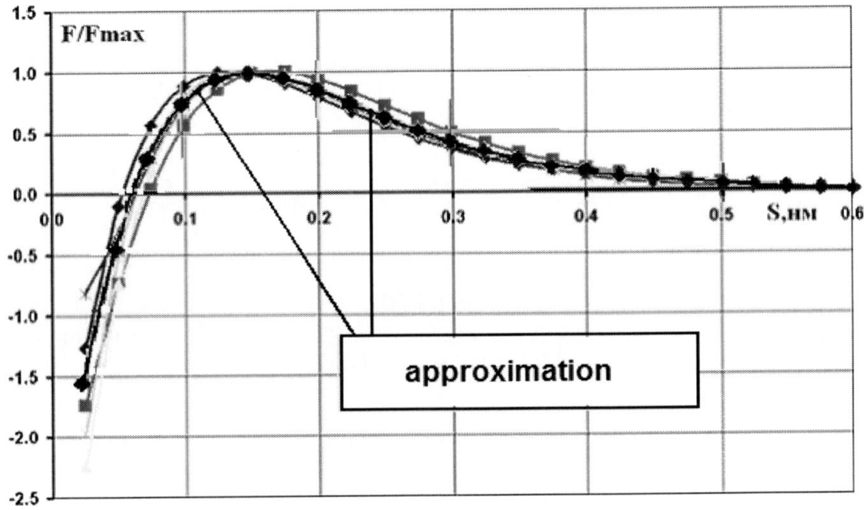

Figure 8. The dependence of the "relative" force $\overline{F}$ of the interaction of the nanoparticles on the distance S [nm] between them.

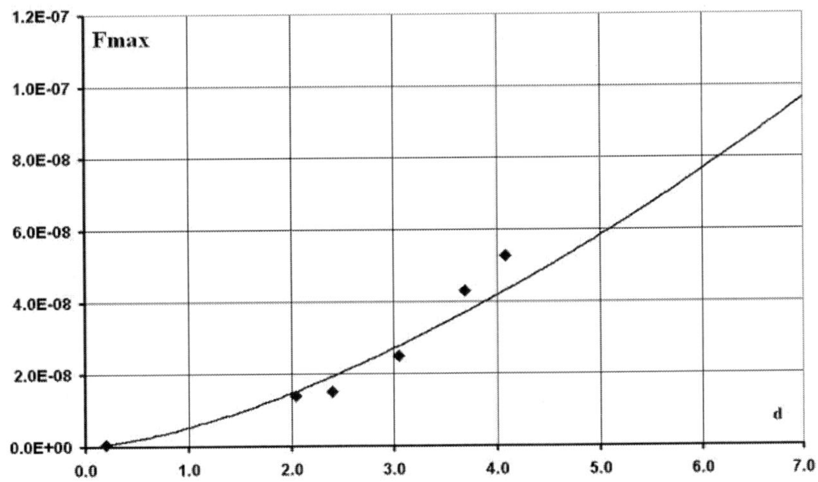

Figure 9. The dependence of the maximal attraction force $F_{\max}$ [N] the nanoparticles on the nanoparticle diameter d [nm].

Using the polynomial approximation of the curve in figure 5 and the power mode approximation of the curve in Figure 6, we obtain

$$\bar{F} = (-1.13S^6 + 3.08S^5 - 3.41S^4 - 0.58S^3 + 0.82S - 0.00335)10^3, \quad (25)$$

$$F_{max} = 0.5 \cdot 10^{-9} \cdot d^{1.499}, \quad (26)$$

$$F = F_{max} \cdot \bar{F}, \quad (27)$$

where d and S are the diameter of the nanoparticles and the distance between them [nm], respectively; F_{max} is the maximal force of the interaction of the nanoparticles [N].

Dependences (25) – (27) were used for the calculation of the nanocomposite ultimate strength for different patterns of nanoparticles' "packing" in the composite (Figure 10).

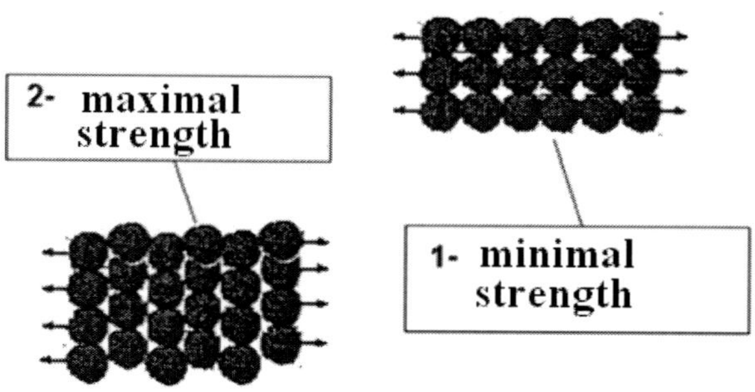

Figure 10. Different types of the nanoparticles' "packing" in the composite.

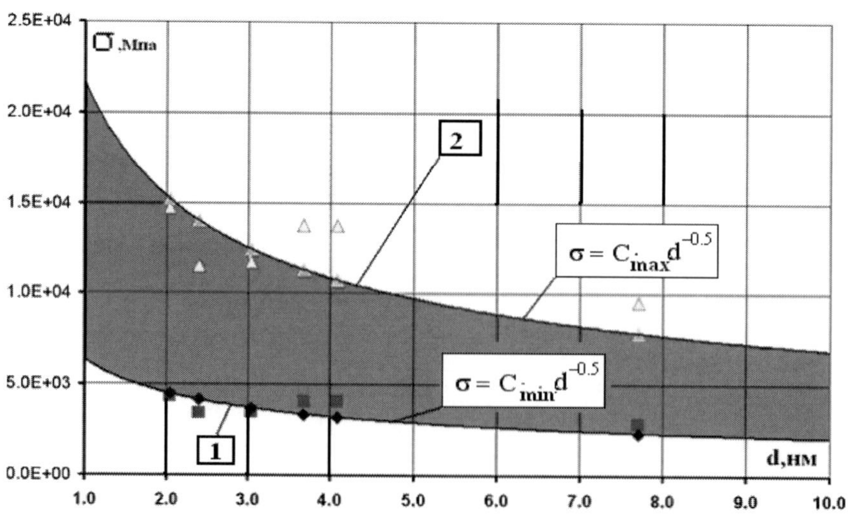

Figure 11. The dependence of the ultimate strength σ [MPa] of the nanocomposite formed by monodisperse nanoparticles on the nanoparticle sizes d [nm].

Figure 11 shows the dependence of the ultimate strength of the nanocomposite formed by monodisperse nanoparticles on the nanoparticle sizes. One can see that with the decrease of the nanoparticle sizes, the ultimate strength of the nanomaterial increases, and vice versa. The calculations have shown that the nanocomposite strength properties are significantly influenced by the nanoparticles' "packing" type in the material. The material strength grows when the packing density of nanoparticles increases. It should be specially noted that the material strength changes in inverse proportion to the nanoparticle diameter in the degree of 0.5, which agrees with the experimentally established law of strength change of nanomaterials (the law by Hall-Petch) [18]:

$$\sigma = C \cdot d^{-0.5}, \tag{28}$$

where $C = C_{\max} = 2.17 \cdot 10^4$ is for the maximal packing density; $C = C_{\min} = 6.4 \cdot 10^3$ is for the minimal packing density.

The electrostatic forces can strongly change force of interaction of nanoparticles. For example, numerical simulation of charged sodium (NaCl) nanoparticles system (Figure 12) has been carried out. Considered ensemble consists of eight separate nanoparticles. The nanoparticles interact due to Van-der-Waals and electrostatic forces.

Results of particles center of masses motion are introduced at Figure 13 representing trajectories of all nanoparticles included into system. It shows the dependence of the modulus of displacement vector $|R|$ on time. One can see that nanoparticle moves intensively at first stage of calculation process. At the end of numerical calculation, all particles have got new stable locations, and the graphs of the radius vector $|R|$ become stationary. However, the nanoparticles continue to "vibrate" even at the final stage of numerical calculations. Nevertheless, despite of "vibration", the system of nanoparticles occupies steady position.

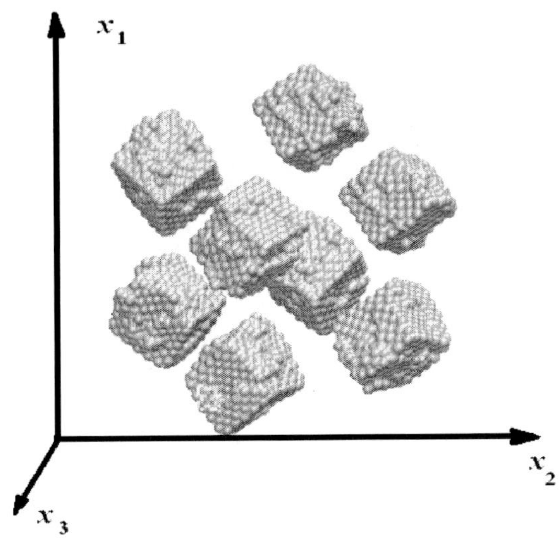

Figure 12. Nanoparticles system consists of eight nanoparticles NaCl.

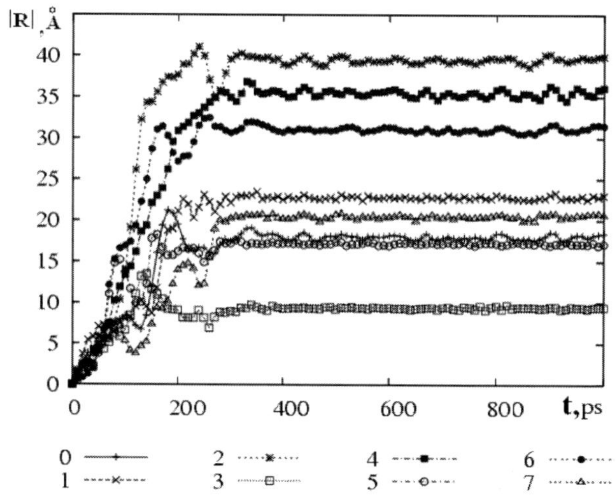

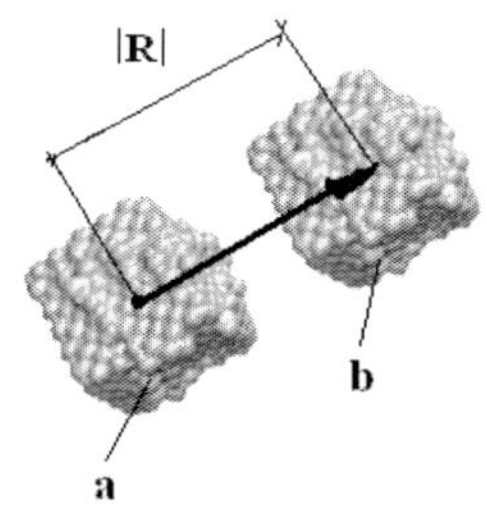

Figure 13. The dependence of nanoparticle centers of masses motion $|R|$ on the time t; a, b--the nanoparticle positions at time 0 and t, accordingly; 1-8 - are the numbers of the nanoparticles

However, one can observe a number of other situations. Let us consider, for example, the self-organization calculation for the system consisting of 125 cubic nanoparticles, the atomic interaction of which is determined by Morse potential (Figure 14).

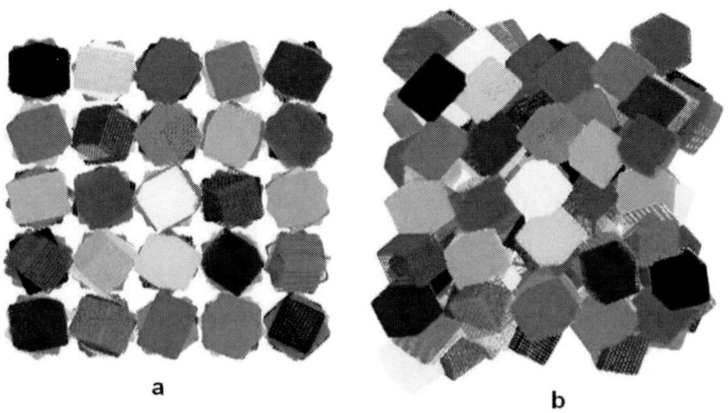

Figure 14. The positions of the 125 cubic nanoparticles: a – initial configuration; b- final configuration of nanoparticles.

As you see, the nanoparticles are moving and rotating in the self-organization process forming the structure with minimal potential energy. Let us consider, for example, the calculation of the self-organization of the system consisting of two cubic nanoparticles, the atomic interaction of which is determined by Morse potential [12]. Figure 15 displays possible mutual positions of these nanoparticles. The positions, where the principal moment of forces is zero, corresponds to pairs of the nanoparticles 2 – 3; 3 – 4; 2 – 5 (Figure 15) and defines the possible positions of their equilibrium.

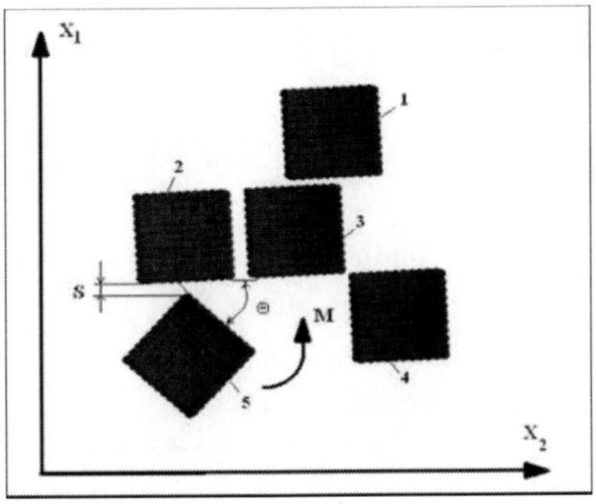

Figure 15. Characteristic positions of the cubic nanoparticles.

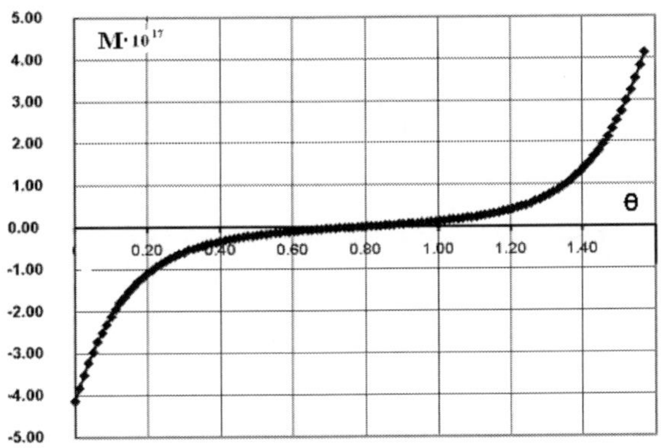

Figure 16. The dependence of the moment M [Nm] of the interaction force between cubic nanoparticles 1 – 3 (see in Figure 9) on the angle of their relative rotation θ [rad].

Figure 16 presents the dependence of the moment of the interaction force between the cubic nanoparticles 1 – 3 (Figure 15) on the angle of their relative rotation. From the plot follows that when the rotation angle of particle 1 relative to particle 3 is $\pi/4$, the force moment of their interaction is zero. At an increase or a decrease in the angle the force moment appears. In the range of $\pi/8<\theta<3\pi/4$ the moment is small. The force moment rapidly grows outside of this range. The distance S between the nanoparticles plays a significant role in establishing their equilibrium. If $S>S_0$ (where S_0 is the distance, where the interaction forces of the nanoparticles are zero), then the particles are attracted to one another. In this case, the sign of the moment corresponds to the sign of the angle θ deviation from $\pi/4$. At $S<S_0$ (the repulsion of the nanoparticles), the sign of the moment is opposite to the sign of the angle deviation. In other words, in the first case, the increase of the angle deviation causes the increase of the moment promoting the movement of the nanoelement in the given direction, and in the second case, the angle deviation causes the increase of the moment hindering the movement of the nanoelement in the given direction. Thus, the first

case corresponds to the unstable equilibrium of nanoparticles, and the second case – to their stable equilibrium. The potential energy change plots for the system of the interaction of two cubic nanoparticles (Figure 17) illustrate the influence of the parameter S. Here, curve 1 corresponds to the condition $S < S_0$ and it has a well-expressed minimum in the $0.3 < \theta < 1.3$ region. At $\theta < 0.3$ and $\theta > 1.3$, the interaction potential energy sharply increases, which leads to the return of the system into the initial equilibrium position. At $S > S_0$ (curves 2 – 5), the potential energy plot has a maximum at the $\theta = 0$ point, which corresponds to the unstable position.

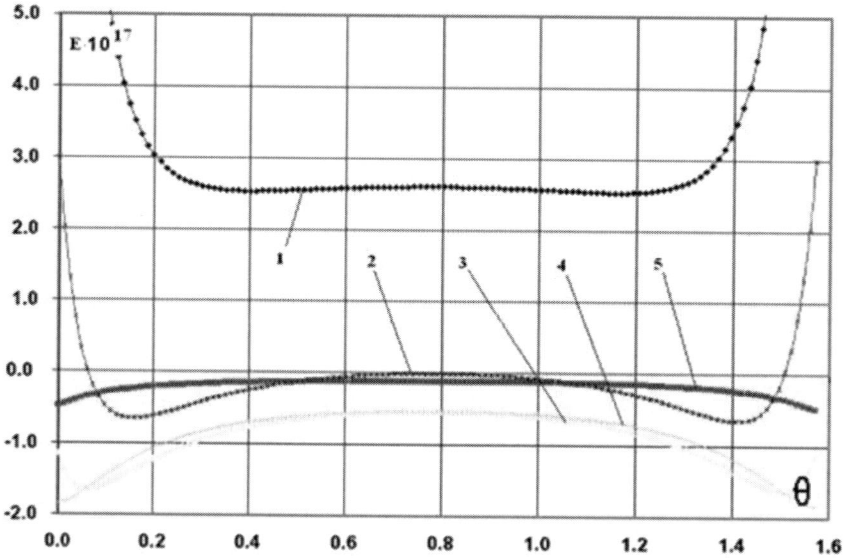

Figure 17. The plots of the change of the potential energy E [Nm] for the interaction of two cubic nanoparticles depending on the angle of their relative rotation θ [rad] and the distance between them (positions of the nanoparticles 1 – 3, figure 9).

The carried out theoretical analysis is confirmed by the works of the scientists from New Jersey University and California University in Berkeley who experimentally found the self-organization of the cubic micro-particles of plumbum zirconate- titanate (PZT) [28]: the ordered groups of cubic micro-crystals from PZT obtained by hydrothermal synthesis formed a flat layer of particles on the air-water interface, where the particle occupied the more stable position corresponding to position 2 – 3 in Figure 15.

Thus, the analysis of the interaction of two cubic nanoparticles has shown that different variants of their final stationary state of equilibrium are possible, in which the principal vectors of forces and moments are zero. However, there are both stable and unstable stationary states of this system: nanoparticle positions 2 – 3 are stable, and positions 3 – 4 and 2 – 5 have limited stability or they are unstable depending on the distance between the nanoparticles.

Note that for the structures consisting of a large number of nanoparticles, there can be a quantity of stable stationary and unstable forms of equilibrium. Accordingly, the stable and unstable nanostructures of composite materials can appear. The search and analysis of the parameters determining the formation of stable nanosystems is an urgent task.

It is necessary to note, that the method offered has restrictions. This is explained by change of the nanoparticles form and accordingly variation of interaction pair potential during nanoparticles coming together at certain conditions.

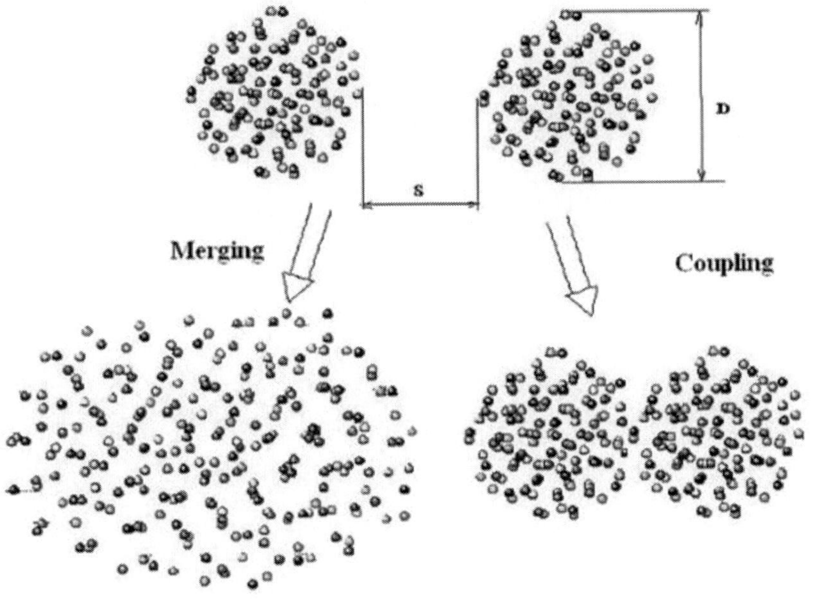

Figure 18. Different type of nanoparticles connection (merging and coupling).

The merge (accretion [4]) of two or several nanoparticles into a single whole is possible (Figure 18). Change of a kind of connection cooperating nanoparticles (merging or coupling in larger particles) depending on the its sizes, it is possible to explain on the basis of the analysis of the energy change graph of connection nanoparticles (Figure 19). From Figure 19 follows, that, though with the size increasing of a particle energy of nanoparticles connection E_{np} grows also, its size in comparison with superficial energy E_S of a particle sharply increases at reduction of the sizes nanoparticles. Hence, for finer particles energy of connection can appear sufficient for destruction of their configuration under action of a mutual attraction and merging in larger particle.

Spatial distribution of particles influences on rate of the forces holding nanostructures, formed from several nanoparticles, also. On Figure 20 the chain nanoparticles, formed is resulted at coupling of three nanoparticles (from 512 atoms everyone), located in the initial moment on one line. Calculations have shown, that in this case nanoparticles form a stable chain. Thus, particles practically do not change the form and cooperate on «small platforms».

In the same figure the result of connection of three nanoparticles, located in the initial moment on a circle and consisting of 256 atoms everyone is submitted. In this case particles incorporate among themselves "densely", contacting on a significant part of the external surface.

Distance between particles at which they are in balance it is much less for the particles collected in group ($L^0_{3np} < L^0_{2np}$) It confirms also the graph of forces from which it is visible, that the maximal force of an attraction between particles in this case (is designated by a continuous line) in some times more, than at an arrangement of particles in a chain (dashed line) $F_{3np} > F_{2np}$.

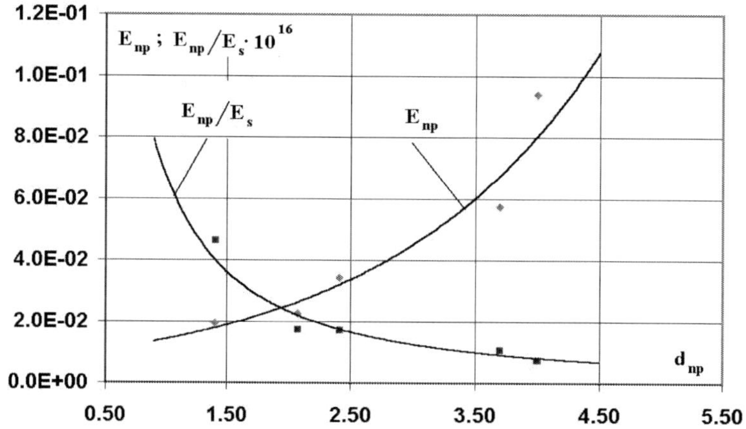

Figure 19. Change of energy of nanoparticles connection E_{np} [Nm] and E_{np} ration to superficial energy E_S depending on nanoparticles diameter d [nm]. Points designate the calculated values. Continuous lines are approximations.

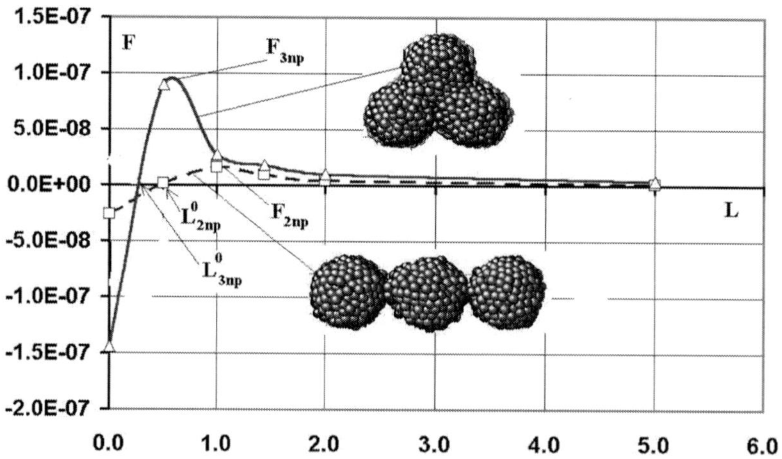

Figure 20. Change of force F [N] of 3 nanoparticles interaction, consisting of 512 atoms everyone, and connected among themselves on a line and on the beams missing under a corner of 120 degrees, accordingly, depending on distance between them L [nm].

Experimental investigation of the spatial structures formed by nanoparticles [4], confirm that nanoparticles gather to compact objects. Thus the internal nuclear structure of the connections area of nanoparticles considerably differs from structure of a free nanoparticle.

Nanoelements kind of interaction depends strongly on the temperature. In Figure 21 shows the picture of the interaction of nanoparticles at different temperatures (Figure 22) It is seen that with increasing temperature the interaction of changes in sequence: coupling (1,2), merging (3,4). With further increase in temperature the nanoparticles dispersed.

Figure 21. Change of nanoparticles connection at increase in temperature.

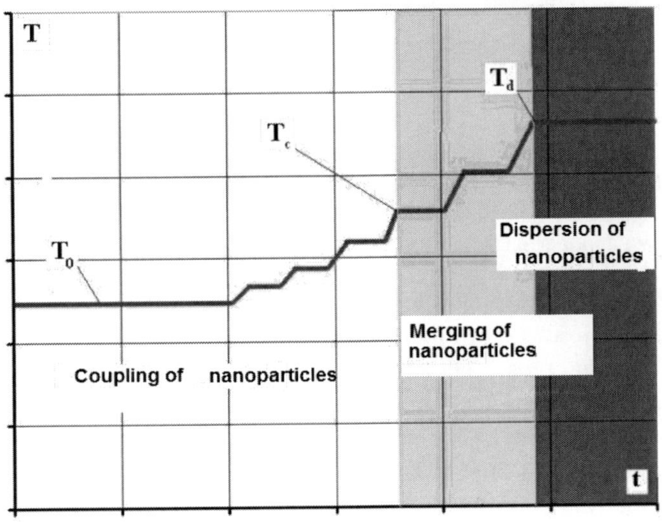

Figure 22. Curve of temperature change.

In the conclusion to the section we will consider problems of dynamics of nanoparticles. The analysis of interaction of nanoparticles among themselves also allows to draw a conclusion on an essential role in this process of energy of initial movement of particles. Various processes at interaction of the nanoparticles, moving with different speed, are observed: the processes of agglomerate formation, formation of larger particles at merge of the smaller size particles, absorption by large particles of the smaller ones, dispersion of particles on separate smaller ones or atoms.

For example, in Figure 23 the interactions of two particles are moving towards each other with different speed are shown. At small speed of moving is formed steady agglomerate (Figure 23).

In a figure Figure 23 (left) is submitted interaction of two particles moving towards each other with the large speed. It is visible, that steady formation in this case is not appearing and the particles collapse.

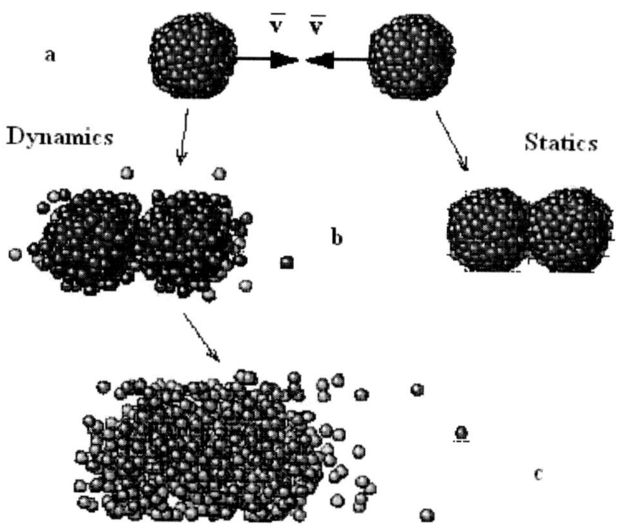

Figure 23. Pictures of dynamic interaction of two nanoparticles: a) an initial configuration nanoparticles; b) nanoparticles at dynamic interaction, c) the "cloud" of atoms formed because of dynamic destruction two nanoparticles.

Nature of interaction of nanoparticles, along with speed of their movement, essentially depends on a ratio of their sizes. In Figure 24 pictures of interaction of two nanoparticles of zinc of the different size are presented. As well as in the previous case of a nanoparticle move from initial situation (1) towards each other. At small initial speed, nanoparticles incorporate at contact and form a steady conglomerate (2).

At increase in speed of movement larger nanoparticle absorbs smaller, and the uniform nanoparticle (3,4) is formed. At further increase in speed of movement of nanoparticles, owing to blow, the smaller particle intensively takes root in big and destroys its.

The given examples show that use of dynamic processes of pressing for formation of nanocomposites demands a right choice of a mode of the loading providing integrity of nanoparticles. At big energy of dynamic loading, instead of a nanocomposite with dispersion corresponding to the initial size of nanoparticles, the nanocomposite with much larger grain that will essentially change properties of a composite can be obtained.

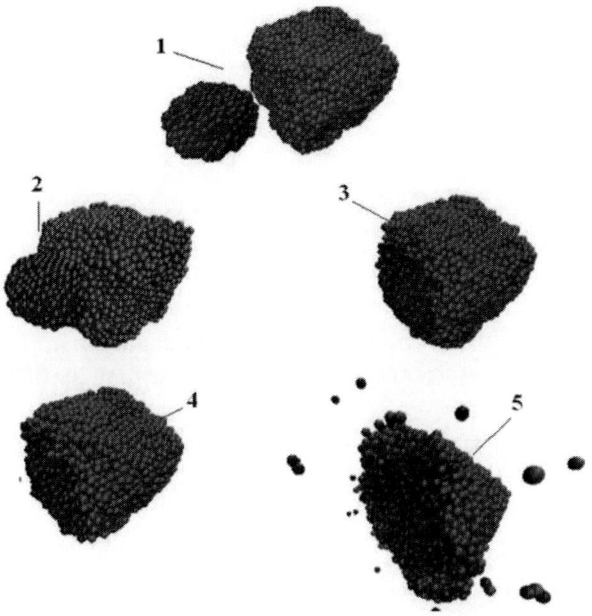

Figure 24. Pictures of interaction of two nanoparticles of zinc: 1-initial configuration of nanoparticles; 2-connection of nanoparticles, 3,4 absorption by a large nanoparticle of a particle of the smaller size, 5-destruction of nanoparticles at blow.

CONCLUSION

In conclusion, the following basic regularities of the nanoparticle formation and self-organization should be noted.

1) The existence of several types of the forms and structures of nanoparticles is possible depending on the thermodynamic conditions.
2) The absence of the crystal nucleus in small nanoparticles.
3) The formation of a single (ideal) crystal nucleus defects on the nucleus surface connected to the amorphous shell.
4) The formation of the polycrystal nucleus with defects distributed among the crystal grains with low atomic density and change interatomic distances. In addition, the grain boundaries are non-equilibrium and contain a great number of grain-boundary defects.
5) When there is an increase in sizes, the structure of nanoparticles is changing from amorphous to roentgen-amorphous and then into the crystalline structure.
6) The formation of the defect structures of different types on the boundaries and the surface of a nanoparticle.
7) The nanoparticle transition from the globe-shaped to the crystal-like shape.
8) The formation of the “regular” and “irregular” shapes of nanoparticles depending on the number of atoms forming the nanoparticle and the relaxation conditions (the rate of cooling, first of all).

9) The structure of a nanoparticle is strained because of the different distances between the atoms inside the nanoparticle and in its surface layers.
10) The systems of nanoparticles can to form stable and unstable nanostructures.

ACKNOWLEDGMENTS

This work was carried out with financial support from the Research Program of the Ural Branch of the Russian Academy of Sciences: the projects 12-P-12-2010, 12-C-1-1004, 12-T-1-1009 and was supported by the grants of Russian Foundation for Basic Research (RFFI) 04-01-96017-r2004ural_a; 05-08-50090-a; 07-01-96015-r_ural_a; 08-08-12082-ofi; 11-03-00571-a.

The author is grateful to his young colleagues doctor A.A. Vakhrushev and doctor A.Yu. Fedotov for active participation in the development of the software complex, calculations and analyzing of numerous calculation results.

The calculations were performed at the Joint Supercomputer Center of the Russian Academy of Sciences.

REFERENCES

[1] Qing-Qing Ni, Yaqin Fu, Masaharu Iwamoto, "*Evaluation of Elastic Modulus of Nano Particles in PMMA/Silica Nanocomposites*" , *Journal of the Society of Materials Science,* Japan, -53,9, 956-961, 2004.

[2] R. S. Ruoff, Nicola M. Pugno, "*Strength of nanostructures*" , Mechanics of the 21st Century. Proceeding of the 21-th international congress of theoretical and applied mechanics. Warsaw: Springer, 303-311,2004.

[3] J. Diao, K Gall., M. L. Dunn, "Atomistic simulation of the structure and elastic properties of gold nanowires" , *Journal of the Mechanics and Physics of Solids,* 52, 9,1935-1962,2004.

[4] R. Dingreville, J. Qu, M. Cherkaoui, "Surface free energy and its effect on the elastic behavior of nano-sized particles, wires and films", *Journal of the Mechanics and Physics of Solids*, 53, 8, 1827-1854, 2004.

[5] H.L. Duan, J. Wang, Z.P. Huang, B.L. Karihaloo, "Size-dependent effective elastic constants of solids containing nano-inhomogeneities with interface stress", *Journal of the Mechanics and Physics of Solids*, 53,7,1574-1596, 2005.

[6] A. I. Gusev and A. A. Rempel, " *Nanocrystalline materials"* Moscow : *Physical mathematical literature*, 2001 (in Russian)

[7] M.R. Hoare, " Structure and dynamics of simple microclusters", *Ach. Chem. Phys.*,40, 49-135,1987.

[8] B.R. Brooks, R.E. Bruccoleri, B.D. Olafson, D.J. States, S. Swaminathan and M. Karplus, "*CHARMM:* A program for macromolecular energy minimization, and dynamics calculations", *J. Comput. Chemistry*, 4, 2, 187-217, 1983.

[9] S. K. Friedlander, "Polymer-like behavior of inorganic nanoparticle chain aggregates ", *Journal of Nanoparticle Research*, 1, 9–15,1999.

[10] M. Grzegorczyk, M. Rybaczuk and K. Maruszewski, " Ballistic aggregation: an alternative approach to modeling of silica sol–gel structures" *Chaos, Solitons and Fractals,* 19, 1003–11, 2004.

[11] E. V. Shevchenko, D. V. Talapin, N. A. Kotov, S. O'Brien and C. B. Murray, "Structural diversity in binary nanoparticle superlattices" *Nature Letter*, 439, 55-59, 2006.

[12] Z.C. Kang and Z. L. Wang, "On Accretion of nanosize carbon spheres", *Journal Physical Chemistry*,100, 5163–65,1996.

[13] I. V. Melikhov and V. E. Bozhevol'nov, "*Variability and self-organization in nanosystems*", *Journal of Nanoparticle Research*, 5, 465–72,2003.

[14] D. Kim and W. Lu, "*Self-organized nanostructures in multi-phase epilayers*" *Nanotechnology*,15, 667–74, 2004.

[15] Kurt E. Geckeler, "*Novel supermolecular nanomaterials: from design to reality*", Proceeding of the 12 Annual International Conference on Composites/Nano Engineering, Tenerife, Spain, August 1-6, CD Rom Edition, 2005.

[16] A.V. Vakhrouchev and A. M. Lipanov, "*A numerical analysis of the rupture of powder materials under the power impact influence*", Computer & Structures ,1/2 (44) 481-86,1992

[17] A.V. Vakhrouchev, "*Modeling of static and dynamic processes of nanoparticles interaction*", CD-ROM Proceeding of the 21-th international congress of theoretical and applied mechanics, ID12054, Warsaw, Poland, 2004.

[18] A.V. Vakhrouchev, "*Simulation of nanoparticles interaction*", Proceeding of the 12 Annual International Conference on Composites/Nano Engineering, Tenerife, Spain, August 1-6, CD Rom Edition, 2005.

[19] Vakhrouchev A.V. "Simulation of nano-elements interactions and self-assembling", Modeling and Simulation in Materials Science and Engineering ,.14, 2006, pp. 975-991

[20] Vakhrouchev A. V. and Lipanov A. M. Numerical Analysis of the Atomic structure and Shape of Metal Nanoparticles //*Computational Mathematics and Mathematical Physics*, 2007, Vol. 47, No. 10, pp. 1702–1711.

[21] Vakhrouchev A.V. Modelling of the Process of Formation and Use of Powder Nanocomposites// Composites with Micro and Nano-Structures. Computational Modeling and Experiments. Computational Methods in Applied Sciences Series. - Barcelona, Spain: Springer Science.- Vol.9.-2008.- pp.107—136.

[22] Vakhrouchev A. V. Modeling of the nanosystems formation by the molecular dynamics, mesodynamics and continuum mechanics methods // *Multidiscipline Modeling in Material and Structures*.-Vol. 5, N. 2, 2009 , pp. 99-118.

[23] Vakhrouchev A.V. Theoretical bases of nanotechnology application to thermal engines and equipment. - Izhevsk, Institute of Applied Mechanics, *Ural Branch of the Russian Academy of Sciences*, 2008.-212p. (In Russian)

[24] Alikin V.N., Vakhrouchev A. V., Golubchikov V.B., Lipanov A.M., Serebrennikov S.Y. Development and investigation of the aerosol nanotechnology. Moscow: Mashinostroenie, 2010.-196p., (in Russian)

[25] W. D. Heerman, "*Computer simulation methods in theoretical physics'* Berlin: Springer-Verlag, 1986

[26] L. Verlet, "Computer "experiments" on classical fluids I. Thermo dynamical properties of Lennard-Jones molecules", *Phys. Rev*., 159, 98-103, 1967.

[27] G. A. Korn and M. T. Korn, "*Mathematical handbook*", New York: McGraw-Hill Book Company, 1968.

[28] Self-organizing of microparticles piezoelectric materials, News of chemistry, datenews.php.htm

In: Quantitative Chemistry, Biochemistry and Biology
Editors: G.E. Zaikov, O.V. Stoyanov et al.
ISBN: 978-1-62948-332-0

Chapter 2

SYNTHESIS OF THERMOPLASTIC BIO-BASED POLYURETHANES ON THE BASE OF VEGETABLE OILS

C. Bueno-Ferrer, N. Burgos and A. Jiménez*
Analytical Chemistry, Nutrition & Food Sciences Department, University of Alicante, Alicante, Spain

ABSTRACT

During the last decade environmental concerns have risen in the modern society and have lead to the search of sustainable alternatives for the global development. Therefore, efforts in research for the new formulations in environmentally-friendly materials are currently increasing. In this sense, the substitution of materials obtained from petrochemical sources by others with renewable origin, such as vegetable oils, is one of the most studied possibilities to reduce the global dependency to fossil fuels. These materials are currently under development and the final goal of these new studies would be the reduction of the industrial carbon footprint and a significant improvement in the sustainability of the current production processes. The versatility and ability to chemical transformation of vegetable oils make them good candidates to be used as polymer matrices after some adequate modifications, giving rise to polymers such as thermoplastic polyurethanes (TPUs). In this context, the main goal of the present work is the proposal of thermoplastic formulations based on the use of vegetable oils as polymer matrices. The study of chemical structures, morphologies, thermal and mechanical behavior of different bio-based TPUs will be presented in this paper.

Keywords: Bio-based polyols, dimmer fatty acids, polyurethanes, hard segment distribution, structure-property relationships

* Telephone: +34-965909660; E-mail: alfjimenez@ua.es.

Introduction

The use of renewable raw materials constitutes a significant contribution to a sustainable development in the plastics production. This strategy is based on the advantages given by Nature synthesis potential and Green Chemistry principles. In this sense, polymers obtained from renewable raw materials have raised some interest in the last years. The development of polymers synthesized from agricultural products, such as starch, cellulose, sugars or lignin has been considerably increased in the last two decades [1]. Amongst all the possible natural sources for polymers, vegetable oils are considered one of the cheapest and abundant in Nature [2]. They can be used as an advantageous chemical platform to polymer synthesis, by their inherent biodegradable condition and low toxicity to humans and the environment. In this context, many efforts are currently going on to propose a great variety of chemical methods to prepare thermoplastics and thermosets based on vegetable oils. This wide range of chemical methods applicable to these natural materials gives rise to many different monomers and polymers with many applications.

Fatty acids are the major chemical entities present in vegetable oils. They are valuable compounds to design specific monomers in the search of polymers with particular properties without any need of important modifications in their native structure. This is an advantageous issue, not only in sustainability terms, but also in industrial applicability and competitiveness in terms of cost and properties [2-5].

Vegetable oils are mainly formed by triglycerols or triglycerides, mainly composed by three fatty acids bonded to a glycerol molecule. Fatty acids constitute 94-96 % of the total triglycerides weight in a vegetable oil and the number of carbon units in their structure is normally between 14 and 22, with zero to three double bonds by fatty acid molecule. The contents in fatty acids in some of the most common vegetable oils are indicated in Table 1.

The use of fatty acids and vegetable oils either in polymer synthesis or as additives comes from some decades, not only by the raising interest in the search for alternatives to fossil fuels but also by the particular chemical characteristics that make them adequate for polymerization processes. Triglycerides are molecules with low reactivity and this fact is a disadvantage in their potential application in polymer synthesis. Nevertheless, the introduction of different functionalities in their reactive sites increases largely the synthetic possibilities of triglycerides [3].

At least three different uses of vegetable oils in polymer formulations can be proposed: (i) as polymer additives (plasticizers, stabilizers, etc); (ii) as building blocks to get polymers from them; and (iii) as units for the thermosets synthesis. Much work on the use of vegetable oils as additives [6-11] and as thermosets precursors [12-18] has been reported, but the development of thermoplastic polymer matrices is still in an early stage of research. Thermoplastics can be easily processed and recycled giving them possibilities in many different applications.

Table 1. Fatty acid distribution in vegetable oils (g fatty acid/100 g oil)

Fatty acid	C:DB	Cotton	Rapeseed	Sunflower	Linseed	Corn	Olive	Palm	Castor	Soybean
Myristic	14:0	0.7	0.1	0.0	0.0	0.1	0.0	1.0	0.0	0.1
Myristoleic	14:1	0.0	0.0	0.0	0.0	0.0	0.0	0.0	0.0	0.0
Palmitic	16:0	21.6	4.1	6.1	5.5	10.9	13.7	44.4	1.5	11.0
Palmitoleic	16:1	0.6	0.3	0.0	0.0	0.2	1.2	0.2	0.0	0.1
Stearic	18:0	2.6	1.8	3.9	3.5	2.0	2.5	4.1	0.5	4.0
Oleic	18:1	18.6	60.9	42.6	19.1	25.4	71.1	39.3	5,0	23.4
Linoleic	18:2	54.4	21.0	46.4	15.3	59.6	10.0	10.0	4.0	53.2
Linolenic	18:3	0.7	8.8	1.0	56.6	1.2	0.6	0.4	0.5	7.8
Ricinoleic	18:1	0.0	0.0	0.0	0.0	0.0	0.0	0.0	87.5	0.0
Arachidic	20:0	0.3	0.7	0.0	0.0	0.4	0.9	0.3	0.0	0.3
Gadoleic	20:1	0.0	1.0	0.0	0.0	0.0	0.0	0.0	0.0	0.0
Eicosadienoic	20:2	0.0	0.0	0.0	0.0	0.0	0.0	0.0	0.0	0.0
Behenic	22:1	0.2	0.3	0.0	0.0	0.1	0.0	0.1	0.0	0.1
Erucic	22:1	0.0	0.7	0.0	0.0	0.0	0.0	0.0	0.0	0.0
Lignoceric	24:0	0.0	0.2	0.0	0.0	0.0	0.0	0.0	0.0	0.0
DB/triglyceride		3.9	3.9	4.7	6.6	4.5	2.8	1.8	2.7	4.6
Iodine index (I)		104-117	91-108	110-143	168-204	107-120	84-86	44-58	82-88	117-143

C, number of carbon atoms; DB, number of C = C double bonds.

The synthesis of thermoplastics from vegetable oils is still in an early stage of the study and development because of the experimental difficulties to be afforded to get reasonable yields in this process. It is known that thermosets obtained from vegetable oils have been largely studied with many reported work [12-22]. This fact is partially due to the own composition of vegetable oils, formed by triglycerides containing different fatty acids with variable number of chain instaurations. Thus, seeds oils are rich in poly-unsaturated fatty acids, giving highly reticulated, rigid and temperature resistant materials [21]. The carbon chains forming the oils can be easily crosslinked by their double bonds. As indicated in Table 1, it can be concluded that most seed oils have fatty acids with 2 or 3 unsaturated bonds as the main component in their lipid profile, except castor, olive and rapeseed oils, which show a mono-unsaturated fatty acid as their main component. Therefore, thermosets can be easily synthesized from oils rich in poly-unsaturated fatty acids, such as those from soya, sunflower or linseed getting polymers with high mechanical and thermal resistance. Castor oil shows high content in ricinoleic acid (87.5 % in total oil weight) and the active site is occupied by an alcohol, while olive and rapeseed oils show a main content in mono-unsaturated oleic acid (71.1 and 60.9 % in total oil weigh, respectively).

It is known that the potential monomers or polymer building blocks should have at least one (in the case of addition polymerization) or two double bonds in their structure (in the case of condensation polymerization) to get thermoplastic materials. Therefore, triglycerides should be modified of functionalized before polymerization. Nevertheless, there are some examples of thermoplastic biomaterials obtained from naturally functionalized castor oil with homogeneous composition and acceptable polymerization yields. The main thermoplastic materials already synthesized from vegetable oils are thermoplastic polyurethanes (TPUs), polyamides (PA), thermoplastic polyesters, polyesteramides and polyanhydrides.

TPUs: Chemistry, Structure and Properties

Polyurethanes are generally synthesized by addition polymerization between a polyalcohol and a poly-isocyanate. This is an exothermic reaction caused by the release of a proton from the alcohol group followed by a general molecular rearrangement by the formation of the urethane bond. [23]. If both reagents are bi-functional linear polyurethanes are obtained, while if functionalities are increased some crosslinked chains are formed, with the formation of reticulated structures. In summary, one of the most common synthesis routes for TPUs consists basically of the reaction of three main components.

1) Polyols with polyester or polyether functionalities with hydroxyl end groups.
2) Di-isocyanate
3) Short-chain diol or diamine used as chain extender.

The clear differences in the structure of all these components are essential to get the final properties of the synthesized thermoplastic polyurethanes. TPUs are block copolymers with $(AB)_n$ segmented structure formed by hard and soft blocks. Soft segments correspond to the elastomeric part of the polymer (polyester chains from the polyol) and they are characterized by a glass transition temperature much lower than ambient, which gives them high flexibility. This is the reason why these parts of the polyurethane are known as the soft block. On the other hand, hard segments are formed by the di-isocyanate and the chain extender forming a rigid structure, mainly formed by the urethane group bonded to aromatic rings. Therefore, some heterogeneity between both blocks in polyurethanes, the hard segment (with high polarity and melting point) and the soft segment (non-polar and low melting point), should be expected, leading to phase separation in the copolymer structure, as presented in Figure 1 [24,25].

However, the phase separation is not complete in TPUs at the molecular scale, and it is possible to find soft segments inside the hard region and vice-versa (Figure 2), as was reported by Tawa et al. [26]. They indicated that urethane groups from neighbour chains could form hydrogen bonds very easily. This inter-molecular bonding leads to the formation of aggregates acting as physical reticulation nodes with crystalline regions dispersed into the soft area in the polymer structure. This would lead to crosslinking with the final result of the increase in the overall rigidity in the TPU. The phase separation between hard and soft segments depends on, among other factors, their affinity, their relative mobility, the chain extender and the isocyanate structural symmetry [25].

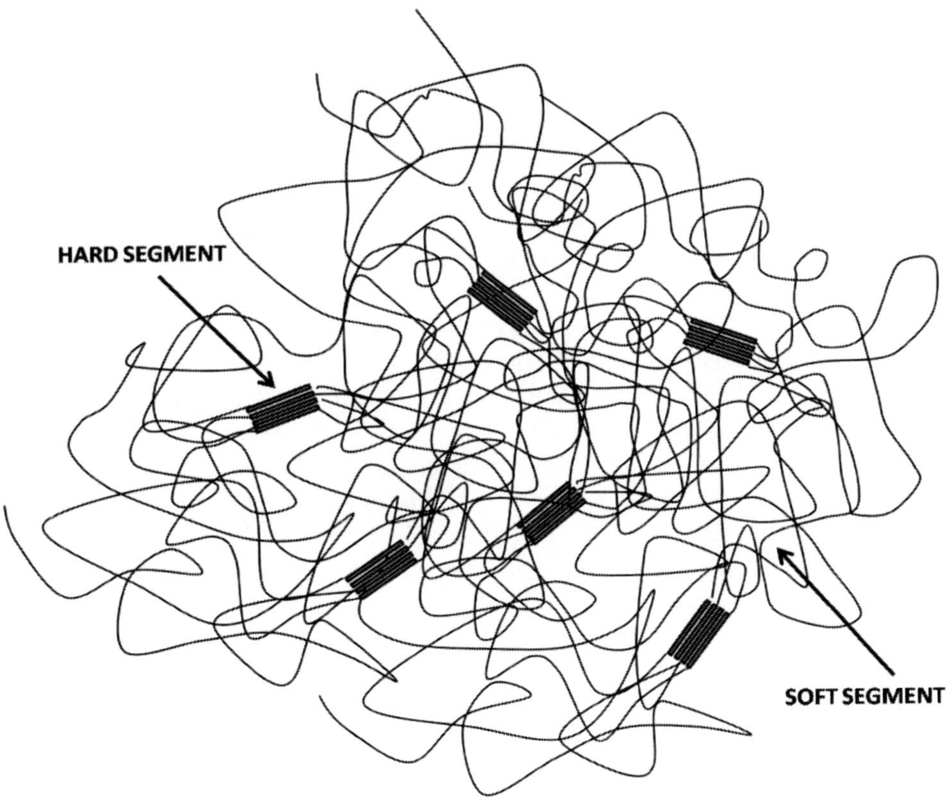

Figure 1. Segmented structure of a thermoplastic polyurethane.

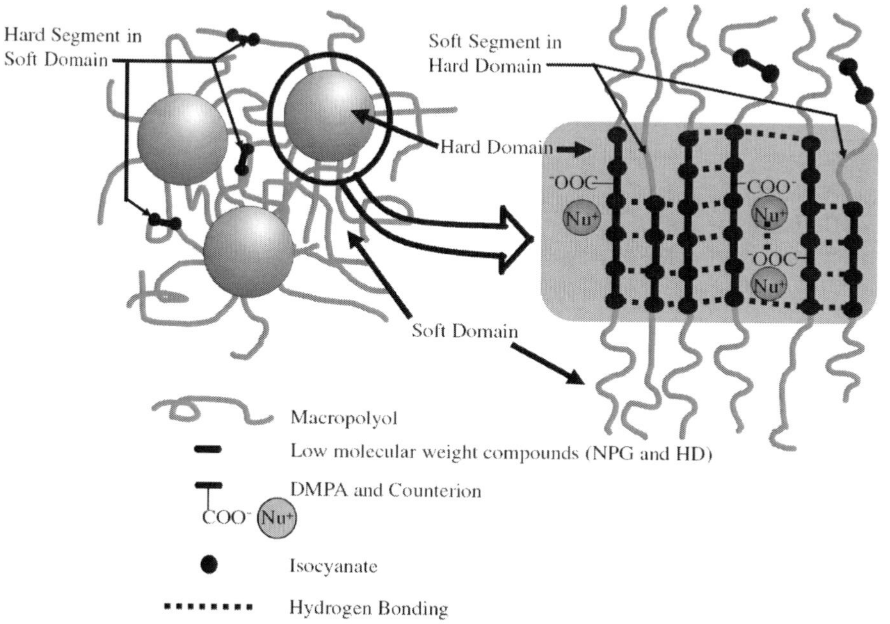

Figure 2. Polyurethane biphasic structure at molecular scale [26], with permission.

TPU structures are generally linear, since the relative amount of hard segments is small. Therefore, the most relevant properties of these polymers are conditioned by the secondary and inter-molecular interactions (mainly Van der Waals) between the soft segments [24,27]. The elastic properties of polyurethanes mainly depend on the polyol chains mobility, which is dependent on their chemical nature and length of the soft segments. The higher the molar mass of the soft segments, the higher tensile strength and elongation to break and consequently the more flexible TPU [24,25].

On the other hand, when the hard segment content is high, the plastic deformation and the polymer general softening are observed after the application of mechanical stresses at high temperature. The thermal stability of polyurethanes is determined by the temperature range where the rigid segments start to melt and consequently their phase separation and their segmented structure. Polyurethanes will show thermoplastic behaviour at higher temperatures than the melting point of the hard regions [28]. Another important feature of the hard/soft ratio in TPUs is the decrease of molar masses and low phase separation in polymers with high values of that ratio, conditioning their rheological and mechanical properties, giving rise to more rigid materials. In addition, the isocyanate structure also influences the final TPU properties. High-volume di-isocyanates lead to polyurethanes with high elastic modulus and tensile strength [25].

The development of TPUs from vegetable oils is the main goal of this study. In the next sections some work performed by our research group for TPUs synthesized from rapeseed oil dimer fatty acids is presented.

EXPERIMENTAL

Materials

The bio-based polyester polyol used in this study was kindly supplied by Croda (Yorkshire, UK) and it is based on dimmer fatty acids from rapeseed oil, with purity higher than 98 % and weight average molar mass (M_w) around 3000 g mol^{-1}. Hydroxyl and acid values are 40 mg KOH g^{-1} and 0.253 mg KOH g^{-1} respectively, as given by the supplier. 4,4'-diphenylmethane di-isocyanate (MDI) was supplied by Brenntag (Rosheim, France). 1,4-butanediol (BDO), dibutylamine, toluene and hydrochloric acid were purchased from Sigma Aldrich (Lyon, France). All reagents were used without any further purification step.

TPU Synthesis

Four different TPUs were prepared with a NCO/OH ratio equal to 1 and increasing HS content (10-40 wt%, named TPU10–TPU40, respectively), by the two-step prepolymer process for TPU polymerization (Table 2). In a first step, the polyol reacted with an excess of MDI (ratio 2:1) for 2 h in a five-necked round-bottom flask having the provision for nitrogen flushing, mechanical stirring and temperature control at 80 ^{0}C. During the synthesis, samples were extracted each 30 min in triplicate and diluted in 20 mL of a standard solution of dibutylamine 0.05 M in toluene for the reaction between the residual di-isocyanate and the amine, to control the NCO consumption during the reaction. Each solution was then stirred at room temperature for 10 h to ensure the complete reaction of NCO groups with dibutylamine. The excess amine was titrated back with standard aqueous HCl 0.05 M solution using bromophenol green as indicator. For each titration, 25 mL of isopropyl alcohol were added to the solution to ensure compatibility between dibutylamine and the HCl solution. It was calculated from these experiments that 56.1 % of NCO groups were consumed at the end of the prepolymer synthesis. This result was used for the addition of the precise amount of diol groups in the processing step. The prepolymer was further melt-blended in a second step by reactive processing with the adequate amount of polyol and chain extender, depending on the HS content targeted in the final TPUs. The prepolymer synthesized in the first step and the calculated amount of polyol for a NCO/OH = 1 ratio were directly introduced in the feeding zone of an internal mixer (counter-rotating mixer Rheocord 9000, Haake, USA) equipped with a pair of high-shear roller-type rotors, at 80 ^{0}C, with a rotation speed of 50 rpm and 15 min processing time. Then, the adequate amount of BDO chain extender was added and the temperature was immediately increased to 180 ^{0}C for 8 min, without any catalyst. After polymerization, all systems were cured overnight in an oven at 70 ^{0}C to ensure the complete reaction of NCO groups, which was further checked by attenuated total reflection Fourier-transform infrared spectroscopy (ATR-FTIR). TPUs were subsequently compression-molded in a hot press at 200 ^{0}C by applying 200 MPa pressure for 5 min and further quenched between two steel plates for 10 min to obtain sheets with 1.5 mm thickness for each system. The expected HS length for each TPU sample was calculated from data in Table 2 by following a protocol already described by Petrovic et al. [29] to determine the average polymerization degree in segmented polyurethanes. This method was applied to the TPUs synthesized in this study resulting in an HS average polymerization degree of 5-7, 7-9, 11-13 and 19-21 units for TPU10, TPU20, TPU30 and TPU40, respectively. These results could give an estimation of the length of HS for each TPU.

Table 2. Calculated hard segment (HS) percentage and reactants amounts for each TPU blend [30]

Sample	HS (wt%)	Polyol (g)	MDI (g)	BDO (g)
TPU10	10	49.5	5.1	0.36
TPU20	20	44.0	9.1	1.94
TPU30	30	38.5	13.0	3.52
TPU40	40	33.0	16.9	5.11

With permission.

MATERIALS CHARACTERIZATION

ATR-FTIR Spectroscopy

ATR-FTIR was used to screen the complete reaction between NCO and OH groups and, consequently, to evaluate the adequate curing of the TPUs. Infrared spectra were collected on a TA Instruments SDT Q600 (Thermo-Nicolet, New Castle, DE, USA) at a resolution of 4 cm^{-1} and 64 scans per run. The ATR accessory was equipped with a Germanium (n = 4) crystal and it was used at a nominal incidence angle of 45^0, yielding 12 interval reflections at the polymer surface.

Thermogravimetric Analysis (TGA)

The four TPU systems as well as the bio-based polyol were analyzed in dynamic mode by using a TGA/SDTA 851e Mettler Toledo (Schwarzenbach, Switzerland) equipment. Approximately 7 mg samples were weighed in alumina pans (70 μL) and they were heated from 30 °C to 700 ^{0}C at 10 ^{0}C min^{-1} under nitrogen atmosphere (flow rate 30 mL min^{-1}). In the case of TPUs, the initial degradation temperature was calculated as the temperature where 5 wt% of the initial mass was lost ($T_{5\%}$).

Differential Scanning Calorimetry (DSC)

Thermal and structural characterization of TPUs and the polyol was carried out by using a TA Instruments Q2000 (New Castle, DE, USA) equipment. Approximately 5 mg of each sample were weighed in aluminum pans (40 μL) and they were subjected to a first heating stage from 30 °C to 240 °C, with a further cooling from 240 °C to -90 °C and a subsequent heating from -90 °C to 240 °C. All steps were carried out at 10 °C min^{-1} under nitrogen (flow rate 50 mL min^{-1}). All tests were performed in duplicate. Glass transition temperatures (T_g) were determined on the second heating scan. T_g of the polyol was determined by using modulated differential scanning calorimetry (MDSC) during a cooling scan from 30 °C to -90 °C at 2 °C min^{-1}, with 60 s period and heat-only mode.

Uniaxial Mechanical Tests

Tensile properties of TPUs were determined with an Instron tensile testing machine (model 4204, USA), at 25 °C and 50 % relative humidity at a rate of 20 mm min^{-1}, using dumbbell specimens (dimensions: 30 x 10 x 1.5 mm^3). For each formulation at least five samples were tested.

RESULTS AND DISCUSSION

ATR-FTIR Spectroscopy

The adequate curing of all TPUs is a key-point prior to the materials characterization, since the presence of residual NCO groups in the final polymer gives an indication of an incomplete synthesis. In this work ATR-FTIR was used to confirm the complete reaction between NCO and OH groups. TPUs spectra are shown in Figure 3 and they could be used to highlight the main structural differences between them. No peak was found at 2270 cm^{-1} (NCO stretching band) suggesting that the reaction was complete in all cases. As expected, the main variations are related to the increasing content in HS and consequently the higher concentration in urethane groups (-NH-CO-O). In Figure 3, vibrations at 3335 and 1550 cm^{-1} corresponded to -NH stretching and bending, respectively. Besides, the peak for the C=O stretching from the urethane group could be observed at ≈1700 cm^{-1} [31-33]. All these bands, assigned to the urethane groups, increased in their intensity from TPU10 to TPU40, with confirmation of the higher concentration in carbamate groups at higher HS contents. Nevertheless, the absorption band at 1735 cm^{-1} was assigned to the C=O group stretching in the polyol, and it was similar for all TPUs, except for TPU10 where this band was broader and almost no discernible from the band at 1700 cm^{-1}, certainly due to the lower content in urethane groups in this material.

Thermogravimetric Analysis (TGA)

The thermal stability of the polyol and all bio-based TPUs were studied by dynamic TGA. Figure 4 shows their mass percentage and derivative curves. The bio-based polyol showed a narrow derivative peak (Figure 4b, left) due to its purity, while Figure 4b (right) clearly shows the derivative curves of TPUs with lower intensity peaks. It is known that degradation of polyurethanes is a complex and multistep process, as observed in Figure 4. An important parameter, the degradation onset, is dependent on the thermal stability of the less thermally stable part on the polyurethane chains [34,35]. Initial degradation temperatures ($T_{5\%}$) of each step were also studied in TPUs and, together with mass loss percentages, allowed to study the differences between samples depending on their HS content. It was noted that the $T_{5\%}$ value of the pure rapeseed oil-based polyol was higher than in the case of TPU systems, as it was expected, but also higher than the $T_{5\%}$ value reported for castor oil [36] and cashew nut shell liquid-based polyols [37]. TPUs also showed higher thermal stability than polymers with similar structures [38]. It has been reported that their first degradation stage is related to urethane bond decomposition into isocyanate and alcohol with possible formation of primary and secondary amines [39,40]. Nevertheless, the complexity of this stage is also related with the HS content. In this way, when this content increases $T_{5\%}$ decreases, making those materials more susceptible to degradation and suggesting that the starting point of degradation takes place predominantly within hard segments.

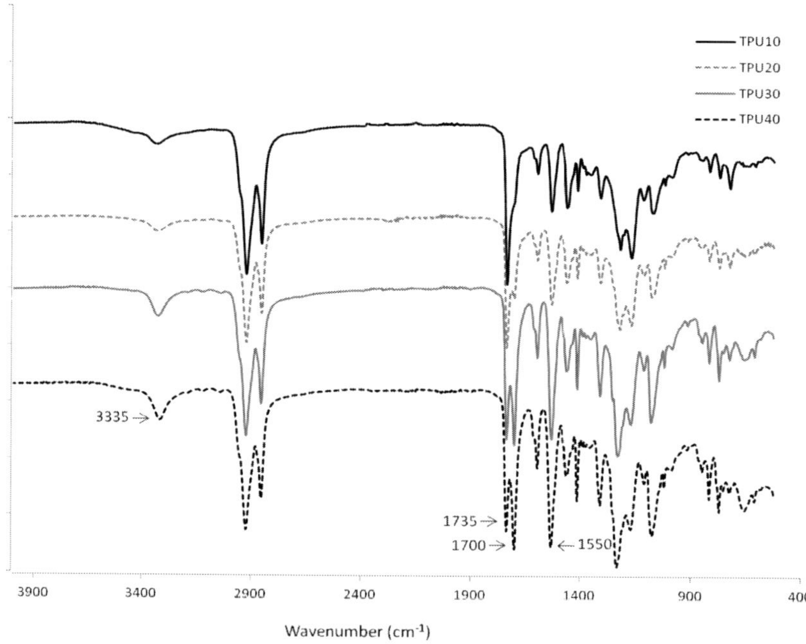

Figure 3. ATR-FTIR spectra of TPU 10-40 wt% of HS and main peak assignments (cm^{-1}) [30], with permission.

Differential Scanning Calorimetry (DSC)

The structure of TPU samples was investigated by DSC while the bio-based polyol was studied both by DSC and MDSC. The main results of this study are shown in Table 3 and Figure 5. This technique is valuable for a precise determination of T_g and from these values it is possible to estimate the real amounts of HS in the unorganized and organized microphases. It has been indicated that HS in TPUs do not fully belong to the hard domains, since some of them can be found in the soft regions and vice versa. Moreover, when MDI is not bonded to the chain extender but to the polyol, presumably in systems with high polyol amount [42], such as in the case of TPU10, this trend is more clearly observed. This phenomenon was also clearly evidenced in the thermal stability of the TPUs, since the initial degradation temperatures ($T_{5\%}$) of TPU30 and TPU40 fell significantly with respect to those obtained for TPU10 and TPU20. This behavior could be associated to the higher HS content. Moreover, it should be mentioned that the mass loss associated with this first degradation stage could be also correlated with the HS content. In samples with higher HS content, such as TPU30 and TPU40, a peak and a shoulder in their derivative curves were observed, both associated with the first stage of the thermal decomposition of urethane bonds (Figure 4b).

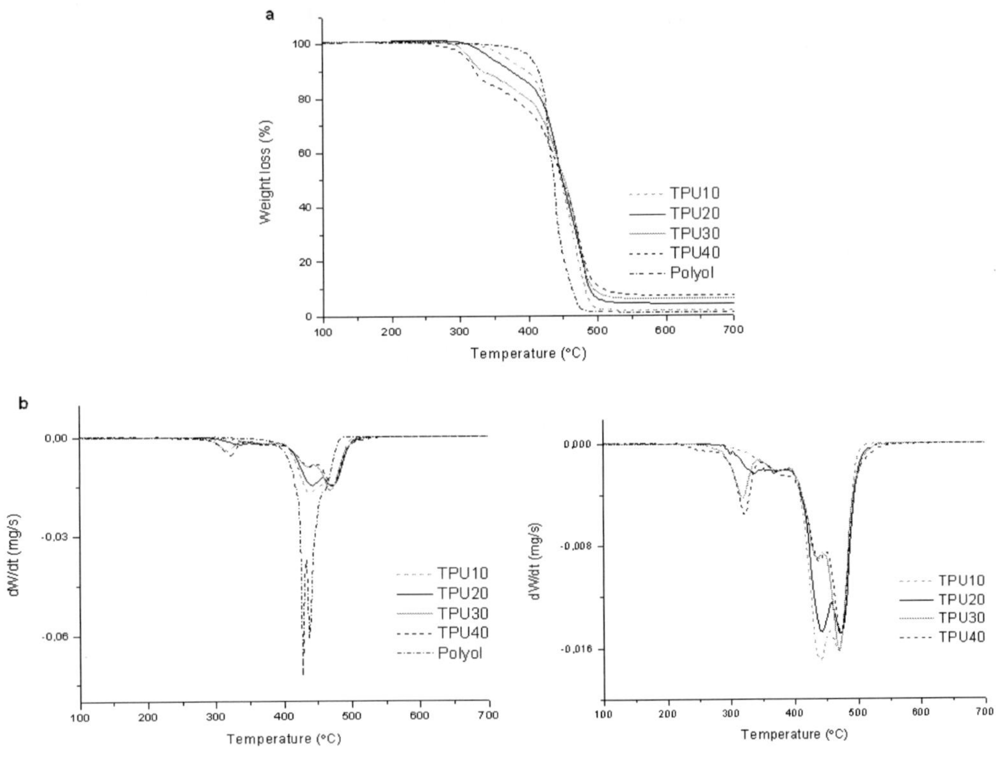

Figure 4. TGA curves for mass percentage (a) and derivative (b) vs. temperature for bio-based polyol and TPUs [41], with permission.

Table 3. Glass transition temperatures (T_g) and hard segment content (%) for TPUs

Sample	T_{gs} SS (°C)	T_{gh} HS (°C)
TPU10	-47.0	---
TPU20	-47.8	122.8
TPU30	-50.0	120.7
TPU40	-51.3	118.1

In a preliminary step the bio-based polyol was studied by conventional DSC but the glass transition was not clearly observed since some melting transitions were superposed in this temperature range, probably associated to the crystalline polymorphism of some oils and fats, such as the case of rapeseed oil. Modulation of DSC data (MDSC) is a powerful tool to separate transitions and to get higher resolutions in particular thermal events. In the case of the polyol, MDSC was used during the cooling cycle and the T_g transition was determined in the reversing phase curve at -61.8 ºC. It was also observed that glass transition temperatures of the SS for all TPUs were higher than that of the polyol (Table 3). T_g values of SS were slightly lower with increasing HS concentrations, as it was reported by Xu et al. [43]. Soft domains have higher mobility when larger HS are present, and this fact could be related with a better microphase separation at high HS content. Besides, the TPUs with lower molar masses and the highest concentrations in HS could contribute to the decrease in T_g values. Higher amount of end chains could result in higher free volume, increasing the mobility of the amorphous phase. Moreover, as it is shown in Figure 5, T_g transition of the MDI-BDO HSs was very difficult to detect due to their stiffness and low mobility [42]. T_g of HS of TPUs slightly increased when their concentration decreased. This result could be attributed to the higher concentration in HS inside the soft domains, as previously explained. Moreover, the intensity of the glass transition of HS is larger at higher HS content but the T_g value is lower, suggesting some interactions between hard and soft phases, that is the low T_g of SS observed for TPU40 which was previously attributed to the microphase separation, could also help to the decrease in T_g of HS in this material due to the higher interaction of soft domains in the hard phase.

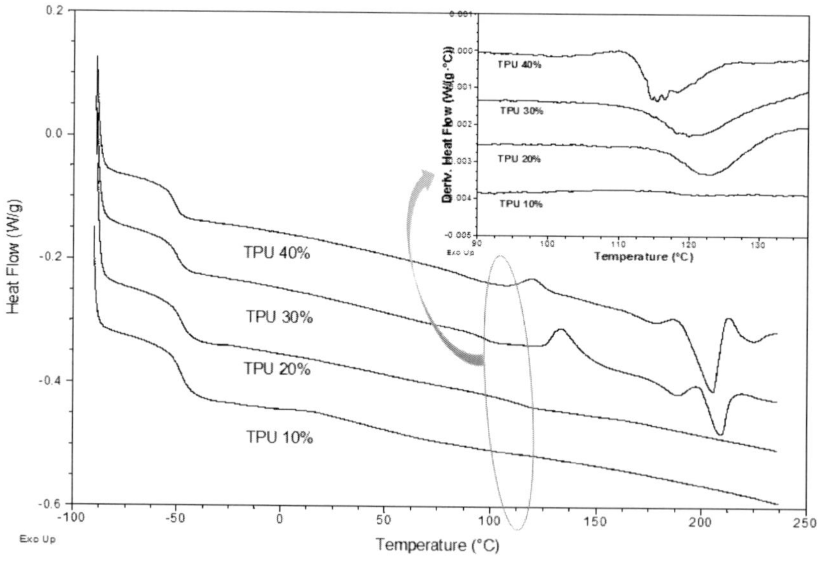

Figure 5. DSC curves from TPU samples and zoom in the zone of T_g of HS [30], with permission.

Uniaxial Mechanical Tests

Tensile properties were determined by uniaxial tensile tests and data are summarized in Table 4 and Figure 6. Results showed that the increase in HS content lead to a brittle material with higher tensile modulus and lower elongation at break (around 25 % for TPU40), as expected, while this material showed lower tensile strength than TPU30 and TPU20. This behavior could be attributed to the fragility provided by the crystalline macro-structures with sizes between 20 and 30 μm that were observed in a previous morphological study [30]. The presence of these macro-structures in TPU40 increased the segregation phase size and may promote points of stress concentration, due to their boundary impingement, which could influence properties such as tensile strength and elongation at break [43]. As expected, TPU40 also exhibited the higher modulus (11.1 MPa), leading to the conclusion that the HS higher crystallinity has a significant effect on the mechanical properties of these TPUs.

Table 4. Uniaxial tensile properties of TPUs [41]

Sample	Tensile strength (MPa)	Elongation at break (%)	Young modulus (MPa)
TPU10	1.3 ± 0.1	> 600	0.7 ± 0.0
TPU20	3.7 ± 0.1	> 600	2.5 ± 0.1
TPU30	5.6 ± 0.3	430 ± 40	8.6 ± 0.5
TPU40	1.5 ± 0.4	25 ± 7	11.1 ± 1.2

With permission.

Figure 6 shows the elastomeric behavior of those TPUs with lower HS content, as demonstrated by their lower moduli. It was also observed in Figure 6 that TPU20 was the only material with a yield point in the range of deformation 250-300 %. This particular behavior, presented as a rubbery region in Figure 6, could be related to a good resistance to the permanent deformation.

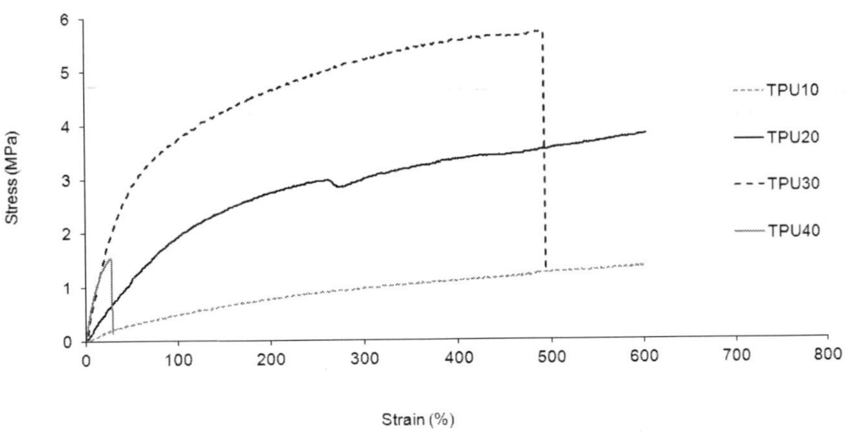

Figure 6. Stress-strain curves of TPUs [41], with permission.

Conclusion

Bio-based TPUs were synthesized from a di-functional dimmer fatty acid-based polyol obtained from rapeseed oil, MDI and BDO at four HS contents, i.e. 10-40 wt%. The polyol characteristics determined the structure and properties of TPUs. FTIR-ATR spectra confirmed that all the isocyanate groups reacted with hydroxyl groups (from polyol or BDO) during the TPUs synthesis. Thermal studies carried out by TGA, DSC and MDSC revealed some interactions between hard and soft domains for all TPUs and a degradation behavior closely linked to their HS concentration. Stress-strain uniaxial tests showed that the increase in HS content in TPUs lead to higher tensile modulus and lower elongation at break. TPU10 and TPU20 showed a strong elastomeric behavior with very high elongation at break (>600 %) and very low elastic modulus.

In summary, TPUs partially synthesized from vegetable oils are very promising materials in good agreement with the current tendency for sustainable development, making them very attractive since they are expected to show specific properties which can be easily tailored by selecting the appropriate HS concentration. These materials could also fulfil many industrial requirements for different fields, such as construction, automotive, textile, adhesive and coatings.

ACKNOWLEDGMENTS

Authors thank the Spanish Ministry of Economy and Competitiveness (MAT2011-28648-C02-01) for financial support.

REFERENCES

[1] G. Lligadas, J. C. Ronda, M. Galià, V. Cádiz. Plant Oils as Platform Chemicals for Polyurethane Synthesis: Current-State-of-the-Art. 2010, *Biomacromolecules,* 11:2825–2835

[2] U. Biermann, W. Friedt, S. Lang, W. Luhs, G. Machmuller, J. O. Metzger, M. R. G Klaas, H. J. Schafer, M. P. Scheider. New Syntheses with Oils and Fats as Renewable Raw Materials for the Chemical Industry. 2000, *Angewandte Chemie International Edition*, 39:2206–2224.

[3] L. Montero de Espinosa, M.A.R. Meier. Plant oils: the perfect renewable resource for polymer science?! 2011, *European Polymer Journal,* 47:837–852

[4] A. Behr, J. P. Gomes. The refinement of renewable resources: new important derivatives of fatty acids and glycerol. 2010, *European Journal of Lipid Science and Technology*, 112(1):31–50.

[5] A. Kockritz, S. N. M. Khot, J. J. Lascala, E. Can, S. S. Morye, G. I. Williams, G. R. Palmese, S. H. Kusefoglu, R. P. Wool. Development and application of triglyceride-based polymers and composites. 2001, *Journal of Applied Polymer Science,* 82(3):703-723.

[6] M.T. Benaniba, N. Belhaneche-Bensemra, G. Gelbard. Stabilizing effect of epoxidized sunflower oil on the thermal degradation of poly(vinyl chloride). 2001, *Polymer Degradation and Stability*, 74:501–505.

[7] M.O. Boussoum, D. Atek, N. Belhaneche-Bensemra. Interactions between poly(vinyl chloride) stabilised with epoxidised sunflower oil and food simulants. 2006, *Polymer Degradation and Stability*, 91:579-584.

[8] J. S. Choi, W. H. Park. Thermal and mechanical properties of Poly(3-hydroxybutyrate-co-3-hydroxyvalerate) plasticized by biodegradable soybean oils. 2003, *Macromolecular Symposium*, 197:65-76.

[9] M.A. Semsarzadeh, M. Mehrabzadeh, S.S. Arabshahi. Mechanical and Thermal Properties of the Plasticized PVC-ESBO. 2005, *Iranian Polymer Journal*, 14(9):769-773.

[10] F. Ali, Y. W. Chang, S. C. Kang, J. Y. Yoon. Thermal, mechanical and rheological properties of poly(lactic acid)/epoxidized soybean oil blends. 2009, *Polymer Bulletin*, 62(1):91-98.

[11] P. Karmalm, T. Hjertberg, A. Jansson, R. Dahl. Thermal stability of poly(vinyl chloride) with epoxidised soybean oil as primary plasticizer. 2009, *Polymer Degradation and Stability*, 94:2275–2281.

[12] M. A. R. Meier, J. O. Metzger, U. S. Schubert. Plant oil renewable resources as green alternatives in polymer science. 2007, *Chemical Society Re*views, 36:1788–1802.

[13] F. S. Güner, Y. Yagci, A. T. Erciyes. Polymers from trygliceride oils. 2006, *Progress in Polymer Science*, 31:633–670.

[14] Y. Lu, R. C. Larock. Novel polymeric materials from vegetable oils and vinyl monomers: Preparation, properties, and applications. 2009, *ChemSusChem,* 2(2):136-147.

[15] S. N. Khot, J. J. Lascala, E. Can, S. S. Morye, G. I. Williams, G. R. Palmese, S. H. Kusefoglu, R. P. Wool. Development and application of triglyceride-based polymers and composites. 2001, *Journal of Applied Polymer Science*, 82(3):703-723.

[16] E. Hablot, D. Zheng, M. Bouquey, L. Averous. Polyurethanes Based on Castor Oil: Kinetics, Chemical, Mechanical and Thermal Properties. 2008, *Macromolecular Materials and Engineering*, 293(11):922-929.

[17] Z. Liu, S. Z. Erhan. "Green" composites and nanocomposites from soybean oil. 2008, *Materials Science and Engineering A*, 483–484:708–711.

[18] K. M. Doll, S. Z. Erhan. The improved synthesis of carbonated soybean oil using supercritical carbon dioxide at a reduced reaction time. 2005, *Green Chemistry*, 7:849–854.

[19] C. Meiorin, M. I. Aranguren, M. A. Mosiewicki. Smart and structural thermosets from the cationic copolymerization of a vegetable oil. 2011, Journal of Applied Polymer Science, 124(6):5071-5078.

[20] X. Kong, T. S. Omonov, J. M.Curtis. The development of canola oil based bio-resins. 2012, *Lipid Technology*, 24(1):7-10.

[21] J. R. Kim, S. Sharma. The development and comparison of bio-thermoset plastics from epoxidized plant oils. 2012, *Industrial Crops and Products*, 36(1):485-499.

[22] R. L. Quirino, J. Woodford, R. C. Larock. Soybean and linseed oil-based composites reinforced with wood flour and wood fibers. 2012, *Journal of Applied Polymer Science* 124(2):1520-1528.

[23] R. B. Seymour, C. E. Carraher. *Polymer Chemistry - An introduction. 2nd Edition.* Marcel Dekker, Basel, Suiza, 1988.

[24] M. Ionescu. Chemistry and Technology of Polyols for Polyurethanes. Rapra Technology Limited, Shropshire, United Kingdom, 2005.

[25] U. Meier-Westhues, *Polyurethanes. Coatings, adhesives and sealants.* Vincentz Network, Hannover, Alemania, 2007.

[26] T. Tawa, S. Ito. The role of hard segments of aqueous polyurethane-urea dispersion in determining the colloidal characteristics and physical properties. 2006, *Polymer Journal*, 38(7):686-693.

[27] R. P. Wool, X. S. Sun. *Bio-Based Polymers and Composites*. Elsevier Academic Press, Burlington, MA, United States, 2005.

[28] S. Yamasaki, D. Nishiguchi, K. Kojio, M. Furukawa. Effects of Polymerization Method on Structure and Properties of Thermoplastic Polyurethanes. 2007, *Journal of Polymer Science B: Polymer Physics*, 45:800-814.

[29] Z. S. Petrovic, M. J. Cevallos, I. Javni, D. W. Schaefer, R. Justice. Soy-oil-based segmented polyurethanes. 2005, *Journal of Polymer Science B*, 43:3178-3190.

[30] C. Bueno-Ferrer, E. Hablot, F. Perrin-Sarazin, M.C. Garrigós, A. Jiménez, L. Avérous. Structure and morphology of new bio-based thermoplastic polyurethanes obtained from dimer fatty acids. 2012, *Macromolecular Materials and Engineering*, 297(8):777-784.

[31] L. Irusta, M. J. Fernandez-Berridi. Aromatic poly(ester–urethanes): effect of the polyol molecular weight on the photochemical behaviour. 2000, *Polymer*, 41:3297-3302.

[32] L. Irusta, J. J. Iruin, G. Mendikute, M. J. Fernández-Berridi. Infrared spectroscopy studies of the self-association of aromatic urethanes. 2005, *Vibrational Spectroscopy*, 39:144-150.

[33] B. B. R. Silva, R. M. C. Santana, M. M. C. Forte. A solventless castor oil. 2010, *International Journal of Adhesion and Adhesives*, 30:559-565.

[34] I. Javni, Z. Petrovic, A. Guo, R. Fuller. Thermal stability of polyurethanes based on vegetable oils. 2000, *Journal of Applied Polymer Science*, 77:1723-1734.

[35] P. Król. Synthesis methods, chemical structures and phase structures of linear polyurethanes. Properties and applications of linear polyurethanes in polyurethane elastomers, copolymers and ionomers. 2007, *Progress in Materials Science*, 52:915-1015.

[36] M. A. Corcuera, L. Rueda, B. Fernandez d'Arlas, A. Arbelaiz, C. Marieta, I. Mondragon, A. Eceiza. Microstructure and properties of polyurethanes derived from castor oil. 2010, *Polymer Degradation and Stability*, 95:2175-2184.

[37] H.P. Bhunia, G.B. Nando, T.K. Chaki, A. Basak, S. Lenka, P.L. Nayak. Synthesis and characterization of polymers from cashewnut shell liquid (CNSL), a renewable resource II. Synthesis of polyurethanes. 1999, *European Polymer Journal*, 35:1381-131.

[38] H. Yeganeh, M. R. Mehdizadeh. Synthesis and properties of isocyanate curable millable polyurethane elastomers based on castor oil as a renewable resource polyol. 2004, *European Polymer Journal*, 40:1233-1238.

[39] E. Hablot, D. Zheng, M. Bouquey, L. Avérous. Polyurethanes based on castor oil: kinetics, chemical, mechanical and thermal properties. 2008, *Macromolecular Materials and Engineering*, 293:922-929.

[40] L. Hojabri, X. Kong, S. S. Narine. Fatty acid-derived diisocyanate and biobased polyurethane produced from vegetable oil: synthesis, polymerization and characterization. 2009, *Biomacromolecules*, 10:884-891.

[41] C. Bueno-Ferrer, E. Hablot, M.C. Garrigós, S. Bocchini, L. Avérous, A. Jiménez. Relationship between morphology, properties and degradation parameters of novative biobased thermoplastic polyurethanes obtained from dimer fatty acids. 2012, *Polymer Degradation and Stability*, 97(10):1964-1969.

[42] K. Bagdi, K. Molnar, B. Pukanszky Jr, B. Pukanszky. Thermal analysis of the structure of segmented polyurethane elastomers.2009, *Journal of Thermal Analysis and Calorimetry*, 98, 825-832.

[43] Xu Y, Petrovic Z, Das S, Wilkes GL. Morphology and properties of thermoplastic polyurethanes with dangling chains in ricinoleate-based soft segments. 2008, *Polymer*, 49:4248-4258.

In: Quantitative Chemistry, Biochemistry and Biology
Editors: G.E. Zaikov, O.V. Stoyanov et al.
ISBN: 978-1-62948-332-0

Chapter 3

CARVACROL AND THYMOL FOR FRESH FOOD PACKAGING

Marina Ramos*[*], *Ana Beltrán, Arancha Valdés, Mercedes A. Peltzer, Alfonso Jiménez and María C. Garrigós

Analytical Chemistry, Nutrition & Food Sciences Department, University of Alicante, Alicante, Spain

ABSTRACT

Active packaging is becoming an emerging food technology to improve quality and safety of food products. One of the most common approaches is based on the release of antioxidant/antimicrobial compounds from the packaging material. In this work an antifungal active packaging system based on the release of carvacrol and thymol was optimized to increase the post-harvest shelf life of fresh strawberries and bread during storage. Thermal properties of the developed packaging material were determined by differential scanning calorimetry (DSC) and thermogravimetric analysis (TGA). Volatile compounds in food samples contained in active packaging systems were monitored by using headspace solid phase microextraction followed by gas chromatography analysis (HS-SPME-GC-MS) at controlled conditions. The obtained results provided evidences that exposure to carvacrol and thymol is an effective way to enlarge the quality of strawberries and bread samples during distribution and sale.

Keywords: Active packaging, carvacrol, thymol, strawberries, bread, thermal properties, volatile profile

1. INTRODUCTION

Active packaging is the one of most relevant approaches to increase the protection and shelf-life of fresh food (Dong Sun Lee, 2008; Singh et al., 2011). Active packaging is broad

* E-mail: marina.ramos@ua.es, Phone: +34965903400 Ext. 1187, Fax: +34965903527.

concept that can be defined as a system in which the product, the packaging material and the environment interact in a positive way to extend food shelf-life (Cutter, 2002; Mastromatteo et al., 2010b). Many different active agents can be incorporated into the packaging material to improve its functionality. The migration of the active compound may be achieved by direct contact between food and the packaging material or through gas phase diffusion from the inner packaging layer to the food surface (Coma, 2008; Conte et al., 2006; Gemili et al., 2009; Mastromatteo et al., 2010b).

Food can be subjected to microbial contamination that is mainly caused by bacteria, yeasts and fungi. Many of these microorganisms can cause undesirable reactions and can deteriorate organoleptic and nutritional properties of foods (Kuorwel et al., 2011b). Different antimicrobial (AM) agents (most of them synthetic) are commonly incorporated directly into the food to diminish food spoilage by microorganisms. But this strategy has several disadvantages such as: the rising consumer refusal for food with synthetic additives; the useless addition of these agents to the food bulk when spoilage occurs primarily on the surface and the undesirable modification of food flavor. Therefore, antimicrobial packaging is a promising method to protect food from microbial contamination with no need of using synthetic additives in food composition (Kuorwel et al., 2013).

Natural AM agents (AMs) have attracted much attention from food and packaging industries by their potential action in food preservation. According to Davidson and Zivanovic, natural AM agents can be classified by their sources in vegetal [herbs and essential oils (EOs)], animals (lysozyme, lactoferrin), microbials (nisin, natamycin) or antimicrobial polymers (chitosan) (Davidson PM, 2003). In these sense, many studies have focused on the AMs present in EOs extracted from plants (basil, thyme, oregano, cinnamon, clove, rosemary) consisting on complex mixtures of different compounds including terpenoids, esters, aldehydes, ketones, acids and alcohols (Dorman and Deans, 2000). These plant EOs are volatile liquids characterized by a strong odor (Bakkali et al., 2008). Extracts derived from herbs and EOs contain many natural compounds such as thymol, linalool and carvacrol with a broad AM range against different pathogenic and spoilage microorganisms including Gram-negative (López et al., 2007b; Suppakul et al., 2011) and Gram-positive species (Gutiérrez et al., 2009; López et al., 2007a); as well as against yeast (Kuorwel et al., 2011a) and molds (Rodriguez-Lafuente et al., 2010). In general, these additives are considered to be safe and they have been classified as GRAS (Generally Recognized As Safe) by the American Food and Drug Administration (Kuorwel et al., 2011b).

Rodriguez et al. studied the addition of EOs to a wax coating in order to develop an antimicrobial active packaging and they assessed their ability to preserve strawberries from microorganism contamination by the release of AMs from the coating (Rodríguez et al., 2007). During this study, there was no direct contact between the EOs and the food product. Therefore, the chemicals responsible for the inhibition of the pathogen growth must have been the natural volatile compounds (eugenol, carvacrol, thymol) present in the headspace packaging. In other study, carvacrol was added to chitosan-based films for active packaging and its antimicrobial efficiency against food spoilage microorganisms was demonstrated by using a headspace chromatographic technique (Kurek et al., 2013). Gutierrez et al used a cinnamon-based active material to increase more than three times the shelf-life of a complex bakery product with minimal changes in the packaging and no additional manipulation steps (Gutiérrez et al., 2009).

In addition to microbial contamination, there are changes in the food macroscopic properties that also induced biochemical reactions and chemical alterations in tissues, such as changes in the volatile profile (Chiralt et al., 2001) and development of undesirable chemicals (i.e. ethanol or acetaldehyde) associated with changes in the respiratory paths (Tovar et al., 2001). Flavor is one of the main factors influencing consumer's food choice (Pozo-Bayón et al., 2006). Volatile compounds are important contributors to flavor and odor of fruits, being aroma a very important quality attribute of strawberries. The flavor of strawberries is comprised of a complex mixture of esters, aldehydes, alcohols, furans and sulfur compounds. Esters are the main headspace volatiles. Methyl esters amount increases with the plant maturation, while ethyl esters amount do not change significantly during the fruit growth (Rizzolo et al., 2007).

In the case of bread, aroma has been largely studied and many methods have been developed to identify the compounds responsible of flavor (Cayot, 2007). More than 540 different compounds have been described in the complex volatile fraction of bread (Ruiz et al., 2003), being alcohols, aldehydes, esters, ketones, acids, pyrazines and pyrrolines the most important quantitatively, but furans, hydrocarbons and lactones were also identified (Poinot et al., 2007).

Solid Phase Microextraction (SPME) has become one of the preferred techniques in aroma analysis, offering solvent free, rapid sampling with low cost and easy preparation. Also, it is sensitive, selective and compatible with low detection limits (Ho et al., 2006). Placed in the sample headspace, SPME is a non-destructive and non-invasive method to evaluate volatile and semi-volatile compounds. In this sense, the extraction of volatile compounds released from a great number of foods has been carried out by using HS-SPME technique (Poinot et al., 2007; Quílez et al., 2006).

Besides the increasing concern in recent years about the use of synthetic polymers due to their poor biodegradability and high permanence in the environment after their use, these materials are still highly competitive by their many advantages including low cost, good processability and excellent mechanical and physical properties. Therefore, the development of antimicrobial packaging materials manufactured from synthetic polymers, such as low-density polyethylene (LDPE), high-density polyethylene (HDPE), polystyrene (PS), polyethylene terephthalate (PET) and polypropylene (PP) is important by their commercial benefits for food packaging (Kuorwel et al., 2011b).

This study focuses on the optimization of antifungal active systems based on the controlled release of carvacrol and thymol from polypropylene (PP) films. The effectiveness of the developed active films was evaluated by studying the headspace volatile composition of two food samples (bread and strawberries) stored at different conditions. This study was carried out by headspace Solid Phase Micro-Extraction (SPME) followed by gas chromatography analysis (HS-SPME-GC-MS). Results were correlated with the antimicrobial activity by visual observation of the fungal growth in the studied food (bread and strawberries). In addition, the effect of the studied additives on the thermal properties of the developed active films was also carried out.

2. EXPERIMENTAL

2.1. Materials

The polymer used in this work was polypropylene (PP) ECOLEN HZ10K (Hellenic Petroleum, Greece), kindly supplied in pellets by Ashland Chemical Hispania (Barcelona, Spain). Melt flow index (MFI) was 3.2 g 10 min^{-1} determined according to ASTM-D1238 standard (230 °C, 2.16 Kg), and density 0.9 g cm^{-3}. Carvacrol (98 %) and thymol (99.5 %) were obtained from Sigma-Aldrich (Madrid, Spain).

Strawberries and sliced bread were purchased from a local market. Damaged, non-uniform, unripe or overripe strawberries were removed and the selected fruits were stored for at least 2 h at 3 °C to ensure their thermal equilibrium. Strawberries were selected for this study due to their rapid post-harvest deterioration, which constitutes a problem on their commercial distribution. Sliced bread was selected due to the increasing consumer demand for fresh bread with long shelf-life.

2.2. Preparation of Active Films

Active films were prepared by melt-blending followed by compression molding by using a method previously reported (Ramos et al., 2012). A Haake Polylab QC mixer (ThermoFischer Scientific, Walham, USA) at 190 °C for 6 min at rotation speed of 50 rpm was used. Both additives were introduced in the mixer once the polymer was already in the melt state to avoid unnecessary losses and to ensure their presence in the final materials. The active antimicrobial films were obtained at 190 °C in a hot press (Carver Inc, Model 3850, USA) for 12 minutes. The average thickness of the films was around 200 μm measured with a Digimatic Micrometer Series 293 MDC-Lite (Mitutoyo, Japan) at five random positions around the film. The final appearance of the films was completely transparent and homogenous.

Two active formulations were prepared: PP containing 8 wt% of thymol (PPT8) and PP with 8 wt% of carvacrol (PPC8). An additional sample without any active compound was also prepared and used as control (PP0).

2.3. Thermal Characterization of Active Films

2.3.1. Thermogravimetric Analysis (TGA)

TGA tests were performed in a TGA/SDTA 851 Mettler Toledo thermal analyzer (Schwarzenbach, Switzerland). Approximately 5 mg samples were weighed in alumina pans (70 μL) and were heated from 30 °C to 700 °C at a heating rate of 10 °C min^{-1} under inert nitrogen atmosphere (flow rate 50 mL min^{-1}).

2.3.2. Differential Scanning Calorimetry (DSC)

2.3.2.1. Determination of Thermal Parameters in Inert Atmosphere

DSC tests were conducted in a TA DSC Q-2000 instrument (New Castle, DE, USA) under inert nitrogen atmosphere. 3 mg samples were introduced in aluminium pans (40μL) and were submitted to the following thermal program: heating from 0 °C to 180 °C at 10 °C min^{-1} (3 min hold), cooling at 10 °C min^{-1} to 0 °C (3 min hold) and heating to 180 °C at 10 °C min^{-1}. The percentage of crystallinity (χ %) for each material was calculated according to the following equation,

$$\chi\ (\%) = \left(\Delta H_m / W \Delta H_m^o\right) \cdot 100 \quad (1)$$

where ΔH_m (J g^{-1}) is the latent heat of fusion of the sample, W is the PP weight fraction in the sample, and ΔH_m^o is the theoretical latent heat of fusion for 100 % crystalline PP, 138 J g^{-1} (Joseph et al., 2003).

2.3.2.2. Evaluation of Oxidation Induction Time

The antioxidant performance of carvacrol and thymol in the developed active films was also studied by DSC by determining their oxidation induction time, OIT (min) (Pospisil et al, 2003; Archodoulaki et al, 2006). The OIT value is defined as the time to the onset of an exothermic oxidation peak in oxidative atmosphere and it was determined by using oxygen and air, as the results obtained can be dependent on the type of atmosphere used for the analysis.

OIT tests were carried out by heating samples at 100 °C min^{-1} under nitrogen (flow rate 50 mL min^{-1}) to the set temperature (200 °C) according to ASTM-D3895-07 Standard. After 5 min, the atmosphere was switched to pure oxygen or air (50 mL min^{-1}). The heat flow was then recorded in isothermal conditions up to the detection of the exothermic peak indicating the beginning of the oxidation reaction. All tests were performed in triplicate for each formulation.

2.4. Study of the Effectiveness of the Active Films to Preserve Perishable Food

2.4.1. Observation of Fungal Growth

The effectiveness of the developed active films was evaluated by putting them in contact with sliced bread and strawberries and further observing the occurrence of fungal growth on food samples with time. For this purpose, food samples were appropriately cut to be placed on the base of disposable polypropylene Petri dishes (inside dimensions: 88 mm diameter x 12 mm high). An additional test was carried out with uncut strawberries which were placed into a polyethylene suitable food container (250 mL, 4 cm high x 13 cm opening diameter) as shown in Figure 1.

Active films were cut with the appropriate dimensions to match the top of the lid of the used containers in order to release the antimicrobial studied agents (carvacrol and thymol) into the packaging headspace. The final containers were then sealed with "Parafilm" in order to avoid losses of volatile compounds and were incubated at 25 ºC and 50% RH in a CM 0/81 climatic chamber (Dycometal, Barcelona, Spain) during 15 days. Food samples stored with the control film (without active compounds) were also studied for comparison.

2.4.2. Headspace Analysis by HS-SPME-GC-MS

Food samples (whole strawberries and sliced bread) were placed into the polyethylene container described in the previous section in direct contact with the PP films and samples were extracted at selected times to determine the headspace composition. Containers were sealed and a PTFE/silicone septum was placed on their top part to allow the insertion of the SPME fibre for volatiles extraction (Figure 1). Samples were then stored in a climatic chamber and tested at different temperatures and days of storage according to the conditions shown in Table 1. In this sense, 25ºC and 4ºC were selected in order to simulate ambient and refrigerated storage conditions, respectively. Three replicates were performed for each food sample and day of study.

HS-SPME analysis of volatile compounds for food samples was performed according to a previously reported method applied to bread samples (Poinot et al., 2008) with slight modifications. Similar conditions to those proposed by G. Blanda et al. Were used in the study with strawberries (Blanda et al., 2009). The SPME fibre used was divinylbenzene/carboxen/polydimethylsiloxane (DVB/CAR/PDMS) 50/30 μm, StableFlex, 1 cm long mounted to an SPME manual holder assembly from Supelco (Bellefonte, PA) (Figure 1). Prior to use, the fibre was conditioned by following the manufacturer's recommendations. The needle of the SPME device was inserted into the container through the septum and the fibre was exposed to the food sample headspace for 30 min at room temperature. The fibre was then retracted into the needle assembly, removed from the container, transferred to the injection port of the GC unit and immediately desorbed.

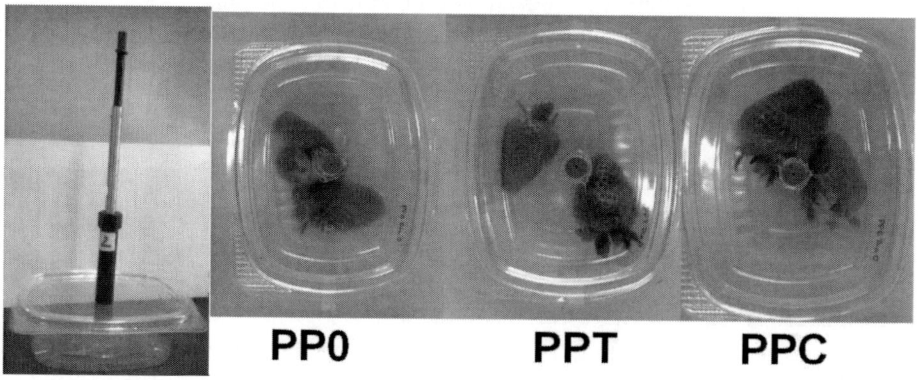

Figure 1. Experimental assembly used for headspace analysis of whole strawberries by HS-SPME.

Table 1. Storage and testing conditions in the headspace study of food by HS-SPME-GC-MS

Food sample	Temperature (ºC)	Days of study					
Slice bread	25	0	2	5	10	15	-
Whole strawberries	25	0	2	4	7	10	-
Whole strawberries	4	0	2	4	7	10	15

An analysis of volatiles produced in the headspace of bread and strawberries packed samples was performed by using a Perkin Elmer TurboMass Gold GC-MS (Boston, MA, USA) equipped with a split/splitless injector and a quadrupole mass spectrometer operating in electronic impact (EI) ionisation mode (70 eV). A SPB-5 capillary column (30 m x 0.25 mm x 0.25 μm; Supelco, Bellefonte, PA, USA) was used. The column temperature was programmed from 40 ºC (hold 10 min) to 120 ºC (hold 1 min) at 5 ºC min^{-1}, to 140 ºC at 2 ºC min^{-1} (hold 0 min) and to 230 ºC at 5 ºC min^{-1} (hold 8 min). Helium was used as carrier gas at a flow rate of 1 mL min^{-1}. Ion source and GC/MS transfer line temperatures were 250 and 270 ºC, respectively. Injector temperature was 270 ºC and time for fibre desorption was fixed at 5 min in the splitless mode (1.5 min splitless-period). After every run, the SPME fibre was conditioned for 30 min at 270 ºC in the injector of the gas chromatograph followed by a blank analysis to avoid carryover of the fibre.

Identification of volatile compounds in strawberries and sliced bread headspace was performed in full scan mode (m/z 30-550). Carvacrol and thymol were identified by a combination of the NIST mass spectral library and gas chromatographic retention times of standard compounds. The rest of volatiles were tentatively identified by their GC/MS spectra. In this sense, the compounds having ≤ 90% similarity with spectra in the NIST library were not taken into consideration. Chromatographic responses of detected volatile compounds (peak area counts) were monitored for comparative measurements of each compound in the studied samples.

3. Results and Discussion

3.1. Thermogravimetric Analysis (TGA)

The effect of carvacrol and thymol on the thermal stability of PP films was studied by TGA under nitrogen atmosphere. The TGA patterns obtained for PP films showed a first degradation step at low temperatures (about 115 ºC) observed only for active films, which was associated to the thermal decomposition of carvacrol and/or thymol and a second step corresponding to the thermal degradation of the polymer matrix (Ramos et al., 2012). The first degradation step observed for active films was considered as an indirect confirmation of the presence of the active compounds (thymol and carvacrol) in the polymer matrix after processing and consequently their ability to act as active agents in these materials as it has been reported by other authors (Persico et al., 2009).

Table 2 summarizes the temperatures obtained for the main degradation step (T_{max}), ascribed to the PP thermal degradation. No significant differences were observed for T_{max} values in all samples. These results showed that the addition of carvacrol and thymol to the polymer matrix did not significantly affect its thermal degradation profile in inert nitrogen atmosphere. However, it would be expectable that a certain amount of carvacrol and thymol would be lost during processing, since materials are submitted to temperatures above the decomposition point of these additives. Therefore, the processing parameters, in particular temperature and time, should be optimized to avoid the excessive evaporation and consequent loss of these additives incorporated to PP (Dobkowski, 2006).

3.2. Differential Scanning Calorimetry (DSC)

3.2.1. Determination of Thermal Parameters in Inert Atmosphere

Four parameters were determined for the thermal characterization of these materials by DSC (Table 2): cold-crystallization temperature, T_{cc} (°C); melting temperature, T_m (°C); crystallization enthalpy, ΔH_c (J g^{-1}) and melting enthalpy, ΔH_m (J g^{-1}). As can be seen in Table 2, melting and cold-crystallization temperatures as well as crystallization enthalpy did not show important differences for all the studied materials. Nevertheless, it should be highlighted that the melting enthalpy of PP0 sample was clearly higher than those obtained for the active materials. In this sense, crystallinity, χ (%), of samples was calculated according to equation (1) to evaluate if the addition of thymol and carvacrol could alter the crystallization behavior of PP. A higher value for χ(%) was determined for the PP0 sample. Therefore, it could be concluded that the PP crystallinity decreases significantly with the addition of thymol and carvacrol. This decrease in crystallinity could be due to the interactions between the polymer matrix and the additive molecules in the PP macromolecular network. A similar effect was reported for PP with the addition of some commercial synthetic antioxidants, such as Irgafos 168 and Irganox 1010 (Alin and Hakkarainen, 2010).

Table 2. TGA, DSC and OIT parameters obtained for all samples

Sample	T_{max} (°C)	T_c (°C)	T_m (°C)	ΔH_c (J g^{-1})	ΔH_m (J g^{-1})	χ (%)	OIT (min) Oxygen[a]	OIT (min) Air[a]
PP0	461	119	161	95.2	99.1	72	0.9 ± 0.3	1.3 ± 0.4
PPC8	462	118	161	89.2	48.1	38	8.5 ± 1.0	20.7 ± 2.8
PPT8	462	115	159	88.9	49.5	39	15.4 ± 1.7	38.8 ± 0.6

[a] mean ± SD (n = 3).

3.2.2. Evaluation of the Oxidation Induction Time (OIT)

The determination of OIT is considered a simple, reliable and fast method for the evaluation of the antioxidants efficiency (Pomerantsev and Rodionova, 2005), corresponding to relative measurements of the materials stability against oxidation. The evaluation of the antioxidant performance of carvacrol and thymol in PP is important since they are supposed not only to play the role of active additives for food, but also to protect the polymer to oxidative degradation during processing and use.

The evaluation of OIT was carried out in two different atmospheres. Air was selected to get a similar situation to the real conditions during materials processing or food shelf-life, while the use of pure oxygen would represent the most aggressive conditions for oxidative degradation. Table 2 shows the results obtained for OIT in both atmospheres.

In both cases it was confirmed the higher efficiency of thymol as an antioxidant when compared to carvacrol. This behavior was also reported by other authors who demonstrated that the antioxidant efficiency of thymol was higher in sunflower oil samples (Yanishlieva et al., 1999). In the case of air atmosphere, as expected, OIT values were higher than those obtained in pure oxygen atmosphere, since the experiment under air is less aggressive to materials (Riga et al., 1998).

In all cases, the increase in OIT values for PP with additives showed the existence of certain antioxidant effect after processing. These results are an additional confirmation that certain amounts of thymol and carvacrol are still remaining in all formulations after processing and they would be able to be released from the material to foodstuff as active additives.

3.3. Study of the Effectiveness of the Active Films to Preserve Perishable Food

3.3.1. Observation Of Fungal Growth

This study was conducted to evaluate the antimicrobial activity of the developed films and their ability to act in active packaging formulations to increase the fresh food shelf-life. It was based on the visual observation of the inhibition of the fungal growth on food samples by the action of the volatile active additives, carvacrol and thymol. In this sense, some studies by other authors showed the effectiveness of these compounds against different fungal strains of particular interest in the food industry (López et al., 2007a).

Figure 2 shows the appearance of sliced strawberries and bread samples at the beginning of the experiment (day 0) and after the observation of microbial growth. Regarding strawberries, satisfactory results were obtained for samples in contact with the films with additives, since no fungal growth was observed until six days of storage. In the case of strawberries in contact with the pure PP film (PP0), a rapid growth of microorganisms was observed at the third day of treatment.

On the other hand, the presence of microorganisms was observed in bread samples in contact with the PP0 film after 13 days of storage, in contrast with bread with the films with additives where no evidence of microbial contamination after 45 days of storage was observed.

However, it was noticed that strawberries lost their organoleptic properties in a few days, even before the visual evidence of fungal growth, when they were cut and stored. For this reason this study was also conducted for uncut strawberries (Figure 3). For the PP0 film microbial growth was observed after 6 days of storage. However, strawberries in contact with the film containing 8 wt% of thymol (PPT8) remained unaltered after 13 days. At this storage time strawberries presented a physical deterioration due to the experimental storage conditions, but it is important to highlight that microbial growth was not observed until the end of the study (15 days).

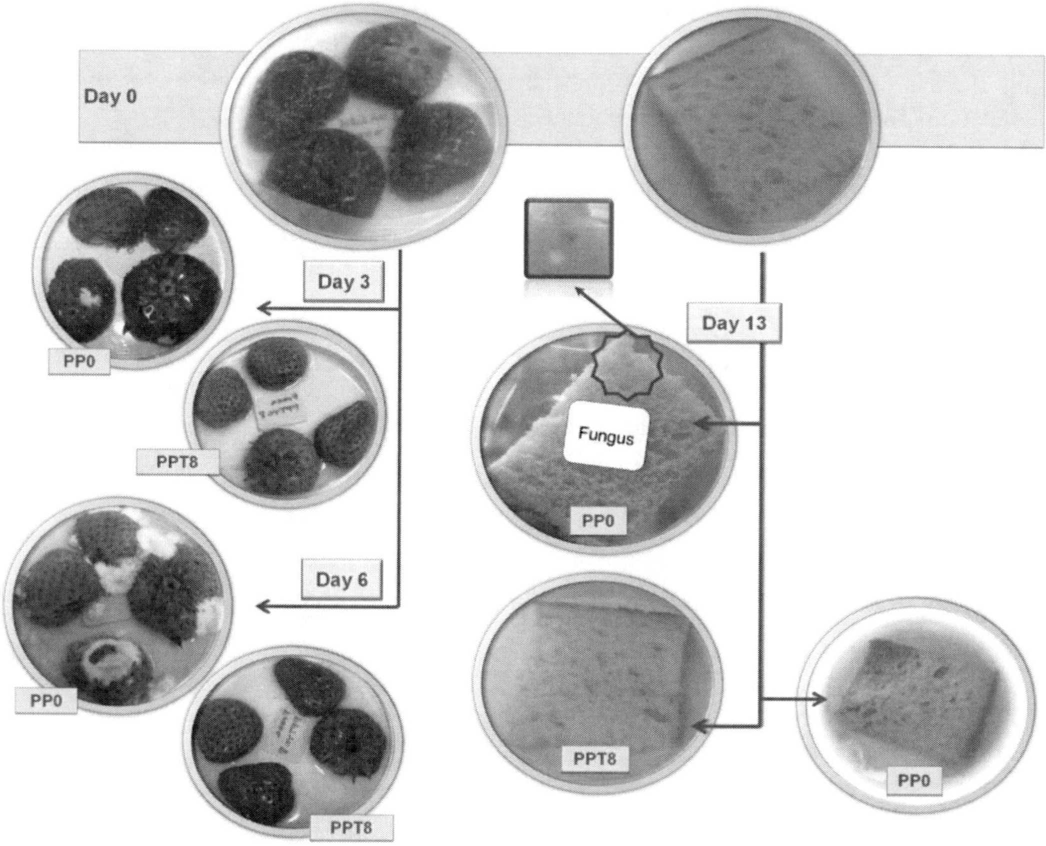

Figure 2. Study of the effectiveness of PPO and active films containing 8 wt% of thymol (PPT8) to preserve cut bread and strawberries by observation of fungal growth.

Similar studies were performed with uncut strawberries by other authors, getting satisfactory results for samples in contact with films treated with essential oils, such as cinnamon, oregano and thyme (Rodríguez et al., 2007). Regarding thyme and oregano essential oils, their antimicrobial activity is due to the high amount of carvacrol and thymol in their composition (Hazzit et al., 2006). Other studies conducted in different fruits and vegetables also demonstrated the effectiveness of the constituents of different essential oils (eugenol, thymol, menthol or eucalyptol) to improve the organoleptic quality of food as well as to reduce the microbial growth, by using also a modified atmosphere (Mastromatteo et al., 2010a).

In conclusion, results obtained from food samples in contact with PP films containing carvacrol and thymol evidenced the effectiveness of these compounds to improve the shelf-life of perishable food, such as strawberries and bread. Accordingly, these results also indicated the potential to use the developed films in active packaging systems to replace the direct addition of preservatives in food formulations.

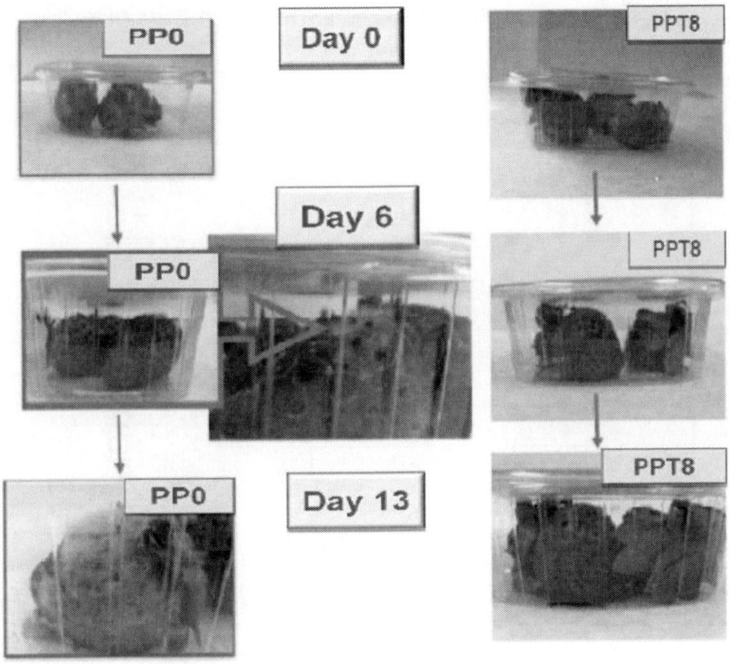

Figure 3. Evaluation of the effectiveness of PPO and active film containing 8 wt% of thymol (PPT8) to preserve uncut strawberries by observation of fungal growth.

3.3.2. Headspace Analysis by HS-SPME-GC-MS

Figure 4 shows the levels of carvacrol, in terms of peak area counts, reached in the headspace of the containers with bread slices after 0, 2, 5, 10 and 15 days of storage at room temperature. As it can be seen, an increase in the amount of carvacrol released from the PP films was observed with time for the bread samples. A high release of carvacrol was observed at 2 days, being released more slowly after 5, 10 and 15 days of storage. This mechanism of controlled release could lead to shelf-life improvement of the stored samples retarding the post-harvest deterioration. This behavior was also observed for strawberries. Regarding the thymol release, a similar trend was shown for both test food samples.

Equilibrium modified atmosphere packaging (EMAP) is the most commonly used packaging technology to reduce the high respiration rate of strawberries. It is known that a suitable atmosphere composition can reduce the respiration rate of fruits and fungal growth with minimal alteration of organoleptic properties (Rizzolo et al, 2007). In this sense, Table 3 shows the compounds identified in the headspace of strawberries with PP0 films after 4 days. One of the most important processes occurring during fruit ripening is the increase in volatiles contributing to fruit aroma and flavor. The major volatiles identified for strawberries stored at room temperature include methyl-isopentanoate, 2-methyl-butylacetate, methyl-hexanoate and hexyl-acetate. The compounds methyl-butanoate and methyl-hexanoate were also found in the headspace composition of refrigerated strawberries at 4 °C. These results are in accordance with those obtained by other authors when studying volatile compounds in the same food samples (Rizzolo et al, 2007; Blanda et al, 2009).

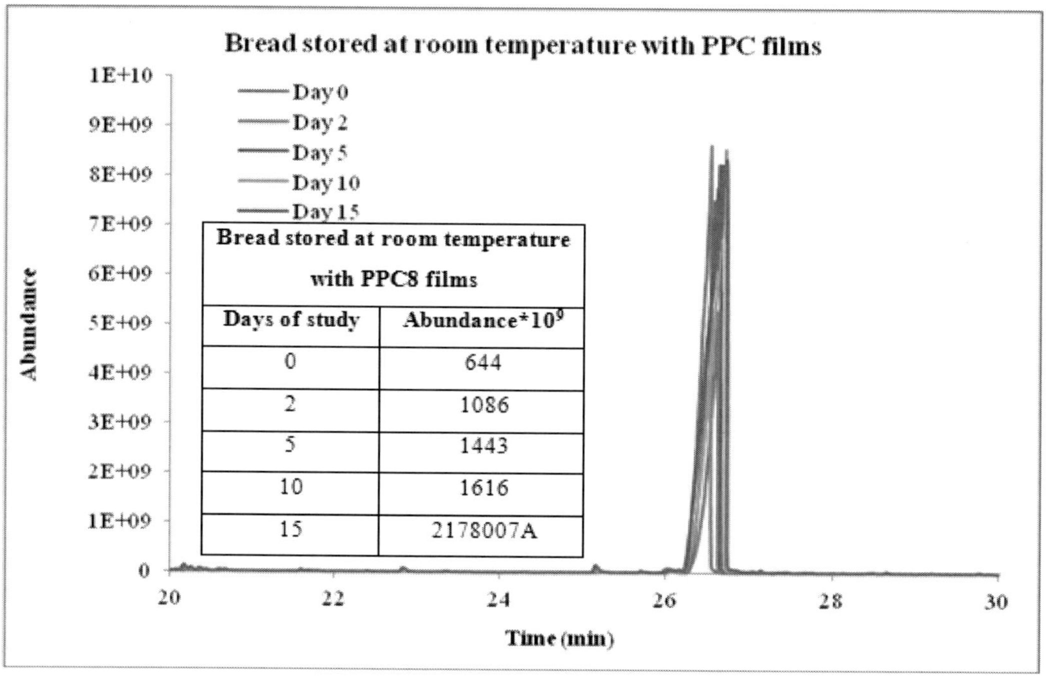

Figure 4. Release of carvacrol in the headspace of bread slices after 0, 2, 5, 10 and 15 days of storage at room temperature.

The addition of thymol and carvacrol to PP films significantly modified the initial atmosphere composition inside packages during the storage of food samples due to their release from the film. This fact could be related to the inhibition of the volatile identified compounds (Table 3) that did not were detected in samples in contact with PPT8 and PPC8 films after 4 days of storage.

Table 3. Identified compounds present in the headspace of food samples packed with PP0 films after 4 days

Bread stored at room temperature		Strawberries stored at room temperature		Strawberries refrigerated at 4 °C	
Time (min)	Compound	Time (min)	Compound	Time (min)	Compound
1.4	Ethanol	3.2	Methyl isopentanoate	1.9	Hexane
5.7	2,4-dimethylheptane	5.8	2-methyl butylacetate	3.1	Methylbutanoate
6.6	2,4-dimethyl heptene	7.4	Methyl hexanoate	3.9	Toluene
7.8	Isononane	10.6	Hexylacetate	6.8	Isopropyl butyrate
13.0	4-methyloctane			12.2	Methyl hexanoate

On the other hand, ethanol was the main volatile found in the headspace of bread in contact with the PP0 film and stored at room temperature for 4 days. This compound was resulting from fermentation and/or lipid oxidation as it has been reported by other authors (Poinot et al, 2007). In this sense, commercial bread samples in contact with PPT8 or PPC8 films after 4 days of storage were characterized by significantly lower amounts of ethanol, suggesting a reduction on the lipid oxidation reactions by the presence of thymol and carvacrol. The improvement on the oxidative stability of bread could be attributed to the release of carvacrol and thymol increasing the shelf-life of bread.

From these results, it can be concluded that the release of both additives from active films to the headspace of the studied packaged foodstuff increased with the storage time, as expected. The volatiles profile obtained by HS-SPME-GC-MS was found to be different for samples in contact with PP0 and those with PPT8 and PPC8, due to the modification of the food headspace composition by the presence of these additives. Therefore, the release of thymol and carvacrol from the active PP films has shown to be effective in maintaining the quality of strawberries and bread during different storage conditions. Finally, it can be concluded that PP films with carvacrol and thymol could be a promising alternative to increase the shelf-life of different foodstuff.

Conclusion

Carvacrol and thymol have demonstrated their potential to be used as active additives in PP films for food packaging applications by their controlled antimicrobial release to foodstuff and also by the possibility to protect food from degradation processes. The addition of carvacrol and thymol did not significantly affect the thermal behavior of PP, but they modified the material crystallinity. PP films containing carvacrol and thymol showed a significant increase in OIT values, indicating that the polymer is well stabilized and a certain amount of these compounds remained in the polymer matrix after processing at relatively high temperatures. In this sense, these additives could be furthermore released from the material playing their role as antimicrobial additives. Therefore, it could be concluded that the addition of antimicrobial additives as carvacrol and thymol at 8 wt% to PP shows potential to improve the food quality and safety.

Acknowledgments

Authors would like to thank Ashland Chemical Hispania for kindly supplying ECOLEN HZ10K PP (Hellenic Petroleum). Marina Ramos would like to thank University of Alicante (Spain) for UAFPU2011-48539721S predoctoral research grant.

References

(2007). ASTM D 3895 - 07. Standard test method for oxidative-induction time of polyolefins by differential scanning calorimetry. *In: Annual book of ASTM standards. Amer. Soc. for Testing & Materials, Philadelphia, PA.*

Alin, J. & Hakkarainen, M., (2010). Type of polypropylene material significantly influences the migration of antioxidants from polymer packaging to food simulants during microwave heating. *Journal of Applied Polymer Science*, 118(2), 1084-1093.

Bakkali, F., Averbeck, S., Averbeck, D. & Idaomar, M., (2008). Biological effects of essential oils - A review. *Food and Chemical Toxicology*, 46(2), 446-475.

Blanda, G., Cerretani, L., Cardinali, A., Barbieri, S., Bendini, A. & Lercker, G., (2009). Osmotic dehydrofreezing of strawberries: Polyphenolic content, volatile profile and consumer acceptance. *LWT - Food Science and Technology*, 42(1), 30-36.

Cayot, N., (2007). Sensory quality of traditional foods. *Food Chemistry*, 101(1), 154-162.

Chiralt, A., Martínez-Navarrete, N., Martínez-Monzó, J., Talens, P., Moraga, G., Ayala, A. & Fito, P., (2001). Changes in mechanical properties throughout osmotic processes: Cryoprotectant effect. *Journal of Food Engineering*, 49(2–3), 129-135.

Coma, V., (2008). Bioactive packaging technologies for extended shelf life of meat-based products. *Meat Science*, 78(1–2), 90-103.

Conte, A., Buonocore, G.G., Bevilacqua, A., Sinigaglia, M. & Del Nobile, M.A., (2006). Immobilization of Lysozyme on Polyvinylalcohol Films for Active Packaging Applications. *Journal of Food Protection*, 69(4), 866-870.

Cutter, C.N., (2002). Microbial control by packaging: A review. *Critical Reviews in Food Science and Nutrition*, 42(2), 151-161.

Davidson PM, Z.S., (2003). *The use of natural antimicrobials. Food preservation techniques.* Boca Raton, Fla.: Woodhead Publishing Limited and CRC Press.

Dobkowski, Z., (2006). Thermal analysis techniques for characterization of polymer materials. *Polymer Degradation and Stability*, 91(3), 488-493.

Dong Sun Lee, K.L.Y., Luciano Piergiovanni., (2008). *Food packaging science and technology*. Boca Raton : CRC Press, c2008

Dorman, H.J.D. & Deans, S.G., (2000). Antimicrobial agents from plants: antibacterial activity of plant volatile oils. *Journal of Applied Microbiology*, 88(2), 308-316.

Gemili, S., Yemenicioǧlu, A. & Altinkaya, S.A., (2009). Development of cellulose acetate based antimicrobial food packaging materials for controlled release of lysozyme. *Journal of Food Engineering*, 90(4), 453-462.

Gutiérrez, L., Escudero, A., Batlle, R.n. & Nerín, C., (2009). Effect of Mixed Antimicrobial Agents and Flavors in Active Packaging Films. *Journal of Agricultural and Food Chemistry*, 57(18), 8564-8571.

Gutiérrez, L., Sánchez, C., Batlle, R. & Nerín, C., (2009). New antimicrobial active package for bakery products. *Trends in Food Science & Technology*, 20(2), 92-99.

Hazzit, M., Baaliouamer, A., Faleiro, M.L. & Miguel, M.G., (2006). Composition of the essential oils of Thymus and Origanum species from Algeria and their antioxidant and antimicrobial activities. *Journal of Agricultural and Food Chemistry*, 54(17), 6314-6321.

Ho, C.W., Wan Aida, W.M., Maskat, M.Y. & Osman, H., (2006). Optimization of headspace solid phase microextraction (HS-SPME) for gas chromatography mass spectrometry (GC-MS) analysis of aroma compound in palm sugar (Arenga pinnata). *Journal of Food Composition and Analysis*, 19(8), 822-830.

Joseph, P.V., Joseph, K., Thomas, S., Pillai, C.K.S., Prasad, V.S., Groeninckx, G. & Sarkissova, M., (2003). The thermal and crystallisation studies of short sisal fibre reinforced polypropylene composites. *Composites Part A: Applied Science and Manufacturing*, 34(3), 253-266.

Kuorwel, K.K., Cran, M.J., Sonneveld, K., Miltz, J. & Bigger, S.W., (2011a). Antimicrobial Activity of Natural Agents against Saccharomyces cerevisiae. *Packaging Technology and Science*, 24(5), 299-307.

Kuorwel, K.K., Cran, M.J., Sonneveld, K., Miltz, J. & Bigger, S.W., (2011b). Essential Oils and Their Principal Constituents as Antimicrobial Agents for Synthetic Packaging Films. *Journal of Food Science*, 76(9), R164-R177.

Kuorwel, K.K., Cran, M.J., Sonneveld, K., Miltz, J. & Bigger, S.W., (2013). Migration of antimicrobial agents from starch-based films into a food simulant. *LWT - Food Science and Technology*, 50(2), 432-438.

Kurek, M., Moundanga, S., Favier, C., Galić, K. & Debeaufort, F., (2013). Antimicrobial efficiency of carvacrol vapour related to mass partition coefficient when incorporated in chitosan based films aimed for active packaging. *Food Control*, 32(1), 168-175.

López, P., Sánchez, C., Batlle, R. & Nerín, C., (2007a). Development of flexible antimicrobial films using essential oils as active agents. *Journal of Agricultural and Food Chemistry*, 55(21), 8814-8824.

López, P., Sánchez, C., Batlle, R. & Nerín, C., (2007b). Vapor-phase activities of cinnamon, thyme, and oregano essential oils and key constituents against foodborne microorganisms. *Journal of Agricultural and Food Chemistry*, 55(11), 4348-4356.

Mastromatteo, M., Conte, A. & Del Nobile, M., (2010a). Combined use of modified atmosphere packaging and natural compounds for food preservation. *Food Engineering Reviews*, 2(1), 28-38.

Mastromatteo, M., Mastromatteo, M., Conte, A. & Del Nobile, M.A., (2010b). Advances in controlled release devices for food packaging applications. *Trends in Food Science & Technology*, 21(12), 591-598.

Persico, P., Ambrogi, V., Carfagna, C., Cerruti, P., Ferrocino, I. & Mauriello, G., (2009). Nanocomposite polymer films containing carvacrol for antimicrobial active packaging. *Polymer Engineering & Science*, 49(7), 1447-1455.

Poinot, P., Arvisenet, G., Grua-Priol, J., Colas, D., Fillonneau, C., Le Bail, A. & Prost, C., (2008). Influence of formulation and process on the aromatic profile and physical characteristics of bread. *Journal of Cereal Science*, 48(3), 686-697.

Poinot, P., Grua-Priol, J., Arvisenet, G., Rannou, C., Semenou, M., Bail, A.L. & Prost, C., (2007). Optimisation of HS-SPME to study representativeness of partially baked bread odorant extracts. *Food Research International*, 40(9), 1170-1184.

Pomerantsev, A.L. & Rodionova, O.Y., (2005). Hard and soft methods for prediction of antioxidants' activity based on the DSC measurements. *Chemometrics and Intelligent Laboratory Systems*, 79(1-2), 73-83.

Pozo-Bayón, M.A., Guichard, E. & Cayot, N., (2006). Flavor Control in Baked Cereal Products. *Food Reviews International*, 22(4), 335-379.

Quílez, J., Ruiz, J.A. & Romero, M.P., (2006). Relationships Between Sensory Flavor Evaluation and Volatile and Nonvolatile Compounds in Commercial Wheat Bread Type Baguette. *Journal of Food Science*, 71(6), S423-S427.

Ramos, M., Jiménez, A., Peltzer, M. & Garrigós, M.C., (2012). Characterization and antimicrobial activity studies of polypropylene films with carvacrol and thymol for active packaging. *Journal of Food Engineering*, 109(3), 513-519.

Riga, A., Collins, R. & Mlachak, G., (1998). Oxidative behavior of polymers by thermogravimetric analysis, differential thermal analysis and pressure differential scanning calorimetry. *Thermochimica Acta*, 324, 135-149.

Rizzolo, A., Gerli, F., Prinzivalli, C., Buratti, S. & Torreggiani, D., (2007). Headspace volatile compounds during osmotic dehydration of strawberries (cv Camarosa): Influence of osmotic solution composition and processing time. *LWT - Food Science and Technology*, 40(3), 529-535.

Rodriguez-Lafuente, A., Nerin, C. & Batlle, R., (2010). Active Paraffin-Based Paper Packaging for Extending the Shelf Life of Cherry Tomatoes. *Journal of Agricultural and Food Chemistry*, 58(11), 6780-6786.

Rodríguez, A., Batlle, R. & Nerín, C., (2007). The use of natural essential oils as antimicrobial solutions in paper packaging. Part II. *Progress in Organic Coatings*, 60(1), 33-38.

Ruiz, J.A., Quilez, J., Mestres, M. & Guasch, J., (2003). Solid-Phase Microextraction Method for Headspace Analysis of Volatile Compounds in Bread Crumb. *Cereal Chemistry Journal*, 80(3), 255-259.

Singh, P., Wani, A.A. & Saengerlaub, S., (2011). Active packaging of food products: Recent trends. *Nutrition and Food Science*, 41(4), 249-260.

Suppakul, P., Sonneveld, K., Bigger, S.W. & Miltz, J., (2011). Diffusion of linalool and methylchavicol from polyethylene-based antimicrobial packaging films. *LWT - Food Science and Technology*, 44(9), 1888-1893.

Tovar, B.z., García, H.S. & Mata, M., (2001). Physiology of pre-cut mango. I. ACC and ACC oxidase activity of slices subjected to osmotic dehydration. *Food Research International*, 34(2–3), 207-215.

Yanishlieva, N.V., Marinova, E.M., Gordon, M.H. & Raneva, V.G., (1999). Antioxidant activity and mechanism of action of thymol and carvacrol in two lipid systems. *Food Chemistry*, 64(1), 59-66.

In: Quantitative Chemistry, Biochemistry and Biology
Editors: G.E. Zaikov, O.V. Stoyanov et al.
ISBN: 978-1-62948-332-0

Chapter 4

SOME PROPERTIES OF SMALL WATER CLUSTERS IN WATER-STARCH SYSTEMS

N.A. Grunina*[*]*, G.I. Tsereteli, T.V. Belopolskaya, O.I. Smirnova and L.O. Rodriguez Castillo

Institute for Physics of Physical Department of St. Petersburg State University, Russia
St. Petersburg, Petrodvorets, Russia

ABSTRACT

The influence of the hydration degree and also the various conformational changes of the starch molecules on the thermodynamic properties of the water clusters forming in the starches of A- and B- types is discussed. It is obtained that the T_m of free water in the starch-water system is always below than T_m of pure water in the wide range of concentrations studied that reflects a high degree of its dispersion. The increasing of water concentrations leads to the increase of water T_m in all starches that means that we deals with the small water clusters whose T_m depends on the size. It is also found that the relation between the values of the melting temperatures of freezable water in the denatured state of the starch (T_m^D) and in the native one (T_m^N) is changing. At denaturation of the starch with free water more than 35% the average size of clusters decreases probably due to the destruction of the largest clusters which are characteristic for the native state in this case. It is found that for the denatured state of the starches studied the dependence of melting heat values (Q_m^D) on humidity has more complicated character than that for the native state. The amount of bound water at denaturation increases while humidity of starch approaches 50%. The maximum possible quantity of bound water for the studied starches of different types in their amorphous state was determined. It is shown that the value of Q_m^D as well Q_m^N at humidity of starch higher than 50% coincides with Q_m of the bulk water. At the same time for $C_{H_2O} < 50\%$ the values Q_m^D in contrast to Q_m^N essentially lower than that for the bulk water.

[*] E-mail: nagrunina@mail.ru.

Keywords: Water cluster, water-starch systems, temperature and heat of water melting, DSC

INTRODUCTION

The interest to the thermal properties of humid biopolymers containing free and bound water maintains over many decades which is connected both with fundamental problems of biophysics and practical applied tasks. The strict classification of water as being either bound or free seems to be a bit oversimplified [1], but nevertheless is widely used up to now [2-6]. The term free corresponds to water which freezes at cooling to low temperatures, and hence melts, and the term bound to that that does not. The quantitative separation of the water of two states depends on the experimental methodic used as well as on the time-scale of experiment.

The complicated hierarchical supermolecular structure of starch leads to the strongly heterogenic distribution of water in its native granules [7]. In respect to the localization degree, some different types of water can be specified in starch granules, mainly, the hydration water incorporated to the crystallite structure, the amount of which in crystals of A- and B-types differs by 5-6 times [8, 9] and the water in the amorphous phase which in its turn is about 70% of the total starch mass [7]. Besides, there is the "micro-capillar" water in starches [10] the content of which is apparently different in the granule growth-rings of different density. These peculiarities of the water states are clearly reflected in the thermal properties of water clusters which are forming in native and amorphous starches.

Numerous calorimetric studies, including ours, of the water state in biopolymers has shown that in humid proteins and DNA [3] as well as in concentrated solutions of carbohydrates of both natural and artificial origin [11] and in starches [12] the small water clusters can be formed the melting temperature (T_m) of which is below 0°C and changes in dependence of water concentration in the system. In other words, T_m clusters of water depends on their size. Influence of surface energy on the energy balance of small particles was discussed, for example, in [13-15]. It should be noted that the dependence of melting temperature of water clusters on its size obtained in different studies has a close character. Anyway, the minimum T_m is observed at the minimum content of free water.

The situation with the dependence of the melting heat Q_m on the cluster size is quite different. Focus only on the data for starches. Different authors report different relation between the Q_m of bulk water and the water clusters in starch-water systems. Naturally, the melting heat of ice (water) in the starch-water system with an excess of water is almost equal to Q_m clean water. About the difference between these values one can speak only for small amounts of free water in the system under study. Whereas some authors claim that the Q_m of bulk water is larger than that of clusters in starches [4, 5], the others are coming to the opposite conclusion [6].

Our previous results on the study of thermal properties of water clusters in the starches of A- and B-types in dependence on humidity have shown that the Q_m of water clusters in these systems in native state does not depend on humidity and coincides with that of the bulk water. However, it has been shown that the denaturation leads to decrease of the obtained Q_m of water at its fixed total content in the sample [12].

The main aim of the present work is the further study of the influence of the hydration degree, on the one hand, and the various conformational changes of the starch molecules, on the other hand, on the thermodynamic properties of the water clusters forming in starch-water systems.The previously obtained data are complimented by the study of one more starch species of A-type, enlarging the range of humidity and usage of different heating rates.

METHODS

Preparation of Starch Samples with Different Humidities

Earlier it was determined that the water concentration in starches under their storage at room temperature is usually about 10÷15% that depends on T_{room} value and humidity of environment. A necessary concentration of water in the samples for different experiments was achieved by two ways. In one case it was achieved either by wetting the samples with water vapour (up to 25%) or by adding water (up to 85%) at T_{room}. In other case from the beginning the aqueous suspensions of starch (85%) were prepared and then they were drying up to a necessary concentration. Prior to the DSC measurements, each sample was kept at room temperature for 1÷2 days. The resulting water content in the sample measured was estimated from its weight loss after drying at 105°C until a constant weight was reached. The total mass of the samples was varied from 50 to 120 mg.

Differential Scanning Calorimetry

DSC measurements were performed with the Differential Scanning Microcalorimeter (Setaram DSC-111, France) with sensitivity of $3 \cdot 10^{-5}$ $J \cdot s^{-1}$. Temperature was controlled with a precision not less than ±0.2 deg over the whole measured interval -60÷120°C. In dependence on the task the different heating rates were used (1; 3; 5 °*C*/min). For the V_{heat} = 5°*C*/min the errors in values of water melting heat in these experiments were about ±5 % in the wide interval humidities from 23÷80 %.

MATERIALS

In this work, along with the starches the thermal properties of which were investigated by us earlier – potato ("Aldrich") and rice ("Sigma"), was studied starch of wheat ("Fluka"). Granules of these starches contain nanocrystalline particles of two types: A- (rice and wheat) and B- (potato), with various amount of crystallohydrate water which can be also considered as a separate type of the bound water.

RESULTS AND DISCUSSION

Using the DSC method we have obtained the temperature dependences of specific heat in the region of the ice-water phase transition for three starches with different types of the crystal lattice in native and denatured states in the range of the hydration degrees 23% ÷ 80%.

Melting Temperature of the Water Clusters in Native and Denatured Starches

Figure 1 shows the experimental data for three starches in the native state. As it is seen, at the increase of the hydration degree the melting curves are shifting to the region of higher temperatures. Whereas all curves are started at temperatures essentially below 0°C (about from – 20°C) and the positions of maximum of the melting curves T_m^N are shifting from -7°C to 6°C. Figure 2, curves 2, 3 show separately the obtained dependences of the T_m of free water in the native and denatured starches on their humidity correspondingly.

In order to understand the obtained results on the behaviour of the T_m of water one should take into account the specifity of determination of temperatures of the first order phase transitions, and the melting points in particular, by the DSC method. In theory, the thermal capacity at such transitions becomes infinite at transition temperature and the melting curve represents δ-function. In the DSC method this δ-function transforms into the melting curve of a finite width [16, 17]. It is firmly established that the T_{max} of the observable curve is the higher the lager is the melting heat Q_m of the transition, the mass of the sample and the heating rate V_{heat}.

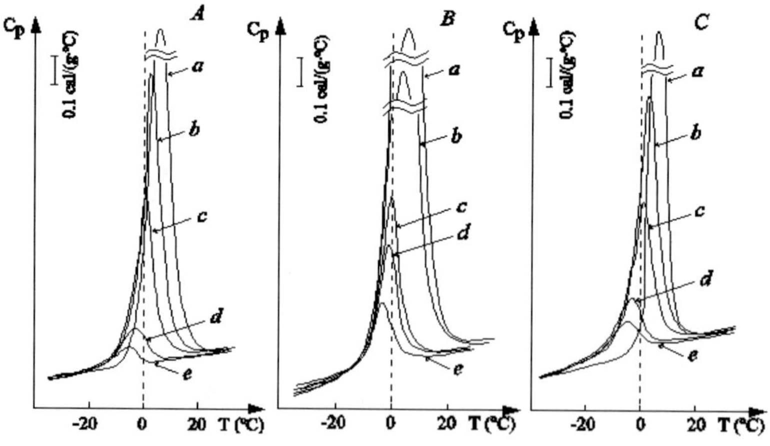

Figure 1. Temperature dependence of the heat capacity in the region of ice-water transition for the native rice (*A*), wheat (*B*) and potato (*C*) starches on humidity of samples:

A – *a* (60%), *b* (35%), *c* (29%), *d* (25.5%), *e* (23.7%);

B – *a* (54%), *b* (45%), *c* (30%), *d* (29%), *e* (26%);

C – *a* (55%), *b* (37%), *c* (32%), *d* (30%), *e* (27.4%). Heating rate is 5 °C/min.

Our studies have shown that at small percentage of the free water and hence at its small mass and small melting heat Q_m the temperature of the melting maximum T_{max} of the free (freezable) water practically (within 1°C) does not depend on the heating rate. So we can consider that obtained under these conditions T_{max} values below 0°C are close to the true melting points of free water in starch-water system (Figure 2). Note, such low values of T_m belong to the water concentrations < 35 %. In our experiments, for an example, 31% of total water in the sample corresponded only to the 7 mg of the mass of free water, i.e. in this range we deal really with small mass of free water.

Further it is necessary to understand how the observed values of T_{max} of the melting curves of the frozen water for samples with high humidity (Figure2. curves *2* and *3*) are corresponding to the true values of T_m of the water clusters. To clear this problem we have compared the observed values of T_{max} with the values of T_m for the samples of pure water with the mass equal to that of the mass of frozen water in the samples studied, using in both measurements the same heating rate 5°/min (Figure2. curves *1*). As it is seen, all the obtained values of T_{max} in the melting curves of the water clusters in the water-starch systems are lower than the corresponding values of T_m for the pure water, which indicates the high degree of water dispersion in all the starches studied.

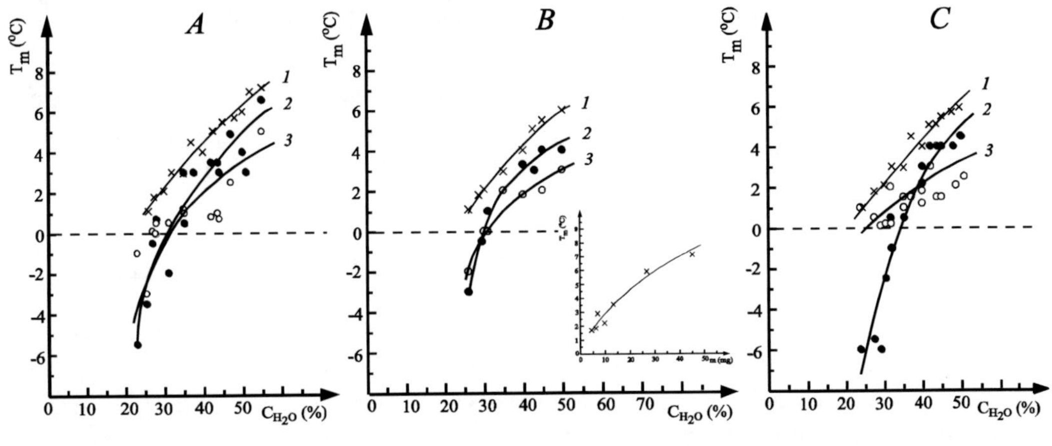

Figure 2. Melting temperature of water for: pure water (1), free water in the native (2) and denatured (3) starches. *A* – rice, *B* – wheat, *C* – potato. On the inset it is shown the dependence T_{max} of the melting curves for the pure water on the mass.

Unfortunately the obtained data does not allow the accurate estimation of the forming water clusters. However, taking into account that the number of water molecules attached to one monomer chain is 3 at the humidity 25% and increases up to 9 at 50% humidity [9] we can suppose that the forming clusters have the nanoscale sizes. As it follows from the obtained data, in the process of further hydration of starch the thermal properties of water clusters converge to that of the bulk water.

Further it was found that the effect of the dependence of T_m of free water on its content in the starch is observed not only for native but for denatured samples of all studied starches (Figure 2, curves 3). In this case the denaturation of starch at all humidities leads to a change of water T_m with respect to the native sample. It was also found that when passing through the degree of hydration of ~30÷35% the relation between the values of the melting temperatures of freezable water in the denatured state, T_m^D , and native state, T_m^N , is changing. This effect is most pronounced for the B-type starch but is less detectable for the A-type. So it can be concluded that during the denaturation the transformation of the water clusters takes place. At small amounts of free water the size of clusters at denaturation increases and at the increase of starch hydration degree the cluster size decreases.

In order to understand the results obtained at humidity higher than ~35% we should take into account the fact that the native granules contain alternating approximately spherical layers of starch with different density. The essential part of the high density layers constitutes the crystallites with the fixed amount of crystallohydrate water, the amount of such water being different for the starches of A- and B-types. It is natural to suppose that the content of free water in the starch layers of low density is higher than in layers of high density and the distribution of water over the volume in the native starch is heterogeneous. At the denaturation at high hydration degrees the destruction of granules brings about more uniform distribution of the water over the starch volume, mostly on the expense destruction of the most large water clusters in the layers of low density. As a result, the average size of clusters is decreasing which leads to decrease of T_m^D in respect to T_m^N. The stronger influence of denaturation on the melting temperatures in B-type starches, as compared to A-type can be attributed to the difference of the granule sizes in these starches. In the potato starch the average size of granules is about 40 μm, in wheat starch is ~20 μm and in the rice starch it is only about 5 μm [7, 18, 19].

As to the region of low humidity (several first %), the changes of the size of nanoclasters at denaturation are caused by the release of the crystallohydrate water the amount of which is different in the starches of A- and B-types which have a different density of packing.

Melting Heat of Water in Starches in their Native and Denatured States

Figure 3 shows the results on the melting heat of freezable water in the native samples for the investigated starches of A- and B- types in the wide range of water concentrations. The presented values are normalized on the total mass of the sample. The obtained results can be approximated by a straight line between the point Q_m=0 cal/g for 23% humidity (corresponding to the maximum content of unfreezable water in native samples of the studied starches) and the point Q_m=80 cal/g at 100% humidity (corresponding to the melting heat of ice). It is seen that the same straight line approximates well all together and the each one sets of data for all three starches. This result provides a convincing proof of the independence of the Q_m of water clusters on the humidity of native starches.

The possibility to approximate the experimental data by the straight line shows that the formation at cooling of the ice matrix of the freezable water does not lead to the freezing of the bound water, i.e. its amount related to the total mass of the starch does not change at variation of the hydration degree. It can be also concluded that the melting heat of the dispersed water in the native starch coincides practically with that of the bulk water at any humidity.

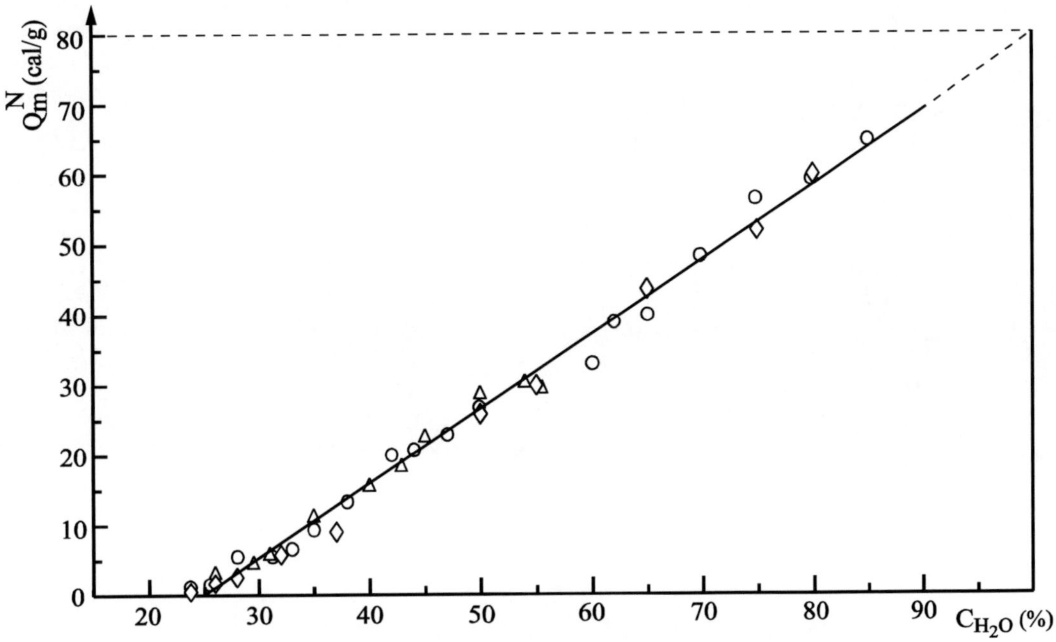

Figure 3. The melting heats of water in the native samples of the rice (o), wheat (Δ) and potato (◊) starches with different content of water.

Melting of the Freezable Water in Denatured State of A- and B- Type Starches

The analysis of the data on the melting heats of water in the studied starches in amorphous state has shown that the melting heats of water in denatured starches (Q_m^D) are smaller than that in native starches (Q_m^N) for the comparable degrees of hydration.

The observed decrease of the melting heat of the free water in the amorphous state of starch indicates the redistribution the relative amounts of free and bound water in the process of thermal destruction of native crystals of all starches studied. As an example, the Figure 4 shows the dependence of Q_m water on humidity of the rice starch in native and denatured states. As it is seen, the obtained dependence of Q_m values on humidity for denatured state has more complicated character than that for the native state that it was also observed in [4-6]. The values of Q_m at humidity higher than 50% are situated on a straight line which also as a line 1 involves the value Q_m of bulk water at 100% humidity. However, the extrapolation of this line to Q_m=0 intercepts the humidity axis at C_{H_2O}=32%. This value exactly indicates the amount of bound water in denatured rice starch with an excess of water. For the wheat and potato starches the corresponding values are 32% and 28%. Therefore, the experimental results show that the amount of bound water at denaturation increases since the total amount of water in the sample is fixed.

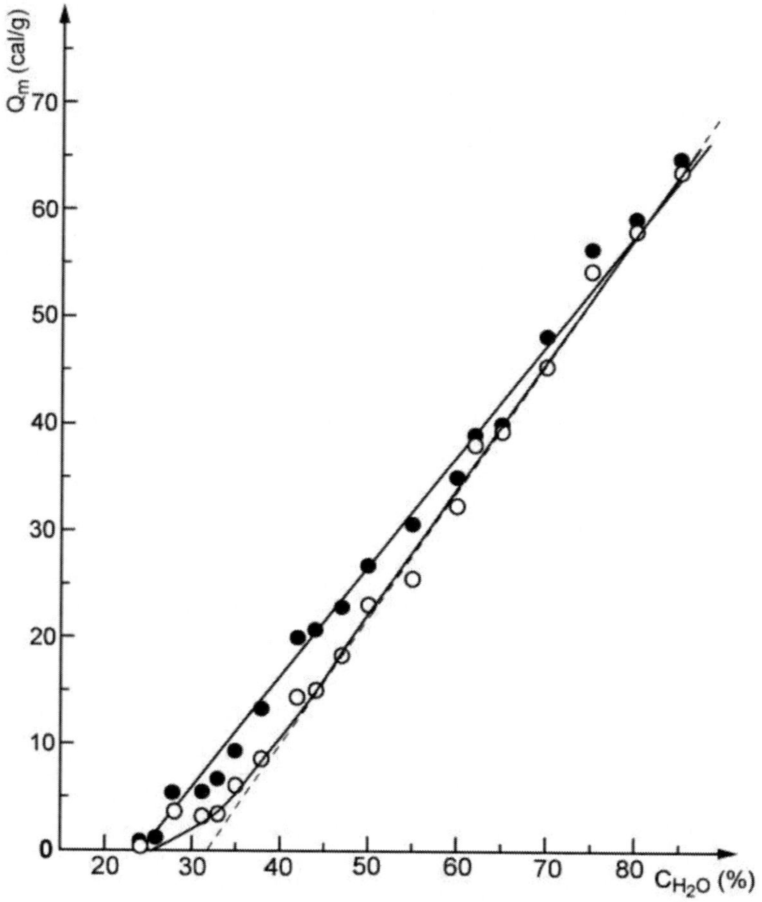

Figure 4. The melting heats of water for the native (•) and denatured (o) samples of the rice dependence on the water content. Dash line is an extrapolation of the Q_m^D curve at high humidities.

In more details the obtained result is demonstrated in Figure 5 where the difference of water melting heats in native and denatured states for all studied starches is shown on a different scale. For denatured samples the equilibrium values of Q_m^D measured after the keeping the samples at room temperature for 24 hours are presented. The maximum difference between Q_m^N and Q_m^D is observed at humidity ~ 40÷50%. The more intense redistribution of free and bound water in A-type starches ($\Delta \cong 6{,}0$ cal/g) as compared to B-type ($\Delta \cong 4{,}0$ cal/g) starches can be attributed to the different content of the crystallohydrate water in crystallites of these two types.

It should be taken into account that the additional (as compared to the native state) binding of water with the starch molecules takes place not only in a new amorphous phase formed from the melted crystallites. The similar process is going also in initial amorphous phase which constitutes approximately 70% of the total starch mass as it is already mentioned. As it is known, in reality the amorphous phase in native granules is not totally amorphous since it is connected with the crystalline phase by common molecular chains. This structural peculiarity of the native samples manifests itself in the ability of granules to swell [20] and in the process of glassification of native starches [21]. Only after the destruction of the granules the amorphous phase became truly amorphous which leads to the additional binding of water and, correspondingly, to the decrease of temperature of glass transition T_g [22-25].

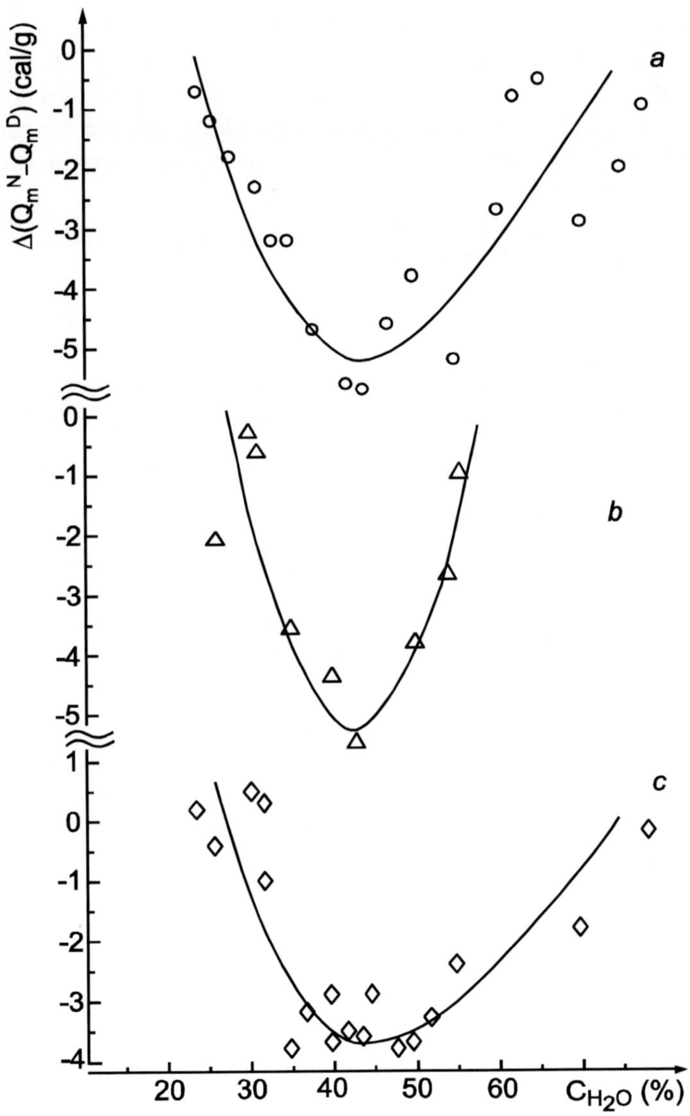

Figure 5. The difference of water melting heats in native and denatured states for the rice (*a*), wheat (*b*) and potato (*c*) starches.

Now we shall try to estimate the relative role of new and initial amorphous phases in the redistribution between free and bound water during the denaturation process. The crystallohydrate water should be considered as a bound water, however it is bound to the starch molecules not in amorphous but in crystalline phase. In starches of B-type the increase of the bound water content in the amorphous phase formed from the melted crystals takes place from 25% to 28% whereas in the A-type starches the corresponding increase is from 5% to 32%. It means that the increase of the bound water content in the layers of "former" dense packing in these starches differs by several ($\approx$ 10) times. At the same time the corresponding increase in the initial amorphous phase is from 23% to 28% for B-type and from 23% to 32% for A-type.

So it can be concluded that the total increase of the ability of starch to bound the water at the transition from native to denatured state is caused both by the destruction of the crystalline phase and by transformation of the properties of initial amorphous phase. The mentioned above quantitative difference in the depth of the minima (Figure5) reflects the different ability of such binding for A- and B-type starches and is caused directly by the difference of their crystal lattices.

In conclusion, let us return to Figure 4. As mentioned above, the line (or a tangent), held in the range of humidity C_{H_2O}>50% hits the point $Q_m = 80$ cal/g at 100% humidity. At the same time, the tangent to this curve for C_{H_2O} <50%, allows to obtain the values of Q_m^D water clusters in water-starch systems significantly lower values of Q_m bulk water. In other words, the obtained results indicate that in denatured starches at low humidity exist the clusters for which not only the melting temperature and the melting heat differ from the corresponding parameters for the bulk water. Both T_m^D, and Q_m^D are lower than that of the bulk water. The similar experimental data for denatured starches which allow us to make the same conclusion reported in [4, 5].

CONCLUSION

1) It is shown that both in native and denatured states of the investigated starches of A- and B-types the melting temperature of water clusters are below than that of the bulk water. As the degree of hydration of the starch increases the T_m of water clusters gradually approaches the T_m of the bulk water.
2) It is found that in the denatured state, in contrast to the native, in all starches the amount of bound water depends on the degree of hydration. It is determined the maximum possible amount of bound water for the studied starches of different types in their amorphous state, which remains unchanged at the concentration of water higher than 50%.
3) It is established that in the process of thermal destruction of native crystals of all starches studied the redistribution the relative amounts of free and bound water takes place. The value of this effect depends on the starch humidity and reaches a maximum around C_{H_2O} =40÷45 %. This effect is more pronounced in the A-type starches (rice, wheat) than in the B-type starch (potato), due to the difference of their crystal lattices.
4) It is obtained that in denatured starches at low humidity exist the clusters for which not only the melting temperature and the melting heat are lower than the corresponding parameters for the bulk water.

REFERENCES

[1] Franks, F., Unfrozen water: yes; unfreezable water: hardly; bound water: certainly not, *Cryo-Lett.*, 1986, *7*, 207.

[2] Water in polymers (ed: S.R. Rowland), Washington, D.C. 1980, p.555.

[3] Mrevlishvili, G.M., Low-temperature Calorimetry of Biological Macromolecules, Metsniyereba, Tbilisi, 1984, p.188.

[4] Tananuwong, K. and Reid, D. S., DSC and NMR relaxation studies of starch-water interactions during gelatinization, *Carbohydr. Polym.*, 2004**,** *58*, 345-358.

[5] Tran, T., Thitipraphunkul, K., Piyachomkwan, K. and Sriroth, K., Effect of chemical modifications and hydrocolloids on cassava starch, functional properties and frozen storage stability, in: Starch Science and Technology (eds: V.P. Yuryev, P. Tomasik, A. Blennow, L.A. Wasserman, G.E. Zaikov), Nova Science Publishers, NY, 2008, pp. 119-128.

[6] S. Suzuki, and S. Kitamura. Unfrozen water in amylosic molecules is dependent on the molecular structures – A differential scanning calorimetric study. *Food Hygrocolloids*, 2008**,** *22*, 862-867.

[7] Eliasson, A.-C., Gudmundsson, M., Starch: physicochemical and functional aspects, in: Carbohydrates in food / Ed. A.-C. Eliasson, Marcel Dekker, Inc., NY, 1996, pp. 431-503.

[8] French, D., Organization of starch granules, in: Starch Chemistry and Technology (eds: R.L. Whister, J.N. BeMiller and E.F. Paschall), Academic Press, Orlando, 1984**,** pp.184-242.

[9] Hizukuri, S., Starch: Analytical Aspects, in: Carbohydrates in food / Ed. A.-C. Eliasson, Marcel Dekker, Inc., NY, 1996, pp. 347-429.

[10] Gabuda, S.P., Rzhavin, A.F., NMR in the crystallohydrates and in the hydrated proteins, Novosibirsk: Nauka, 1978, 160 (in Russia).

[11] Slade, L. and Levine, H, A food polymer science approach to structure-property relationships in aqueous food systems: non-equilibrium behavior of carbohydrate-water systems, in: Water Relationships in Foods (eds: H.Levine and L.Slade), Plenum Press, NY, 1991, pp. 29-101.

[12] Belopolskaya, T.V., Tsereteli G.I., Grunina N.A., Smirnova, O.I., Rodriguez, L.O., DSC study of water melting in starches with different type of crystal lattice // The Abstracts of XVIII International Starch Convention, Cracow, 2010, pp. 59-60.

[13] Hill, T.L., Thermodynamics of small system, *Nano Letters*, 2001, *1*, N3, 111-112.

[14] Wunderlich, B., Macromolecular physics: *Crystal melting*, Vol. 3, NY-London, 1980, p. 363.

[15] Pavlov, V.A., *Ph.D. Theses*, Skt. Petersburg State Univ., 2008, p. 125.

[16] Bershtein, V.A., Egorov, V.M., Differential scanning calorimetry of polymer: physics, chemistry, analysis, technology, N.Y.-London-Toronto-Sydney-Singapore, 1994, p. 253.

[17] Godovsky, Yu. K., Thermophysical methods in study of polymers. *Khimiya, Moscow*, 1976, p. 215 (in russian).

[18] Donald, A. M., Understanding starch structure and functionality, in: Starch in food: Structure, function and applications / Ed. A.-C. Eliasson, Woodhead Publishing Limited, Cambridge, 2004, pp. 154-182.

[19] Andreev, N. R., The basis of native starches production. Pishepromizdat, Moscow, 2001, p. 272 (in russian).

[20] Tester, R. F. and Morrison, W.R., Swelling and gelatinization of cereal starches. I. Effects of amylopectin, amylase, and lipids, *Cereal Chem.*, 1990, *67*, 551-557.

[21] Slide, L. and Levine, H., Non-equilibrium melting of native granular starch: Part I. Temperature location of the glass transition associated with gelatinization of A-type cereal starches, *Carbohydr. Polym.*, 1988, *8*, 183-203.

[22] Zeleznak, K., Hoseney, R., The glass transition of starch, *Cereal Chem.*, 1987, *64*, 121-124.

[23] Bizot, H., Bail, P. Le, Leroux, B., Davy, J., Roger, P., Buleon, A., Calorimetric evaluation of the glass transition n hydrated linear and branched polyanhydrogtucose compounds, *Carbohydr. Polym.*, 1997, *32*, 33-50.

[24] .Belopolskaya, T.V., Tsereteli, G.I., Grunina, N.A., Smirnova, O.I., Calorimetric Study of The Native and Postdenatured Structures in Starches With Different Degree of Hydration, J. *Thermal.Analysis & Calorimetry*, 2008, *92*, 677-682.

[25] Tsereteli, G.I., Belopolskaya, T.V., Grunina, N.A., Dehydrated Native Biopolymers – an Unique Representative of Glassy Systems, *J. Thermal.Analysis & Calorimetry*, 2008, *92*, 711-716.

In: Quantitative Chemistry, Biochemistry and Biology
Editors: G.E. Zaikov, O.V. Stoyanov et al.
ISBN: 978-1-62948-332-0

Chapter 5

SOLUBILITY OF ALUMINUM POLYHYDROXOCHLORIDE IN THE WATER-ACETONE SOLUTION

V.F. Kablov, N.U. Bykadorov, O.K. Zhokhova, M.P. Spiridonova and D.A. Provotorova *

Volzhsky Polytechnical Institute, branch of Federal State Budgetary Educational Institution of Higher Professional Education, Volgograd State Technical University, Volzhsky, Volgograd Region, Russia

ABSTRACT

The investigation of aluminum polyhydroxochloride (APHC) solubility in the water-acetone system for extracting it in the solid state has been conducted. In addition, the water has been removed from initial liquid APHC without thermal heating for a short time. The optimal conditions for achieving high extent of APHC transition from the liquid into the solid state have been determined.

Keywords: Aluminum polyhydroxochloride, solvent, acetone, water-acetone fraction

INTRODUCTION

One of the main objectives of technically advanced countries in the world is the rational use of water resources of the planet and protecting them from pollution. The solution of this problem is particularly achieved by using coagulation-flocculation treatment methods for natural and waste waters. Nowadays, the number of studies related to the investigation of a mechanism of obtaining polymers with flocculating properties and their characterization has increased. A range of real-world manufacturing operations based on the application of these polymers is very wide. They are concentrating and sludge decantation, waste water treatment

* E-mail: Olga_VPI@mail.ru.

and conditioning, paper and hardboard manufacturing, mineral processing, etc. All these processes use electrolytes and polyelectrolytes as coagulating and flocculating agents.

The need to expand a range of flocculating agents is caused by large differences in structures of the systems treated with coagulants and flocculants.

Aluminum polyhydroxochloride (APHC) is widely used as a coagulant in waste water treatment and conditioning [1, 2], for obtaining modified coagulants based on it [3, 4], to reduce water permeability of oil wells, and in other areas. However, commercial coagulants based on the main aluminum chlorides are mostly liquids that make a number of inconveniences during its exploitation. Liquid APHC has a high corrosion activity because of free hydrochloric acid formed by APHC hydrolysis. It requires application of special reusable package and causes many problems when delivering it to the area of consumption.

To exclude these disadvantages, it is appropriate to transform the coagulant in a more convenient solid form. There are several ways to transform aluminum hydroxochloride from the liquid into the solid state. For example, it is dewatering by thermal heating [6], addition of various salts (NaCl, $Al_2(SO_4)_3$, $MgCl_2$, etc.) [7,8], and introducing amine containing compounds in a system, such as hexamethylenetetraamine [9]. Nevertheless, the first method consume much power as the initial liquid aluminum hydroxochloride has moisture content of 54-70 %, and various salt additives increase the cost of a solid product. Besides, all the moisture, which is in a bound state, stays in a finished product that increases transportation expenses.

In this work the solid APHC was obtained using a solvent replacement technique. Acetone was chosen as a solvent due to its complete miscibility with water in all proportions, and APHC is slightly soluble in it.

Experimental Part

The subject of research was highly basic APHC [10] (TC 2152-312-05763450-2001) in the form of a water solution with a basic compound content of 54 % (polymer $Al_2(OH)_5Cl$) that is 10.6 % wt by aluminum and 5.5 % wt by chloride ions. Analytical grade acetone was used as a solvent (GOST 2603-79).

The solid product was obtained by mixing the acetone solution with APHC in weight ratios of 1: (1.03 - 4.11) at room temperature. The acetone : water weight ratio in the APHC solution was 1: (0.11 - 0.45) at that.

The aluminum content was determined using complexometric titration, and chloride ions were detected by argentometric titration.

Results and Discussion

For determining the optimal mode of APHC transition from a solution to the solid state there were conducted studies with different proportions of APHC : acetone. For example, 20 ml (15.8 g) of acetone was added to the APHC solution containing 10.8 g of a basic compound. A precipitate was separated and dried. The weight of the solid phase was 4.2 g, i.e., only 38.9 % of APHC transformed into the precipitate, the rest was in the solution. An incomplete separation of APHC occurred due to the mass ratio of acetone : water in the solution was equal to 1:0.58. In this case the effect of water prevails. Water is more polar substance and APHC has high solubility in the system.

When increasing the amount of acetone introduced up to 50 ml (39.5 g) to the initial weight of the APHC solution (20 g), the mass of formed precipitate was 9.6 g or 88.9 % which is far more than in the previous experiment. Here, the acetone : water proportion was 1 : 0,233 by mass. So, the amount of added acetone has a considerable influence on the extent of APHC transition from the liquid to the solid state.

During the experiment it was noticed that the filtration process slowed down or even stopped. It is connected with evaporation of acetone during filtration, as acetone has the low boiling temperature (56,24 ^{0}C), but the relative water concentration in a dispersion phase, in turn, increases. As a result, the precipitate dissolves and transforms into a high viscosity gel which bridges the pores of a filter.

In the investigation it was also considered an influence of a sequence of component introduction to the system. That is why in the following set of experiments an order of component mixing was changed: different amounts of APHC were added to 100 ml of acetone. The results are in Table 1.

Table 1. Influence of a ratio in the acetone: water system on forming the solid APHC

No.	Amount of APHC solution (g), added to 100 ml of acetone	Water weight (g), introduced with APHC	Acetone : water ratio, weight parts	Note
1	19,2	8,8	1 : 0,11	98,6% of solid APHC is formed
2	29,6	13,6	1 : 0,17	95,7% of solid APHC is formed
3	46,1	21,2	1 : 0,27	88,7% of solid APHC is formed
4	62,6	28,8	1 : 0,36	The system divided on two phases: 54% precipitate and gel
5	76,9	35,4	1 : 0,45	The system divided on two phases: 27% precipitate and high viscosity gel having much adhesion to glass and filtering material

As seen in Table 1, when the acetone : water ratio from 1:0,11 to 1:0,27 is achieved (experiments 1-3) the solid APHC precipitates and then is well filtered. It determines a high extent of APHC transition from the liquid into the solid state.

When the acetone: water ratio is from 1:0,36 to 1:0,45, the amount of formed precipitate reduces dramatically, and adhesion to glass and filtering material increases. It shows that the water concentration exceeded the acetone concentration and APHC solubility significantly decreased, therefore APHC remains in the solution forming APHC gel of high viscosity. In other words, when the water-acetone ratio is higher than 1:0,27, much of APHC (50 – 80%) remains in the solution and is carried along with the filtrate. The reason can be the fact that acetone evaporates faster than water, and the water concentration increases in the heterogeneous system; The APHC precipitate begins to dissolve and then disappears. As a result, the APHC gel of high viscosity is formed.

So, if a ratio of 1: 0,30 is achieved, there occurs a necessity of the APHC-acetone-water system regeneration for acetone reusing. For this purpose a fractional distillation of the filtrate containing acetone, water and partially APHC was conducted. In this set of experiments a filtrate with the acetone: water ratio of 1: 0,55 was chosen. Table 2 shows the results of 60 ml supernatant fluid distillation.

Table 2. Results of the water-acetone filtrate fractional distillation

No.	Temperature, ^{0}C	Amount of distillate, ml	Residual, ml
1	58 – 59	35	25
2	61 - 62	20	5

At the boiling point of acetone (56,24^{0}C) there is no distilling. At temperatures of 58–59 ^{0}C (Table 2) acetone is distilled with partial water content (azeotrope), the amount of water-acetone fraction was 58.3% of the initial volume at that. According to the reference data [11], in this interval a mixture of acetone and water containing 83.4% acetone is distilled.

At temperatures of 61–62 ^{0}C 33,3 % of the water-acetone mixture with the acetone content of 53,7 – 58,0 % is distilled. The residual (water and APHC) in the volume was 8,4%.

The obtained water-acetone fractions were tested for determining a possibility of extracting APHC from the solution. Fraction № 1 proved to precipitate APHC with less quantative ratio than at using pure acetone. Fraction № 2 extracts the precipitate worse than fraction № 1, and the gelling process is observed because of large water content in the solution. Fraction № 3 does not precipitate APHC from the solution as it does not include acetone.

In order to define a possibility of changing APHC contents when it is extracted from the solution with acetone, a chemical analysis of different fractions was conducted. The results are given in Table 3.

Table 3. Percentage of Al^{3+} : Cl^- in different fractions

Fraction	Component content, % wt		Percentage of Al^{3+} : Cl^- in a fraction
	Al^{3+}	Cl^-	
APHC initial solution	**10,60**	**5,50**	**1 : 0,519**
Solid APHC, extracted with acetone	22,10	11,50	1 : 0,520
Filtrate	0,1	0,056	1 : 0,560

As seen in Table 3, percentage of Al^{3+} and Cl^- ions in the liquid and solid states of APHC is the same, i.e., acetone treatment does not change APHC chemical composition. The filtrate analysis showed some increase of chloride-ions compared to their initial concentration and concentration in the solid product. It may be connected with partial transition of hydrochloric acid from the APHC solution into the filtrate.

CONCLUSION

Thus, the results of research showed the possibility of APHC transition from the liquid into the solid state by its interaction with acetone.

The process is carried out without any power consumption for a short time.

With a help of analytical control methods it was proved, that the chemical composition was not changed when the solid APHC was extracted from the solution by the proposed method.

The obtained product having not high hygroscopicity, due to the absence of additional inorganic salts (in contrast to the proposed in [7, 8]), enables its long-term storage in adverse conditions.

REFERENCES

[1] Method of natural and waste water purification from suspended particles: Patent RU № 2174104 / I.A. Novakov, N.U. Bykadorov, S.S. Radchenko, O.K. Zhokhova, A.I. Parkhomenko. – 2001. Bul. № 27.

[2] I.A. Novakov, N.U. Bykadorov, S.S. Radchenko, O.K. Zhokhova, E.E. Utkina, N.V. Zyuzina. Study of obtaining dispersion in aluminum hydroxochloride synthesis // Chemistry and Technology of Element Organic Monomers and Polymers: proceedings. – Volgograd, 1998. – p. 176.

[3] Method of obtaining basic aluminum chlorides (variants): patent RU № 2083495 / I.A. Novakov, N.U. Bykadorov, S.S. Radchenko, Yu. N. Kargin, V.F. Mokhov, O.K. Zhokhova. – 1997. Bul. № 19.

[4] Method of obtaining water soluble reagent for natural and waste water purification and phase separation: patent RU № 2174105 / I.A. Novakov, N.U. Bykadorov, S.S. Radchenko, O.P. Otchenashev, O.K. Zhokhova – 2001. Bul. № 27.

[5] N.U. Bykadorov, V.F. Kablov, S.N. Bondarenko, D.A. Kondrutsky, O.K. Zhokhova. Study of water segregation process in alumosilicate processing with a blend of polyacrylamide and aluminum hydroxochloride // *International Journal of Applied and Fundamental Research*. – 2010. - № 2. – p. 32.

[6] Alcohol soluble basic aluminum chlorides and method of making same: patent US № 3904741 / John L. Jones et al. Filed: Oct. 26, 1970. Publ. 09.09.1975.

[7] Method of obtaining solid aluminum chloride containing coagulant (variants): patent RU № 2122973 / I.A. Novakov, N.U. Bykadorov, S.S. Radchenko, O.K. Zhokhova, E.E. Utkina. – 1998. Bul. № 34.

[8] S.S. Radchenko, N.U. Bykadorov, I.A. Novakov, O.K. Zhokhova, E.E. Utkina, O.P. Otchenashev, F.S. Radchenko. On a structure formation of highly basic aluminum hydroxochloride in concentrated solutions and new compositions of coagulants based on it // *Russian Journal of Applied Chemistry*. – 2002. – Vol. 75. Issue 4. – p. 529.

[9] Method of obtaining solid aluminum chloride containing coagulant: patent RU № 2255898 / N.U. Bykadorov, I.A. Novakov, V.F. Kablov, S.S. Radchenko, O.K. Zhokhova, D.A. Kondrutsky. – 2005. Bul. № 19.

[10] S.S. Radchenko, N.U. Bykadorov, I.A. Novakov, O.K. Zhokhova, E.E. Utkina, O.P. Otchenashev. Study of precipitation process of fine dispersions with coagulants based on aluminum hydroxochloride with different viscosity characteristics // Chemistry and Technology of Element Organic Monomers and Polymers: proceedings. – Volgograd, 1999. – p. 142.

[11] A. Gordon, R. Ford. Satellite chemist. Physical and chemical properties, methods, bibliography. – Moscow: *Mir,* 1976. – 541 p.

In: Quantitative Chemistry, Biochemistry and Biology
Editors: G.E. Zaikov, O.V. Stoyanov et al.
ISBN: 978-1-62948-332-0

Chapter 6

DIVERSIFIED FENOMENA OF ENTROPY

G.A. Korablev[1*], N.G. Petrova[2], R.G. Korablev[1], A.K. Osipov[1] and G.E. Zaikov[3]

[1] Izhevsk State Agricultural Academy, Izhevsk, Udmurt Republic
[2] Ministry of Informatization and Communication of the Udmurt Republic
[3] Institute of Biochemical Physics, Russian Academy of Science, Moscow, Russia

ABSTRACT

Various demonstrations of entropy have been discussed, including spatial-energy interactions of atom-molecular systems.

Similarly to the ideas of thermodynamics on the static entropy it is proposed to apply the concept of business quality entropy with the help of which it is possible to evaluate the critical limits of consolidated superpower business structures. The probable formation processes of a business structure are correlated via the entropy of random variables in it.

Keywords: Thermodynamic probability, statistic entropy, business entropy, entropy of a random variable, spatial-energy parameter

1. INTRODUCTION

The idea of entropy appeared based on the second law of thermodynamics and ideas of the adduced quantity of heat.

The adduced quantity of heat in isothermal process is the relation between the quantity of heat Q obtained by the system and temperature T of heat-release body.

The entropy is the function S of the system state whose differential in the elementary reversible process equals the relation between the infinitely little quantity of heat transferred to the systems and its absolute temperature:

* Email: KorablevGA@mail.ru.

$$dS = {}^{\delta Q}/_{T}$$

Using such heat-physical definition we can calculate only the difference between entropies. The entropy itself can only be found with the accuracy to the constant summand (integration constant).

In statistic thermodynamics the entropy of the equilibrious system equals the logarithm of the probability of its definite macrostate:

$$S = k \ln W, \tag{1}$$

where W – number of available states of the system or degree of the degradation of microstates; k – Boltzmann's constant.

Or:

$$W = e^{S/k} \tag{2}$$

These correlations are general assertions of macroscopic character, they do not contain any references to the structure elements of the systems considered and they are completely independent from microscopic models [1].

Therefore the application and consideration of these laws can result in a large number of consequences which are most fruitfully used in statistic thermodynamics.

In statistic thermodynamics the entropy is a function of the system state which helps estimating the process directions and possible changes in them.

At any spontaneous changes in the isolated system the entropy always increases:

$$\Delta S > 0$$

The sense of the second law of thermodynamics comes down to the following:

The nature tends from the less probable states to more probable ones. The most probable is the uniform distribution of molecules through the entire volume. From the macrophysical point, these processes consist in equalizing the density, temperature, pressure and chemical potentials, and the main characteristic of the process is the thermodynamic probability W.

In actual processes in the isolated system the entropy growth is inevitable – disorder and chaos increase in the system, the quality of internal energy goes down.

The thermodynamic probability equals the number of microstates corresponding to the given macrostate.

Since the system degradation degree is not connected with the physical features of the systems, the entropy statistic concept can also have other applications and demonstrations (apart from statistic thermodynamics).

"It is clear that out of the systems completely different by their physical state, the entropy can be the same if their number of possible microstates corresponding to one macroparameter (whatever parameter it is) coincide. Therefore the idea of entropy can be used in various fields. The increasing self-organization of human society ... leads to the increase in entropy and disorder in the environment that is demonstrated, in particular, by a large number of disposal sites all over the earth" [2].

In this research we are trying to apply the concept of entropy to the efficiency of business structures and evaluation of the degree of spatial-energy interactions.

2. ENTROPY OF BUSINESS STRUCTURE AGGREGATION

The main properties of capitalistic system providing its economic advantages are: 1) effective competition, and 2) maximal personal interest of each worker.

But on different economy concentration levels these *ab initio* features function and demonstrate themselves differently. Their greatest efficiency corresponds to small business – when the number of company staff is minimal, the personal interest is stronger and competitive struggle for survival is more active. With companies and productions increase the number of staff goes up, the role of each person gradually decreases, the competition slackens as new opportunities for coordinated actions of various business structures appear. The quality of economic relations in business goes down that is entropy increases. Such process is mostly vivid in monostructures at the largest enterprises of large business (syndicates and cartels).

The concept of thermodynamic probability as a number of microstates corresponding to the given macrostate can be modified as applicable to the processes of economic interactions that directly depend on the parameters of business structures.

A separate business structure can be taken as the system macrostate, and as the number of microstates – number of its workers (N) which is the number of the available most probable states of the given business structure. Thus it is supposed that such number of workers of the business structure is the analog of thermodynamic probability as applicable to the processes of economic interactions in business.

Therefore it can be accepted that the total entropy of business quality consists of two entropies characterizing: 1) decrease in the competition efficiency (S_1) and 2) decrease in the personal interest of each worker (S_2), i.e.:. S_1 is proportional to the number of workers in the company: , and S_2 has a complex dependence not only on the number of workers in the company but also on the efficiency of its management. It is inversely proportional to the personal interest of each worker. Therefore it can be accepted that , where – coefficient of personal interest of each worker.

By analogy with Boltzmann's equation (1) we have:

$$S = (S_1 + S_2) \sim \left[\ln N + \ln\left(\frac{1}{\gamma}\right)\right] \sim \ln\left(\frac{N}{\gamma}\right)$$

or

$$S = k \ln\left(\frac{N}{\gamma}\right),$$

where k – proportionality coefficient;

Here N shows how many times the given business structure is larger than the reference small business structure, at which N = 1, i.e. this value does not have the name.

For non-thermodynamic systems we take k = 1. Therefore:

$$S = \ln\left(\frac{N}{\gamma}\right) \tag{3}$$

In Table 1 you can see the approximate calculations of business entropy by the equation (3) for three main levels of business: small, medium and large. At the same time it is supposed that number N corresponds to some average value from the most probable values. You can see more details in [3].

When calculating the coefficient of personal interest it is considered that it can change from 1 (one self-employed worker) to zero (0), if such worker is a deprived slave, and for larger companies it is accepted as .

Table 1. Entropy growth with the business increase

Structure parameters	**Business**	
	Small	**Average**
$N_1 - N_2$	10 – 50	100 – 1000
	0.9 – 0.8	0.6 – 0.4
S	2.408 – 4.135	5.116 – 7.824
	3.271	6.470

Despite the rather approximate accuracy of such averaged calculations, we can make quite a reliable conclusion on the fact that business entropy, with the aggregation of its structures, sharply increases during the transition from the medium to large business as the quality of business processes decreases. The application of more accurate initial data allows obtaining specific values of business entropy, above which the process of economic relations can reach a critical level.

Actually: "New properties of the system can also accompany the process of its degradation. Under the development we often understand the increase in the system (entropy decrease), but as applicable to the company under order we more often understand regulated hierarchy which cannot be considered as the development attribute. ... Other development criteria – transition of the system to less probable states and diversity increase including the diversity of possible (potential) states of the systems" [4].

Therefore if crisis phenomena are often followed by the aggregation of business structures – nevertheless this "cannot be considered the development attribute".

And the diversity of business systems is expressed in small and medium business. Therefore the optimal criteria of the more qualitative business are defined by the maximum value of their entropy: S = 6.47 (in relative units).

In live systems the entropy growth is compensated via the negative entropy (negoentropy) which is formed through the interaction with the environment. That is a live system is an open one. And business cannot be an isolated system for a long period without the exchange process and interactions with the environment. The role of the external system diminishing the increase in the business entropy must be fulfilled, for example, by the corresponding state and public structures functionally separated from business. The demonopolization of the largest economic structures carried out from the "top" in the evolution way can be the inevitable process here.

But the increase in the personal interest of each worker is defined not only by the parameters of business systems but also depends on the overall arrangement of these processes by the employer. For instance, Ford managed to find such ways of work organization which sharply increased the personal interest of all its employees.

In thermodynamics it is considered that the uncontrollable entropy growth results in the stop of any macrochanges in the systems, i.e. to their death. Therefore the search of methods of increasing the uncontrollable growth of the entropy in large business is topical. At the same time, the entropy critical figures mainly refer to large business. A simple cut-down of the number of its employees cannot give an actual result of entropy decrease. Thus the decrease in the number of workers by 10% results in diminishing their entropy only by 0.6% and this is inevitably followed by the common negative unemployment phenomena.

Therefore for such supermonostructures controlled neither by the state nor by the society the demonopolization without optimization (i.e. without decreasing the total number of employees) is more actual to diminish the business entropy.

3. Entropy of an Elementary Business Structure

In the process of a new business structure formation we go through seeking and recruiting the personnel, at the same time, the number of personnel of which should correspond to the most probability of this process and for the given system is N_0. Here the key role is played by the probability of random values of this process. The similar picture is also characteristic for informative events: "It appears that for the information characteristics it is also possible to introduce the entropy notion. In the information theory we introduce such a value called the random value entropy:

$$H = \sum_n P_n \log_2(P_n^{-1}) \qquad (4)$$

Here: value H equals the number of binary digits required for difference (record) of the allowed value of the random value x [2]";

P_n – probability of the appearance of each given record of the random value.

Following the equation (4), the entropy of a random value is proportional to the total of probabilities and inversely proportional to the logarithm of their probabilities.

For the characteristic of continuous random value we use the function of probability distribution density [5,6]:

$$y = f(x) = P(\Delta x_i)/\Delta x_i,$$

where – probability of the random value ingress into the interval of its values.

In its sense, the random value entropy is reversely proportional to this function:

$$S_0 \sim {}^1/_y$$

or

$$S_0 \sim \Delta x_i [P(\Delta x)]^{-1} \tag{5}$$

Modifying the equations (4) and (5) and transferring from the binary system to the decimal one, we can have:

$$S_0 = \Delta x_i \ln[P(\Delta x)_i]^{-1} \tag{6}$$

Let us consider the application of the equation (6) to the distribution of the probabilities of random processes during the formation of the personnel of an elementary business structure regarding the most probable value N_0.

Each interval of the random values relatively to N_0 can be more or less than its value and equals by the module:

$$\Delta N = |N_0 - N_i|$$

For the probability we take the value:

$$P(\Delta x_i) = \frac{N_0}{N_0 + \Delta N}$$

Then the equation (6) looks as follows:

$$S_0 = \Delta N_i \ln \left[\left(\frac{N_0}{N_0 + \Delta N_i} \right) \right]^{-1} \tag{7}$$

The calculations of some points of graphic dependence S_0 on by the equation (7) at =20 result in Figure 1.

From the calculations and Figure 1 it is seen how the entropy goes up with the deviation of the amount of personnel N from the optimally acceptable value N_0. At the same time, not only those events at which are irrational, but also those, at which . Thus, the non-optimal increase in the bureaucratic apparatus, which seems to lighten the management work, actually results in such business entropy increase.

Thus, the given technique of entropy evaluation in a separate business structure allows establishing the norms of acceptable deviations from the most probable value of random processes in it.

The use of business entropy idea allows evaluating the quality of business processes, in particular, for the obtaining of the values of their critical parameters.

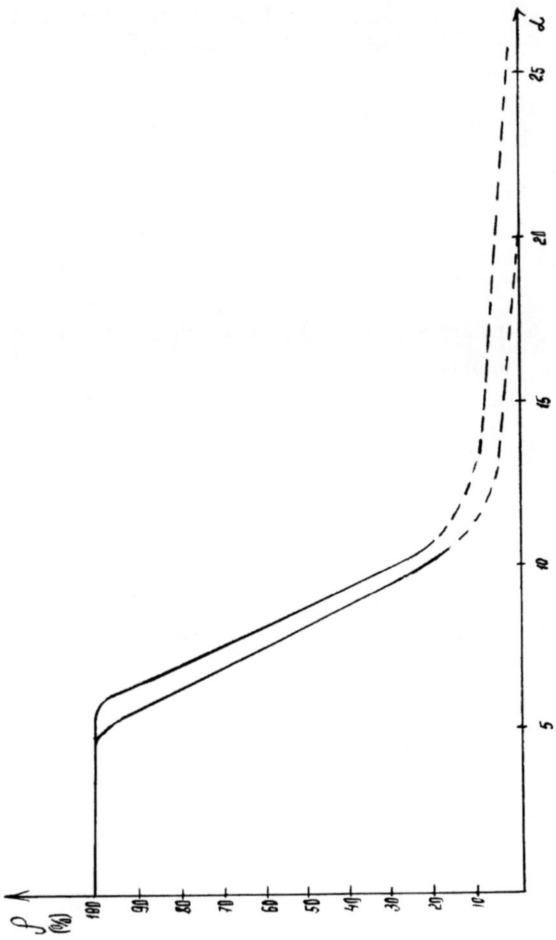

Figure 1. Entropy of random processes in the formation of the given business structure personnel.

4. Entropy of the Degree of Spatial-Energy Interactions

The idea of spatial-energy parameter (P-parameter) which is the complex characteristic of the most important atomic values responsible for interatomic interactions and having the direct bond with the atom electron density is introduced based on the modified Lagrangian equation for the relative motion of two interacting material points [7].

The value of the relative difference of P-parameters of interacting atoms-components – the structural interaction coefficient is used as the main numerical characteristic of structural interactions in condensed media:

$$\alpha = \frac{P_1 - P_2}{(P_1 + P_2)/2} 100\% \quad (8)$$

Applying the reliable experimental data we obtain the nomogram of structural interaction degree dependence () on coefficient , the same for a wide range of structures (Figure 2). This approach gives the possibility to evaluate the degree and direction of the structural interactions of phase formation, isomorphism and solubility processes in multiple systems, including molecular ones.

Such nomogram can be demonstrated [7] as a logarithmic dependence:

$$\alpha = \beta \ln(\rho^{-1}), \tag{9}$$

where coefficient β – the constant value for the given class of structures. From the average value β can structurally change mainly within ± 5%. Thus coefficient α is reversely proportional to the logarithm of the degree of structural interactions and therefore can be characterized as the entropy of spatial-energy interactions of atomic-molecular structures.

Actually the more is ρ, the more probable is the formation of stable ordered structures (e.g. the formation of solid solutions), i.e. the less is the process entropy. But also the less is coefficient α. Moreover, the values of coefficient α for a high degree of structural interactions (up to 7%) additively coincide with the entropy values for small and medium business. Such regularity in the value changes is also preserved further: for the limited degree of structural interactions coefficient α rises sharply and similarly does the value S during the transition from the medium to large business.

Comparing the nomogram (Figure 2) with the data from the Table 1, we can see the additivity of business entropy values (S) with the values of the coefficient of spatial-energy interactions (α).

Therefore as applicable to business processes the idea of business quality is similar to the concept of structural interaction degree (ρ). All this allows approximately defining the critical values of these parameters. Thus, at ... the value is , that corresponds to the number of business structures in the range between 10000 and 100000 workers (in the average about 55000).

Equation (9) does not have the complete analogy with Boltzmann's equation (1) as in this case not absolute but only relative values of the corresponding characteristics of the interacting structures are compared which can be expressed in percent. This refers not only to coefficient α but also to the comparative evaluation of structural interaction degree (ρ), for example – the percent of atom content of the given element in the solid solution relatively to the total number of atoms.

Therefore in equation (9) coefficient k = 1.

CONCLUSION

The relative difference of spatial-energy parameters of interacting structures can be the numerical characteristic of the entropy of is process:

ABOUT THE AUTHORS

1) Grigory Andreevich Korablev – Professor, Doctor of Technical Science, Head of the Physics Department of Izhevsk State Agricultural Academy, e-mail: korablevga@mail.ru
2) Natalia Grigorievna Petrova – specialist-expert of the Department of Information Security and Communication of the Ministry of Informatization and Communication of the Udmurt Republic, e-mail: biakaa@mail.ru
3) Roman Grigorievich Korablev – PhD student of Izhevsk State Agricultural Academy.
4) Anatoly Konstantinovich Osipov – Professor, Doctor of Science in Economics, Head of the Department of Management and Law of Izhevsk State Agricultural Academy.
5) Gennady Efremovich Zaikov – Doctor of Chemical Science, Professor of N.M. Emmanuel Institute of Biochemical Physics, RAS, e-mail: chembio@sky.chph.ras.ru

REFERENCES

[1] F. Reif. Statistic physics. M.: Nauka, 1972, 352 p.
[2] L.A. Gribov, N.I. Prokopyeva. Basics of physics. M.: Vysshaya shkola, 1992, 430 p.
[3] A.K. Osipov. Challenges and ways of effective management of a small public catering business. *Bulletin of Izhevsk State Agricultural Academy,* 2010, №1, p. 29-34.
[4] T.Yu. Ivanova, V.I. Prikhodko, Cybernetic-synergetic approach in management theory // *Management in Russia and abroad*, №5, 2004, p. 132-137.
[5] I.V. Khrusheva. Probability theory. Lan Publishers, 2009, 304 p.
[6] N.M. Liventsev. Course of physics. Lan Publishers, 2012, 672 p.
[7] Korablev G.A. Spatial-Energy Principles of Complex Structures Formation // Brill Academic Publishers and VSP, Netherlands, 2005, 426pp. (Monograph).

In: Quantitative Chemistry, Biochemistry and Biology
Editors: G.E. Zaikov, O.V. Stoyanov et al.
ISBN: 978-1-62948-332-0

Chapter 7

Quantitative Calculation of Spatial-Energy Interactions Entropy

G.A. Korablev[1*], N.G. Petrova[2], R.G. Korablev[1] and G.E. Zaikov[3]
[1] Izhevsk State Agricultural Academy, Izhevsk, Udmurt Republic
[2] Ministry of Informatization and Communication of the Udmurt Republic
[3] Institute of Biochemical Physics, Russian Academy of Science, Moscow, Russia

Abstract

Similar to the ideas of thermodynamics on the static entropy, the concept of the entropy of spatial-energy interactions is used. The variability of entropy demonstrations is discussed.

Keywords: Thermodynamic probability, entropy, statistic thermodynamics, spatial-energy parameter

1. Introduction

The idea of entropy appeared based on the second law of thermodynamics and ideas of the adduced quantity of heat.

In statistic thermodynamics the entropy of isolated and equilibrious system equals the logarithm of the probability of its definite macrostate:

$$S = k \ln W, \tag{1}$$

where W – number of available states of the system or degree of the degradation of microstates; k – Boltzmann's constant.

* Email: KorablevGA@mail.ru.

Or:

$$W = e^{S/k}$$

These correlations are general assertions of macroscopic character, they do not contain any references to the structure elements of the systems considered and they are completely independent from microscopic models [1].

Therefore the application and consideration of these laws can result in a large number of consequences which are most fruitfully used in statistic thermodynamics.

In statistic thermodynamics the entropy is a function of the system state which helps estimating the process directions and possible changes in them.

At any spontaneous changes in the isolated system the entropy always increases:

$$\Delta S > 0$$

The sense of the second law of thermodynamics comes down to the following:

The nature tends from the less probable states to more probable ones. The most probable is the uniform distribution of molecules through the entire volume. From the macrophysical point, these processes consist in equalizing the density, temperature, pressure and chemical potentials, and the main characteristic of the process is the thermodynamic probability W.

In actual processes in the isolated system the entropy growth is inevitable – disorder and chaos increase in the system, the quality of internal energy goes down.

In actual processes in the isolated system the entropy growth is inevitable – disorder and chaos increase in the system, the quality of internal energy goes down.

The thermodynamic probability equals the number of microstates corresponding to the given macrostate.

Since the system degradation degree is not connected with the physical features of the systems, the entropy statistic concept can also have other applications and demonstrations (apart from statistic thermodynamics).

"It is clear that out of the systems completely different by their physical state, the entropy can be the same if their numbers of possible microstates corresponding to one macroparameter (whatever parameter it is) coincide. Therefore the idea of entropy can be used in various fields. The increasing self-organization of human society … leads to the increase in entropy and disorder in the environment that is demonstrated, in particular, by a large number of disposal sites all over the earth" [2].

In this research we are trying to apply the concept of entropy to the efficiency of business structures and evaluation of the degree of spatial-energy interactions.

2. Spatial-Energy Parameter

The comparison of multiple regularities of physical chemical and biological processes allows assuming that in many cases the principle of adding the reciprocals of volume energies or kinetic parameters of the interacting structures is fulfilled.

Lagrangian equation for the relative motion of two interacting material points with masses m_1 and m_2 in coordinate x looks as follows:

$$m_r x'' = -\frac{\partial U}{\partial x}, \tag{2}$$

where

$$\frac{1}{m_r} = \frac{1}{m_1} + \frac{1}{m_2} \tag{2a}$$

where U – mutual potential energy of material points; m_r – reduced mass.

This equation can be brought to the following view:

$$\frac{1}{\Delta U} \approx \frac{1}{\Delta U_1} + \frac{1}{\Delta U_2} \tag{3}$$

in which the resultant energy characteristic of the interaction system of two material points is found by the principle of adding the reciprocals of initial energies of interacting subsystems.

"Electron with mass m moving near the proton with mass M is equivalent to the particle with the mass " [3].

Therefore modifying the equation (3) we can assume that the energy of valent orbitals of the atom (responsible for interatomic interactions) can be calculated [4] by the principle of adding the reciprocals of some initial energy components based on the following equations:

$$\frac{1}{q^2/r_i} + \frac{1}{W_i n_i} = \frac{1}{P_E} \tag{4}$$

or

$$\frac{1}{P_0} + \frac{1}{(Wrn)_i}; \tag{5}$$

$$P_E = \frac{P_0}{r_i} \tag{6}$$

where: W_i – orbital energy of electrons [5]; r_i – orbital radius of i orbital [6]; $q=Z^*/n^*$ – by [7,8], n_i – number of electrons of the given orbital, Z^* and n^* – nucleus effective charge and effective main quantum number, r – bond dimensional characteristics.

The value P_O is called spatial-energy parameter (SEP) and the value P_E – effective P-parameter (effective SEP). The effective SEP has a physical sense of some averaged energy of valent electrons in the atom and is a direct characteristic of the electron density in the atom at the given distance from the nucleus (r_i).

During the formation of solid solution and in other structural equilibrium-exchange interactions there must be the same electron density in the contact points of atoms-components. This process is followed by the re-distribution of electron density between valent zones of both particles and transition of some electrons from external spheres into the adjacent ones.

It is obvious that with the proximity of electron densities in free atoms-components, the transfer processes between the boundary atoms of particles are minimal that contributes to a new structure formation. Thus, the task of degree evaluation of such structural interactions in many cases comes down to the comparative assessment of electron densities of valent electrons in free atoms (on averaged orbitals) participating in the process by the equations:

$$\alpha = \frac{P_0'/r_i' - P_0''/r_i''}{(P_0'/r_i' + P_0''/r_i'')/2} \cdot 100\% \qquad (7)$$

or

$$\alpha = \frac{P_S' - P_S''}{P_S' + P_S''} \cdot 200\% \qquad (8)$$

where P_S – structural parameter is found by the equation:

$$\frac{1}{P_S} = \frac{1}{N_1 P_E'} + \frac{1}{N_2 P_E''} + \cdots \qquad (9)$$

where N_1 and N_2 – number of homogeneous atoms in subsystems.

The nomogram of the dependence of structural interaction degree (ρ) on coefficient α the same for a wide range of structures is made up by all the data obtained. (Fig. 1).

This approach gives the possibility to evaluate the degree and direction of the structural interactions of phase formation, isomorphism and solubility processes in multiple systems, including molecular ones.

Such nomogram can be demonstrated as a linear logarithmic dependence:

$$\alpha = \beta \ln(\rho^{-1}) \qquad (10)$$

where coefficient β – the constant value for the given class of structures. From the average value β can structurally change mainly within ± 5%. Thus coefficient α is inversely proportional to the logarithm of the degree of structural interactions and therefore can be characterized as the entropy of spatial-energy interactions of atomic-molecular structures.

Actually the more is ρ, the more probable is the formation of stable ordered structures (e.g. the formation of solid solutions), i.e. the less is the process entropy. But also the less is coefficient α. That is the relative difference of spatial-energy parameters of interacting structures can be the quantitative characteristic of this process entropy:

$$\alpha = S = \beta \ln(\rho^{-1}) \qquad (10a)$$

Equation (10) does not have the complete analogy with Boltzmann's equation (1) as in this case not absolute but only relative values of the corresponding characteristics of the interacting structures are compared which can be expressed in percent. This refers not only to coefficient α but also to the comparative evaluation of structural interaction degree (ρ), for example – the percent of atom content of the given element in the solid solution relatively to the total number of atoms.

Therefore in equation (10) coefficient k = 1.

3. Entropy of Business Structure Aggregation

The main properties of business systems providing their economic advantages are: 1) effective competition, and 2) maximal personal interest of each worker.

But on different economy concentration levels these *ab initio* features function and demonstrate themselves differently. Their greatest efficiency corresponds to small business – when the number of company staff is minimal, the personal interest is stronger and competitive struggle for survival is more active. With companies and productions increase the number of staff goes up, the role of each person gradually decreases, the competition slackens as new opportunities for coordinated actions of various business structures appear. The quality of economic relations in business goes down that is entropy increases. Such process is mostly vivid in monostructures at the largest enterprises of large business (syndicates and cartels).

The concept of thermodynamic probability as a number of microstates corresponding to the given macrostate can be modified as applicable to the processes of economic interactions that directly depend on the parameters of business structures.

A separate business structure can be taken as the system macrostate, and as the number of microstates – number of its workers (N) which is the number of the available most probable states of the given business structure. Thus it is supposed that such number of workers of the business structure is the analog of thermodynamic probability as applicable to the processes of economic interactions in business.

Therefore it can be accepted that the total entropy of business quality consists of two entropies characterizing: 1) decrease in the competition efficiency (S_1) and 2) decrease in the personal interest of each worker (S_2), i.e.:. S_1 is proportional to the number of workers in the company: , and S_2 has a complex dependence not only on the number of workers in the company but also on the efficiency of its management. It is inversely proportional to the personal interest of each worker. Therefore it can be accepted that, where – coefficient of personal interest of each worker.

By analogy with Boltzmann's equation (1) we have:

$$S = (S_1 + S_2) \sim \left[\ln N + \ln\left(\frac{1}{\gamma}\right)\right] \sim \ln\left(\frac{N}{\gamma}\right)$$

or,

$$S = k\ln\left(\frac{N}{\gamma}\right)$$

where k – proportionality coefficient;

Here N shows how many times the given business structure is larger than the reference small business structure, at which N = 1, i.e. this value does not have the name.

For non-thermodynamic systems we take k = 1. Therefore:

$$S = \ln\left(\frac{N}{\gamma}\right) \tag{11}$$

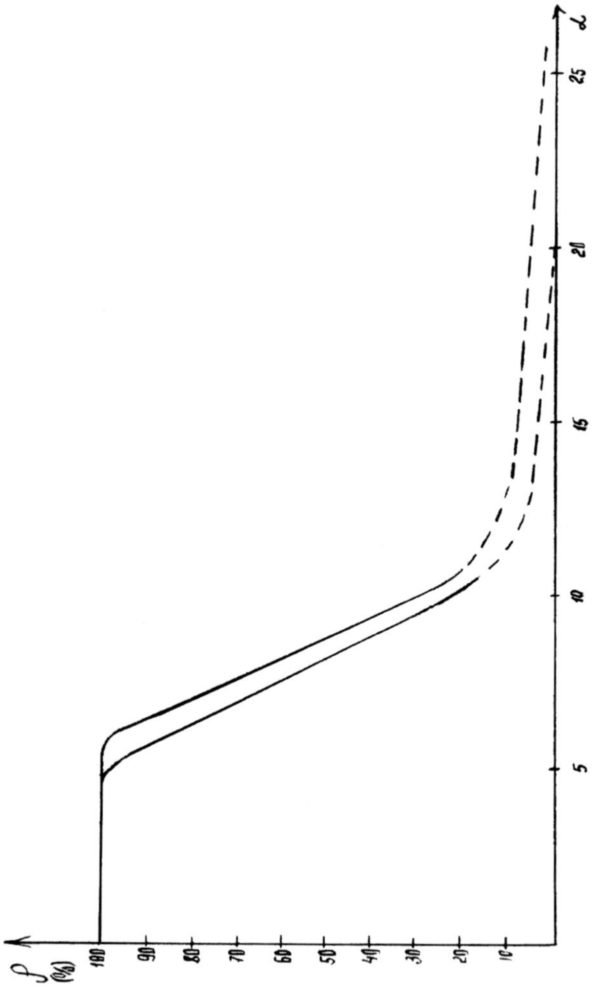

Figure 1.

In Table 1 you can see the approximate calculations of business entropy by the equation (11) for three main levels of business: small, medium and large. At the same time it is supposed that number N corresponds to some average value from the most probable values. You can see more details in [9].

When calculating the coefficient of personal interest it is considered that it can change from 1 (one self-employed worker) to zero (0), if such worker is a deprived slave, and for larger companies it is accepted as γ=0,1–0,01.

Table 1. Entropy growth with the business structure increase

Structure parameters	Business		
	Small	**Average**	**Large**
$N_1 - N_2$	10 – 50	100 – 1000	10000 – 100000
	0.9 – 0.8	0.6 – 0.4	0.1 – 0.01
S	2.408 – 4.135	5.116 – 7.824	11.513 – 16.118
	3.271	6.470	13.816

Despite of the rather approximative accuracy of such averaged calculations, we can make quite a reliable conclusion on the fact that business entropy, with the aggregation of its structures, sharply increases during the transition from the medium to large business as the quality of business processes decreases. The application of more accurate initial data allows obtaining specific values of business entropy, above which the process of economic relations can reach a critical level.

Comparing the nomogram (Fig. 1) with the data from the Table 1, we can see the additivity of business entropy values (S) with the values of the coefficient of spatial-energy interactions (α).

Moreover, the values of coefficient α for a high degree of structural interactions (up to 7%) coincide with the entropy values for small and medium business. Such regularity in the value changes is also preserved further: for the limited degree of structural interactions coefficient α rises sharply and similarly does the value S during the transition from the medium to large business.

All this confirms the assertion that the entropy concept has a multiple demonstration in various research areas.

Conclusion

The concept of entropy of spatial-energy interactions can be used similar to the ideas of thermodynamics on the statistic entropy.

About the Authors

1) Grigory Andreevich Korablev – Professor, Doctor of Technical Science, Head of the Physics Department of Izhevsk State Agricultural Academy, e-mail: korablevga@mail.ru
2) Natalia Grigorievna Petrova – specialist-expert of the Department of Information Security and Communication of the Ministry of Informatization and Communication of the Udmurt Republic, e-mail: biakaa@mail.ru
3) Roman Grigorievich Korablev – PhD student of Izhevsk State Agricultural Academy.
4) Gennady Efremovich Zaikov – Doctor of Chemical Science, Professor of N.M. Emmanuel Institute of Biochemical Physics, RAS, e-mail: chembio@sky.chph.ras.ru

REFERENCES

[1] F. Reif. *Statistic physics*. M.: Nauka, 1972, 352 p.
[2] L.A. Gribov, N.I. Prokopyeva. Basics of physics. M.: Vysshaya shkola, 1992, 430 p.
[3] Eyring H., Walter J., Kimball J. *Quantum chemistry* // M., I.L., 1948, 528 p.
[4] Korablev G.A. Spatial-Energy Principles of Complex Structures Formation // Brill Academic Publishers and VSP, Netherlands, 2005, 426pp. (Monograph).
[5] Fischer C.F. Average-Energy of Configuration Hartree-Fock Results for the Atoms Helium to Radon // Atomic Data, 1972, № 4, 301-399.
[6] Waber J.T., Cromer D.T. Orbital Radii of Atoms and lons // *J.Chem. Phys*, 1965, V 42,-№ 12, 4116-4123.
[7] Clementi E., Raimondi D.L. Atomic Screening constants from S.C.F. Functions, 1 // *J.Chem. Phys*., 1963, v.38, №11, 2686-2689.
[8] Clementi E., Raimondi D.L. Atomic Screening constants from S.C.F. Functions, 1 // *J. Chem. Phys*., 1967, V.47, № 4, 1300-1307.
[9] A.K. Osipov. Challenges and ways of effective management of a small public catering business. Bulletin of Izhevsk State Agricultural Academy, 2010, №1, p. 29-34.

In: Quantitative Chemistry, Biochemistry and Biology
Editors: G.E. Zaikov, O.V. Stoyanov et al.
ISBN: 978-1-62948-332-0

Chapter 8

BACTERIAL PLASMIDS: CLONING OF MULTI-COPY CPG MOTIFS

A.S. Korovashkina, S.V. Kvach and A.I. Zinchenko*
Institute of Microbiology, National Academy of Sciences,
Minsk, Belarus

ABSTRACT

Two plasmids with different copy numbers of unmethylated CpG dinucleotides in particular sequence contexts (CpG motifs) were constructed. The plasmid pCpG-KH11 carries 96 copies of CpG motif GTCGTT, with the latter being the most effective for stimulating the human immune system in accordance with the literature and data. The plasmid pCpG-KM11 contains 120 repetitions of GACGTT-motif found to enhance immune effects in mice, fish, etc. The sequence of inserted fragments was verified by PCR and restriction enzyme digestion.

It is suggested that constructed plasmids may be used to improve protein and DNA vaccines and to treat cancer, infectious and allergic diseases.

Keywords: CpG motifs, polymerase chain reaction, recombinant CpG-enriched plasmid, *Escherichia coli*

INTRODUCTION

Unmethylated CpG dinucleotides in particular sequence contexts (CpG motifs), which are present in bacterial (but not mammalian) DNA stimulate innate host defense mechanisms by activating Toll-like receptor 9 (TLR9) signaling pathways [1–3]. Synthetic oligodeoxynucleotides (ODNs) containing CpG motifs have been shown to act as potent adjuvants that can induce T helper-1 (Th-1)-type immune responses by mimicking the effects of bacterial CpG DNA [4–7]. Synthetic CpG ODNs can improve immune responses to a

* E-mail: zinch@mbio.bas-net.by; fax: +375(172)64-47-66.

variety of coadministered antigens, and therefore represent a promising new approach for improving vaccine efficacy [8–12]. Recently, synthetic CpG ODNs have been shown to enhance the immunogenicity of various microbial and tumor antigens in humans [13–16].

ODNs with natural phosphodiester bonds (PO-ODNs) are susceptible to nuclease degradation in cells [17]. Therefore, in order to provide resistance to nuclease activity and ensure efficient uptake of CpG ODNs by cells, phosphorothioate backbone-modified CpG ODNs containing sulfur substituted for the non-bridging oxygen atoms (PS-ODNs) were developed [18, 19]. However, several laboratories have shown that severe side effects occurred in PS-ODNs-treated mice [20–23]. Therefore, production of a potent immunostimulatory CpG PO-ODN that does not cause side effects is highly desirable for inducing a well-controlled immune response.

Another way of CpG DNA production is construction of CpG-enriched plasmids with subsequent cloning in *Escherichia coli* cells. It is known that CpG motifs present in the backbone of DNA vaccines play an important role in the development of immune responses, and increasing the number of the motifs can enhance immune responses to various antigens [24–28]. Compared with synthetic ODNs, plasmids have more stable chemical properties and are economical in scaled-up production. Moreover, such CpG-containing plasmids are non-toxic for organisms as they lack unnatural phosphorothioate linkages. Hence, CpG-enriched plasmids may prove to be more useful than protein or peptide vaccine adjuvants.

Taking all aspects into consideration, the aim of the research was the construction of recombinant CpG-enriched plasmids on the basis of a multi-copy vector.

MATERIALS AND METHODS

To construct multi-copy CpG fragments, two partially overlapping primers containing four CpG motifs were chemically synthesized (CpG-F and CpG-R). *Hind*III restriction sites were inserted at 5′ ends of the primers (the nucleobases are in italic).

Primers for cloning of **GTCGTT**-motif:

CpG-F-human –
5′-*AGCTTC***GTCGTT**TT**GTCGTT**TT**GTCGTT**TT**GTCGTT***A*-3′;
CpG-R-human –
5′-*AGCTT***AACGAC**AA**AACGAC**AA**AACGAC**AA**AACGAC**G*A*-3′.

Primers for cloning of **GACGTT**-motif:

CpG-F-mouse –
5′-*AGCTTC***GACGTT**TT**GACGTT**TT**GACGTT**TT**GACGTT***A*-3′;
CpG-R-mouse –
5′-*AGCTT***AACGTC**AA**AACGTC**AA**AACGTC**AA**AACGTC**G*A*-3′.

To anneal the primers 50 μL of the reaction mixture containing 500 pmol of each primer, 20 mM Tris-HCl (pH 7.5), 10 mM $MgCl_2$ and 50 mM NaCl was heated at 95°C and slowly cooled down to 4°C. The products obtained were double stranded DNA molecules with sticky

3' and 5' complementary ends. After phosphorylation of 5' ends with T4 polynucleotide kinase (Sileks, Russia) the products were ligated to each other. The fragments were separated using agarose gel electrophoresis and the ones with the highest size (900–1300 base pairs) where isolated and purified with MinElute Gel Extraction Kit (Qiagen, Germany) according to manufacturer recommendations.

Purified double-stranded CpG polynucleotides were ligated into the vector pXcmkn12 (Cloning Vector Collection, Japan) previously digested with *Hind*III. A 10 μL aliquot of the ligation mixture was used to transform *E. coli* DH5α competent cells (Novagen, USA). Cells where grown on Luria-Bertani (LB) agar plates with 100 μg/mL ampicillin.

Several obtained colonies were screened by a polymerase chain reaction (PCR) for plasmid pXcmkn-12 enriched with CpG motifs using primers terminating the cloned fragments. Amplification by PCR was performed in a reaction mixture (30 μL) containing crude cell lysate as a DNA template, 5 pmol of each synthetic primer (5'-GTAAAACGACGGCCAGT-3' and 5'-CAGGAAACAGCTATGAC-3'), four deoxyribonucleoside triphosphates (0.02 mM each), $MgCl_2$ (2 мM), $(NH_4)_2SO_4$ (17 mM), 67 mM Tris-HCl buffer (pH 8.3), 0.02% Tween 20 and 1 U of *Taq* DNA polymerase (Promega, USA). The reaction was conducted for 30 cycles (30 s at 95°C, 30 s at 55°C and 1 min at 72°C). Amplification products were analyzed using agarose gel electrophoresis.

Plasmids with the longest inserted fragments (designated pCpG-KH11 and pCpG-KM11) were chosen for subsequent manipulations. *E. coli* DH5α cells containing target plasmids were grown at 37°C with orbital shaking at 200 rpm in 250 mL Erlenmeyer flasks with 50 mL of ZYM505 culture medium: tryptone – 1.0%; yeast extract – 0.5%; glycerol – 0.5%; glucose – 0.05%; Na_2HPO_4 – 0.025 M; KH_2PO_4 – 0.025 M; NH_4Cl – 0.05 M; Na_2SO_4 – 0.005 M; ampicillin – 150 μg/mL in deionized water adjusted to pH 7.0 with KOH [29]. After 16 h of growth cells were harvested by centrifugation for 10 min at 12,000 g, washed once with TE buffer (10 mM Tris-HCl + 1 mM EDTA; pH 8.0) and slurried in the same buffer. Plasmids were isolated from obtained cell biomass using alkaline lysis method [30].

Restriction analysis of constructed plasmids was performed by their incubation with mixture of restriction nucleases *Nde*I and *EcoR*I. The presence of multi-copy CpG motifs in cloned fragments was affirmed by incubation of restriction products with *Hind*III restrictase.

The obtained plasmids pCpG-KH11 and pCpG-KM11 were introduced into *E. coli* strain BLR(DE3). The transformed cells were grown at 37°C with orbital shaking at 200 rpm in 250 mL Erlenmeyer flasks containing 50 mL of ZYM505 culture medium described above. Cultivation continued during 14–16 h. The obtained biomass was used for plasmid isolation by alkaline lysis method.

Results and Discussion

Many new vaccines under development consist of rationally designed recombinant proteins that are relatively poor immunogens unless combined with potent adjuvants. There is only one adjuvant in common use in the U.S., aluminum phosphate or hydroxide (e.g. alum). This adjuvant, however, has significant limitations, particularly regarding generation of strong cell-mediated (T-cell) immune responses. In recent years, a novel adjuvant, JVRS-100, composed of cationic liposome DNA complexes (CLDC) has been evaluated for immune enhancing activity. The JVRS-100 adjuvant has been shown to elicit robust immune responses compared to CpG ODN and alum adjuvants, and it efficiently enhances both humoral and cellular immune responses [31]. Guan et al. [32] and Guo et al. [33] demonstrated that plasmids containing CpG motifs could function as immune potentiators to porcine reproductive and respiratory syndrome killed virus vaccines. In view of these scientific data, the aim of the work was the preparation of recombinant CpG-enriched plasmids for further priority use as vaccine adjuvants.

Our work resulted in construction of two plasmids containing multi-copy CpG motifs. The plasmid pCpG-KH11 carries multiple repetitions of CpG motif GTCGTT. The cloned motif is known to be the most effective in stimulation of the human immune system. The plasmid pCpG-KM11 contains a series of GACGTT motif found to enhance immune effects in mice and other small-size animals. Schematic illustration of the different steps involved in the plasmids construction is shown in Figure 1.

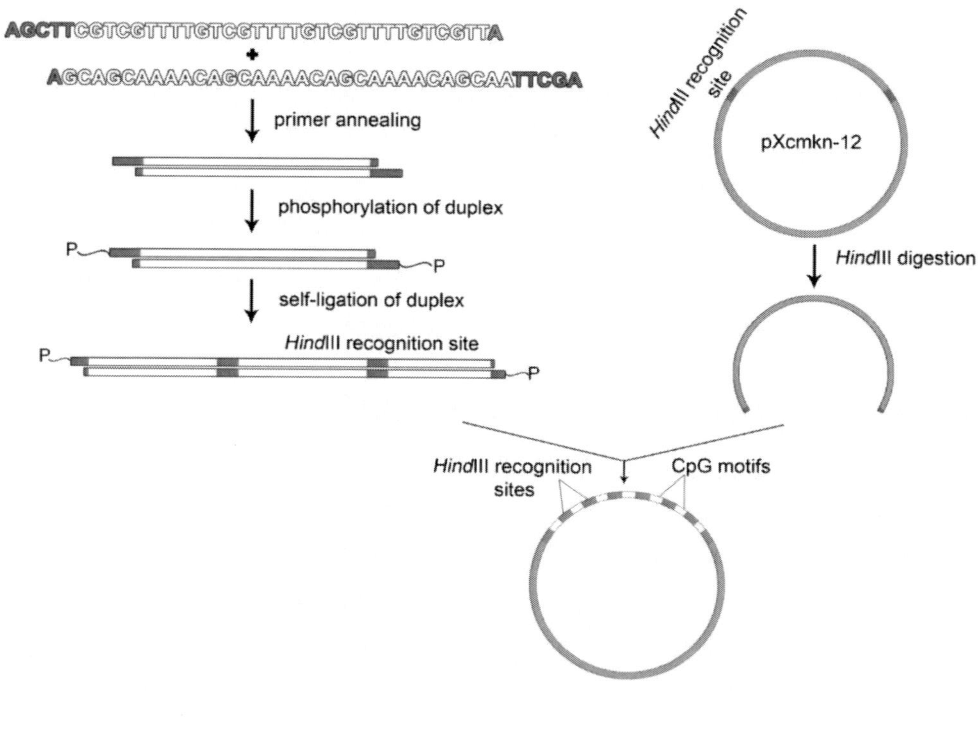

Figure 1. Schematic illustration of the different steps involved in the construction of the recombinant plasmid pCpG-KH11 containing multi-copy CpG motifs.

Characterization of constructed plasmids was performed by restriction analysis. Purified plasmids were incubated with mixture of restriction endonucleases *Nde*I and *Eco*RI which recognition sites terminated the cloned fragments. The length of cloned fragments according to electrophoregram (Figure 1) was 960 base pairs (for GTCGTT motif) and 1200 base pairs (for GACGTT motif).

The multi-copy CpG fragment consists of tandem repetitions containing 4 reiterations of CpG-motifs separated by *Hind*III restriction sites. To confirm structure of cloned fragments digestion products were additionally incubated with *Hind*III restrictase. Scheme of restriction analysis is shown in Figure 2. Electrophoregram of pCpG-KH11 and pCpG-KM11 restriction analysis is shown in Figure 3.

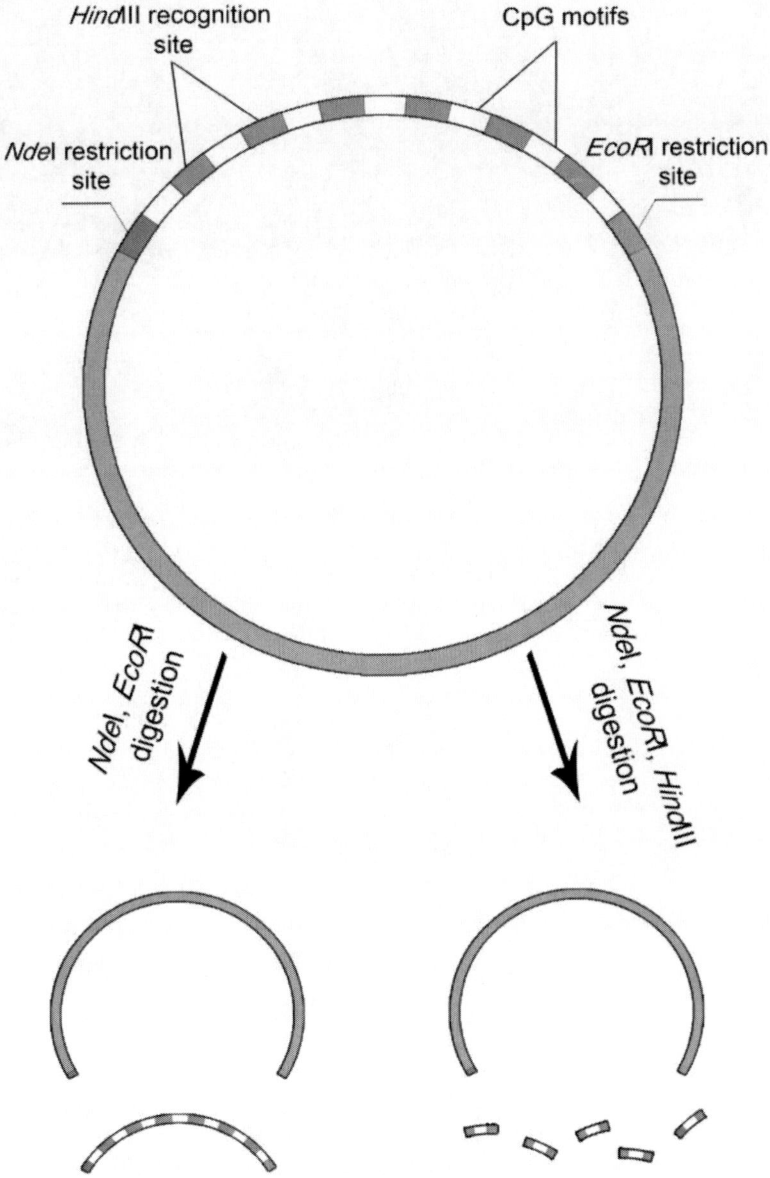

Figure 2. Scheme of restriction analysis of the plasmids containing multi-copy CpG motifs.

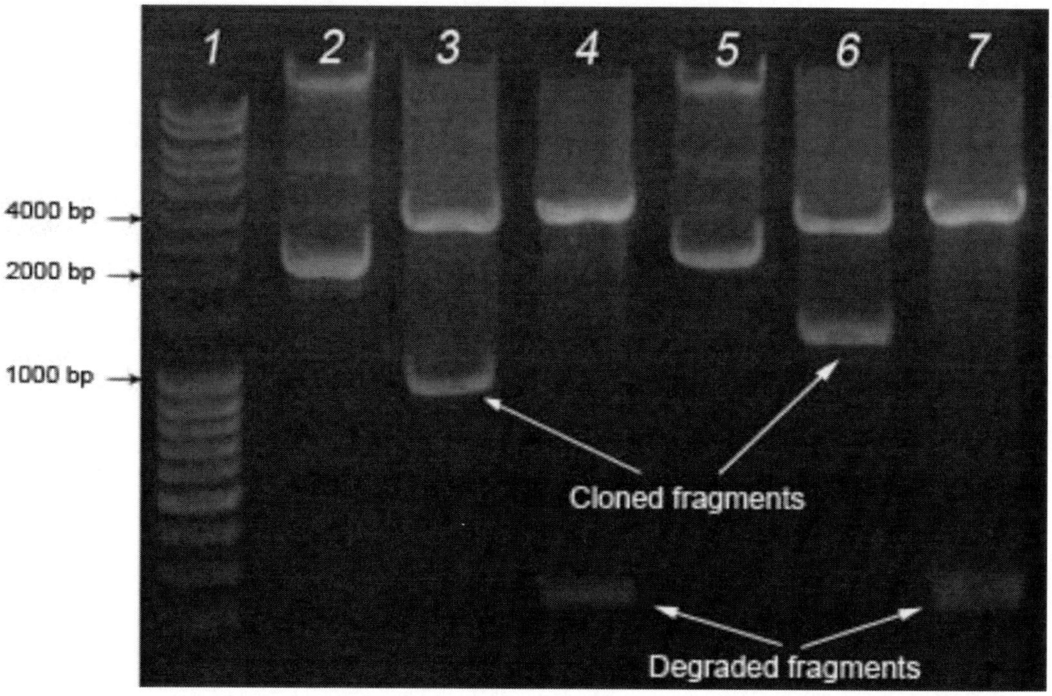

Figure 3. Electrophoregram of restriction analysis of constructed plasmids. Line 1 is DNA molecular size markers; line 2 is purified plasmid pCpG-KH11; line 3 is pCpG-KH11 digestion products after treatment with *Nde*I and *EcoR*I; line 4 is pCpG-KH11 digestion products after treatment with *Nde*I, *EcoR*I and *Hind*III; line 5 is purified plasmid pCpG-KM11; line 6 is pCpG-KM11 digestion products after treatment with *Nde*I and *EcoR*I; line 7 is pCpG-KM11 digestion products after treatment with *Nde*I, *EcoR*I and *Hind*III.

The number of cloned CpG motifs was calculated by the following formula:

$$X = \frac{D1}{40} \times 4,$$

where:

$D1$ – length of the cloned fragment identified after *Nde*I and *Eco*RI digestion (bp);
40 – length of 1 tandem repetition of CpG motifs (bp);
4 – number of CpG motifs in 1 tandem repetition.

By our estimates, the plasmid pCpG-KH11 carries 96 copies of GTCGTT motif and the plasmid pCpG-KM11 contains 120 repetitions of GACGTT motif.

The resultant recombinant plasmids pCpG-KH11 and pCpG-KM11 were introduced into *E. coli* strain BLR(DE3). The transformed cells were grown at 37°C with orbital shaking at 200 rpm in 250 mL Erlenmeyer flasks containing 50 mLof ZYM505 culture medium. The recovered cell biomass was used for isolation and purification of target plasmids. It was shown that the plasmids were produced at a relatively high level (5.5–6.0 mg/L), so that the selected recombinant strains can be successfully used for biotechnological production of CpG-enriched plasmids.

CONCLUSION

The completed work resulted in construction of the plasmids pCpG-KH11 carrying 96 copies of CpG motif GTCGTT and pCpG-KM11 with 120 repetitions of GACGTT motif. The cloned motifs are considered to be the most effective in stimulation of the human and mice immune system, respectively. Using methods of genetic engineering, strain *E. coli* CpG-KH11 producing plasmid pCpG-KH-11 and strain *E. coli* NBM-11 generating plasmid pCpG-KM11 have been obtained. The strains are potent sources of CpG-enriched plasmids and they make it possible to produce 5.5–6.0 mg of target plasmid DNA per 1 L of culture medium.

It is suggested that the constructed strains may be used in biotechnological production of pharmaceutically valuable CpG DNA.

ACKNOWLEDGMENTS

The work was supported by the grant No. 3.07 from the Belarus State Research Program «Fundamental Basics of Biotechnologies».

REFERENCES

[1] A. M. Krieg, A. K. Yi, S. Matson, T. J. Waldschmidt, G. A. Bishop, R. Teasdale, G. A. Koretzky, D. M. Klinman: *Nature*, 374 (6522), 546 (1995).

[2] H. Hemmi, O. Takeuchi, T. Kawai, T. Kaisho, S. Sato, H. Sanjo, M. Matsumoto, K. Hoshino, H. Wagner, K. Takeda, S. Akira: *Nature*, 408 (6813), 740 (2000).

[3] Y. M. Murad, T. M. Clay: Biodrugs, 23 (6), 361 (2009).

[4] G. B. Lipford, M. Bauer, C. Blank, R. Reiter, H. Wagner, K. Heeg: *Eur. J. Immunol.*, 27 (9), 2340 (1997).

[5] D. F. Hoft, C. S. Eickhoff, O. K. Giddings, J. R. C. Vasconcelos, M. M. Rodrigues: *J. Immunol.*, 179 (10), 6889 (2007).

[6] V. A. Conforti, D. M. De Avila, N. S. Cummings, R. Zanella, K. J. Wells, H. Ulker, J. J. Reeves: *Vaccine*, 26 (7), 907 (2008).

[7] Y. Goldfarb, M. Benish, E. Rosenne, R. Melamed, B. Levi, A. Glasner, S. Ben-Eliyahu: *J. Immunother.*, 32 (3), 280 (2009).
8. D. Verthelyi, R. T. Kenney, R. A. Seder, A. A. Gam, B. Friedag, D. M. Klinman: *J. Immunol.*, 168, 1659 (2002).

[8] C. Cheng, I. Bettahi, M. I. Cruz-Fisher, S. Pal, P. Jain, Z. Jia, J. Holmgren, A. M. Harandi, L. M. De La Maza: *Vaccine*, 27 (44), 6239 (2009).

[9] 10 M. H. Huang, S. C. Lin, C. H. Hsiao, H. J. Chao, H. R. Yang, Et Al.: *Plos One*, 5 (8), E12279 (2010).

[10] 11. F. Steinhagen, T. Kinjo, C. Bode, D. M. Klinman: *Vaccine*, 29 (17), 3341 (2011).

[11] *12.* R. Rappuoli, C. W. Mandl, S. Black, E. De Gregorio*: Nat. Rev. Immunol., 11 (12), 865 (2011).*

[12] 13. C. L. Cooper, J. B. Angel, I. Seguin, H. L. Davis, D. W. Cameron: *Clin. Infect. Dis.*, 46 (8), 1310 (2008).

[13] 14. F. Qian, K. M. Rausch, O. Muratova, H. Zhou, G. Song, A. Diouf: *Vaccine*, 26 (20), 2521 (2008).

[14] 15. R. D. Ellis, G. E. Mullen, M. Pierce, R. D. Ellis, G. E. Mullen, M. Pierce, L. B. Martin, K. Miura, M. P. Fay, C. A. Long, D. Shaffer, A. Saul, L.H. Miller, A. P. Durbin: *Vaccine*, 27 (31), 4104 (2009).

[15] 16. J. Karbach, S. Gnjatic, A. Bender, A. Neumann, E. Weidmann, J. Yuan, C. A. Ferrara, E. Hoffmann, L. J. Old, N. K. Altorki, E. Jager: *Int. J. Cancer.*, 126 (4), 909 (2010).

[16] 17. Q. Zhao, S. Matson, C. J. Herrera, E. Fisher, H. Yu, A. M. Krieg: *Antisense Res. Dev.*, 3, 53 (1993).

[17] 18. C. A. Stein, C. Subasinghe, K. Shinozuka, J. S. Cohen: *Nucleic Acid Res.*, 16, 3209 (1988).

[18] 19. Q. Zhao, T. Waldschmidt, E. Fisher, C. J. Herrera, A. M. Krieg: *Blood*, 84, 3660 (1994).

[19] 20. T. Sparwasser, L. Hultner, E. S. Koch, A .Luz, G. B. Lipford, H. Wagner: *J. Immunol.*, 162, 2368 (1999).

[20] 21. G. M. Deng, I. M. Nilsson, M. Verdrengh, L. V. Collins, A. Tarkowski: *Nat. Med.*, 5, 702 (1999).

[21] 22. M. Heikenwalder, M. Polymenidou, T. Junt, C. Sigurdson, H. Wagner, S. Akira, R. Zinkernagel, A. Aguzzi: *Nat. Med.*, 10, 187 (2004).

[22] 23. D. Kim, J. W. Rhee, S. Kwon, W. J. Sohn, Y. Lee, D. W. Kim, D. S. Kim, H. J. Kwon: *Biochem. Biophys. Res. Commun.*, 379, 362 (2009).

[23] 24. Y. Kojima, K. Q. Xin, T. Ooki, K. Hamajima, T. Oikawa, K. Shinoda, T. Ozaki, Y. Hoshino, N. Jounai, M. Nakazawa, D. Klinman, K. Okuda: *Vaccine*, 20 (23-24), 2857 (2002).

[24] 25. R. A. Pontarollo, L. A. Babiuk, R. Hecker, S. Van Drunen Littel-Van Den Hurk: *J. Gen. Virol.*, 83, 2973 (2002).

[25] 26. C. Coban, K. J. Ishii, M. Gursel, D. M. Klinman, N. Kumar: *J. Leukoc. Biol.*, 78 (3), 647 (2005).

[26] 27. L. Liu, X. Zhou, H. Liu, L. Xiang, Z. Yuan: *Immunology*, 115 (2), 223 (2005).

[27] 28. Z. Chen, J. Cao, X. Liao, J. Ke, S. Zhu, P. Zhao, Z. Qi: *Viral Immunol.*, 24 (3), 199 (2011).

[28] 29. F.W. Studier: *Protein Expr. Purif.*, 41 (1), 207 (2005).

[29] 30. J. Sambrook, E. F. Fritsch, T. Maniatis: Molecular Cloning. A Laboratory Manual. Cold Spring Harbor Laboratory Press, Cold Spring Harbor, New York, 1989. 2222 P.

[30] 31. S. Chang, J. Warner, L. Liang, J. Fairman: *Biologicals*, 37 (3), 141 (2009).

[31] 32. Z. Quan, Z. G. Qin, W. Zhen, X. Z. Feng, J. Hong, Z. H. Fei: *Vet. Immunol. Immunopathol.*, 136 (3-4), 257 (2010).

[32] 33. X. Guo, Q. Zhang, S. Hou, G. Zhai, H. Zhu, J. M. Sanchez-Vizcaino: *Vet. Immunol. Immunopathol.*, 144 (3-4), 405 (2011).

In: Quantitative Chemistry, Biochemistry and Biology
Editors: G.E. Zaikov, O.V. Stoyanov et al.
ISBN: 978-1-62948-332-0

Chapter 9

USING FULL-LENGTH DIGUANYLATE CYCLASE OF *THERMOTOGA MARITIME* FOR ENZYMATIC SYNTHESIS OF 3′,5′-CYCLIC DIGUANYLATE

A. S. Korovashkina, S. V. Kvach, M. V. Kvach, A. N. Rymko and A. I. Zinchenko*

Institute of Microbiology, National Academy of Sciences, Minsk, Belarus

ABSTRACT

Using methods of genetic engineering, two strains producing recombinant full-length diguanylate cyclase (DGC) of *Thermotoga maritima* with native and mutant allosteric sites have been constructed. The obtained strains have several-fold higher DGC-producing level than those described in literature. It has been shown that target enzymes are produced substantially in the form of active inclusion bodies. 7-fold increase in activity of the *T. maritima* DGC was caused by mutation in allosteric site. The possibility of applying full-length DGC of *T. maritima* in the form of inclusion bodies for synthesis of 3',5'-cyclic diguanylate has been originally demonstrated.

Keywords: c-di-GMP, diguanylate cyclase, enzymatic synthesis, allosteric site mutation, inclusion bodies, *Thermotoga maritima*, *Escherichia coli*

INTRODUCTION

The 3',5'-cyclic diguanylate (c-di-GMP) is the second messenger found in bacteria for intracellular signal transduction [1]. Accumulating evidence suggests that c-di-GMP regulates a wide variety of bacterial processes such as biofilm formation, motility, and virulence expression [2]. Moreover, recent studies have demonstrated that exogenous c-di-GMP treatment inhibits adhesive *Staphylococcus aureus* cell-to-cell interactions and biofilm

* E-mail: zinch@mbio.bas-net.by; fax: +375(172)64-47-66.

formation [3] and that it is effective also in a murine model of mastitis infection [4]. These findings emphasize its prominent antimicrobial prospects. It was demonstrated that c-di-GMP stimulates the immune system to prevent bacterial infections and is therefore evaluated as a potential vaccine adjuvant [5, 6]. It was also reported that c-di-GMP inhibits cancer cell proliferation *in vitro* and could therefore be used as an antitumor therapeutic agent [7].

Characterization of the macromolecules involved in c-di-GMP signaling, including the enzymes and receptors, implies the use of c-di-GMP as substrate or ligand for biochemical assay. The c-di-GMP can also serve as starting material for the synthesis of c-di-GMP analogues to be used as chemical probes or inhibitors [8]. It appears therefore that large quantities of this cyclic dinucleotide are required. Currently c-di-GMP is mainly produced by chemical methods suffering from such limitations as complexity, high cost and low yield of the end product [9]. On the other hand, the enzymatic production of c-di-GMP using diguanylate cyclase (DGC) domain-containing proteins involves only a single condensation of two guanosine-5'-triphosphate (GTP) molecules [10]. The reaction scheme is shown in Figure 1.

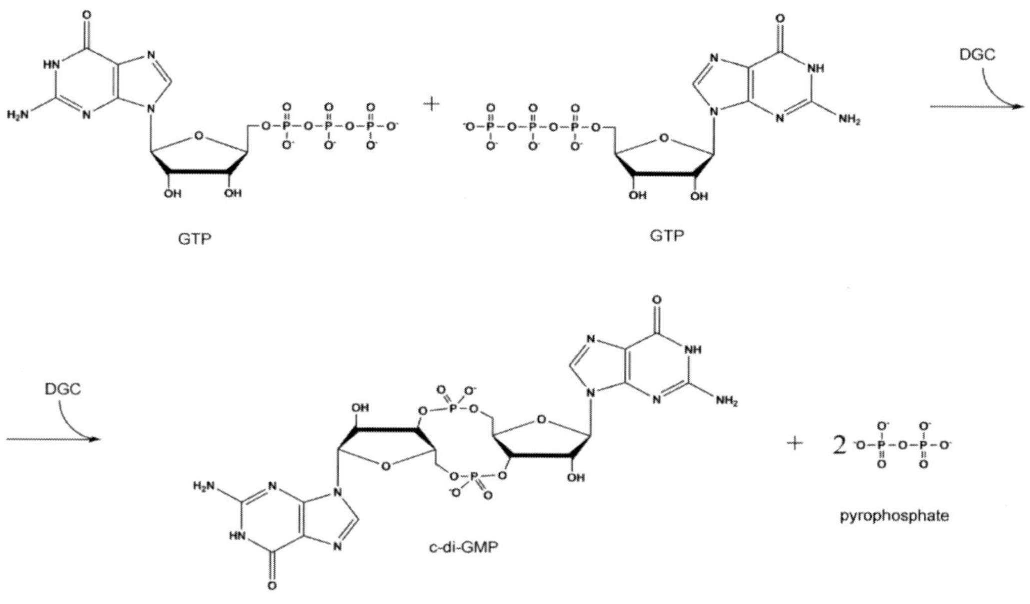

Figure 1. Synthesis of of c-di-GMP from GTP by DGC.

An enzymatic large-scale synthesis of c-di-GMP from GTP has been reported for DGC of *Gluconacetobacter xylinum* [11], DGC of *Escherichia coli* [12] and a DGC fragment of *Thermotoga maritima* [8]. Mutations were introduced in allosteric sites of DGC of *Gluconacetobacter xylinum* and *T. maritima* to increase yield of target product. However, these proteins are produced at relatively low level (10–15 mg/L of culture broth) or require complex and labor-consuming procedure of refolding. Furthermore, the mesophilic proteins of *Gluconacetobacter xylinum* and *E. coli* are rather unstable in solution.

Full-length DGC of *T. maritima* (TM1788, GenBank) comprises catalytic GGDEF domain and hydrophobic domain of unknown function [8]. It was shown earlier that the synthesis of c-di-GMP from two molecules of GTP requires the DGC protein to function in its dimeric form [1]. Previously used DGC fragment of *T. maritima* containing only catalytic domain in the reaction mixture corresponds to monomeric form and synthesis of c-di-GMP is achieved by the formation of transient dimer by the protein in the solution [8].

Considering the above mentioned aspects, we presumed that application of full-length DGC from *T. maritima* comprising both catalytic and hydrophobic domains would increase the reaction rate due to formation of stable DGC dimers.

The aim of the research was the construction and characterization of *E. coli* strains producing full-length DGC of *T. maritima* with native and mutant allosteric sites and verification of c-di-GMP synthesis using the obtained enzymes.

MATERIALS AND METHODS

The DGC-encoding gene (*TM1788*) was amplified by PCR from the chromosomal DNA of *T. maritima* using the following synthetic oligonucleotide primers: DgcfullF 5′-ATTGTG *CATATG*AAAGTTTCCGGGGGGGAAGT-3′ and DgcfullR 5′-ATCTAG*CTCGAG*TG AGA GACTGAAATAAGGCAC-3′. *Nde*I (DgcfullF) and *Xho*I (DgcfullR) restriction sites were inserted at 5′ ends of the primers (the bases are in italic). Point mutation R158A was introduced by megaprimer method [13] with primers: RXXD-f 5'- ACAGG GTT**G** **C**AAGAAGCGATGTGGTGGC-3′ and RXXD-r 5′-ATCGCTTCT**TGC**AACCCTGT CCAG AATCC-3′ (mutation is in bold).

Amplification was performed using 1 U of High-Fidelity DNA Polymerase (Fermentas, Lithuania) and a PTC-200 GeneAmp PCR system (Bio-Rad, USA). The resulting products were isolated with Wizard SV gel and PCR clean-up system (Promega, USA), digested with *Nde*I and *Xho*I restriction endonucleases and ligated into the pET24b+ (Novagen, USA) expression vector, which had been restricted with the same enzymes.

As a result plasmids bearing the gene of full-length DGC of *T. maritima* with native (pETdgc-w7) and mutated (pETdgcR158A) allosteric sites were constructed. The recombinant plasmids were introduced into *E. coli* strain BL21(DE3). The transformed cells were grown at 37°C with orbital shaking at 200 rpm in 250 mL Erlenmeyer flasks containing 50 mL of ZYM505 culture medium [14]: tryptone – 1.0%; yeast extract – 0.5%; glycerol – 0.5%; glucose – 0.05%; α-D-lactose – 0.2%; Na_2HPO_4 – 0.025 M; KH_2PO_4 – 0.025 M; NH_4Cl – 0.05 M; Na_2SO_4 – 0.005 M; kanamycin – 50 μg/mL in deionized water adjusted to pH 7.0 with KOH. Protein synthesis was triggered by autoinduction. Cells were harvested by centrifugation for 10 min at 12,000 g, washed once with buffer A containing 50 mM Tris-HCl (pH 8.0) and 100 mM KCl and then slurried in the same buffer. Cells sonication was performed using Sonifier-450 (Branson, USA) with following program: output – 0.05 kW; temperature – 4°C; duration – 600 pulses 0.5 s each. Obtained crude cell lysate was centrifuged during 30 min at 21,000 *g*.

The precipitate containing inclusion bodies was washed with 1 M urea solution containing 2% Triton X-100. After centrifugation the precipitate was washed with buffer A to remove urea and Triton X-100 traces and resuspended in the same buffer.

Inclusion bodies concentration measurement was determined by modified biuret method. In brief, suspension containing inclusion bodies was boiled during 5 min with 1% sodium dodecyl sulfate (SDS). Further measurement was performed according to standard protocol [15].

Analysis of protein composition was conducted with SDS polyacrylamide gel electrophoresis (SDS-PAGE). Purity and molecular weight of obtained proteins was evaluated using program TotalLab TL120 (Nonlinear Dynamics, USA).

DGC activity assay was carried out in 100 μL reaction mixture containing 1 μL freshly produced DGC, 1 mM GTP, 50 mM Tris-HCl (pH 8.0), 5 mM $MgCl_2$ and 250 mM NaCl. The reaction mixture was incubated at 55°C and aliquots at different time points were taken. The samples were loaded onto an TSK DEAE-3SW HPLC-column (LKB-Bromma, Sweden) and eluted at a flow rate 1 mL/min by linear gradient from 3 mM KCl+3 mM KH_2PO_4 (pH 3.3) to 360 mM KCl+360 mM KH_2PO_4 (pH 4.9) for 25 min. The amount of enzyme converting 1 μmol of substrate per minute was assumed as 1 enzyme activity unit.

Results and Discussion

The large-scale enzymatic synthesis of c-di-GMP using DGC of *Gluconacetobacter xylinum* [11], *E. coli* [12] and DGC fragment of *T. maritima* [8] has previously been reported. The most promising from the described enzymes is the fragment of DGC from hyperthermophilic microorganism *T. maritima* due to its high thermostability [8]. However, this enzyme lacks a hydrophobic domain existing in the full-length protein and therefore can't form a stable dimeric structure in the solution, which is necessary for c-di-GMP synthesis. We supposed that full-length DGC of *T. maritima* comprising both catalytic and hydrophobic domains would possess higher enzymatic activity due to formation of stable dimers in the reaction mixture.

Full-length DGC of *T. maritima* with native allosteric site (DGC^{W}) and DGC carrying allosteric site mutation R158A (DGC^{R158A}) were obtained in the current work to investigate the influence of allosteric inhibition on the enzyme activity.

At the initial stage, using methods of genetic engineering, two strains producing DGC of *T. maritima* with native (*E. coli* pTmadgc-w7) and mutant (*E. coli* pTmadgc) allosteric site have been constructed. Autoinduction of target proteins was performed in ZYM505 medium. Analysis of protein composition of cell lysate and its soluble and insoluble fractions was conducted with SDS-PAGE (Figure 2).

The Figure 2 shows that the target proteins are produced mainly in the form of inclusion bodies. In the soluble fractions, the proteins with the molecular weight corresponding to full-length DGC (29 kDa) were not detected. This could be explained by the presence of a hydrophobic domain at the N-terminus of full-length DGC leading to its irreversible aggregation [8].

Expression level of DGC^{R158A} (6%) calculated with program TotalLab TL120 was approximately 4-fold lower than that of DGC^{W} (23%). This might be due to higher enzymatic activity of DGC^{R158A} and its increased toxicity for cells.

Earlier inclusion bodies were regarded as inactive protein aggregates and to obtain the soluble protein with natural activity labor-consuming and inefficient procedure of refolding had to be arranged [16]. The recent studies have shown, however, that certain proteins possess enzymatic activity in the form of inclusion bodies [17, 18].

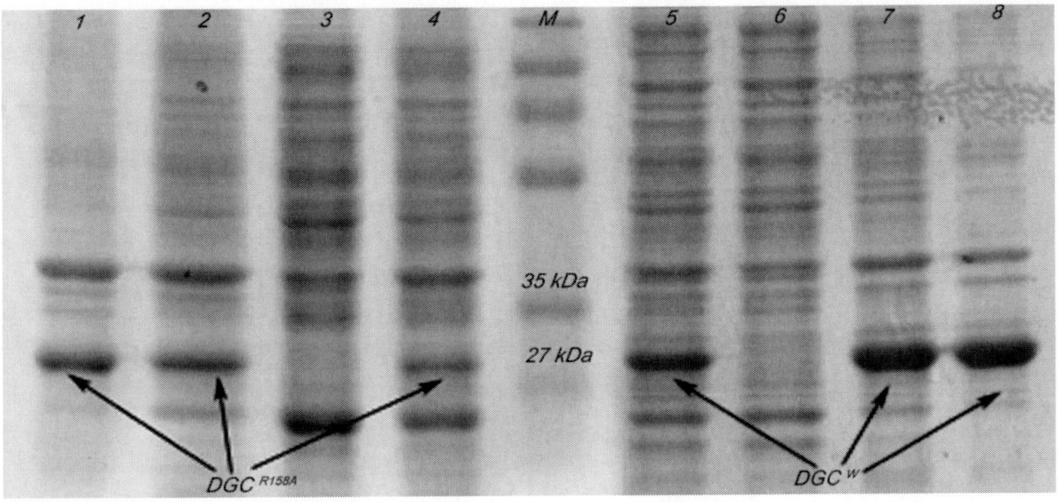

Figure 2. Electrophoregram of protein composition of E. coli pTmadgc and E. coli pTmadgc-w7 cells. Line M, protein molecular size markers; lines 4 and 5, protein compositions of E. coli pTmadgc и E. coli pTmadgc-w7 cells after autoinduction; lines 3 and 6, protein compositions of soluble fraction of E. coli pTmadgc (3) and E. coli pTmadgc-w7 (6) cell lysate; lines 2 and 7, protein compositions of insoluble fraction of E. coli pTmadgc (2) and E. coli pTmadgc-w7 (7) cell lysate; lines 1 and 8, purified DGC^{R158A} and DGC^{W}, respectively.

Taking it into account, we analyzed enzymatic activities of both soluble and insoluble fractions of cell lysate. No DGC-activity was revealed in the soluble fractions, while the activity values 4.23 U/g protein for DGC^{R158A} and 1.84 U/g protein for DGC^{W} were recorded in the insoluble fractions. However, GTP was cleaved to guanosine-5'-diphosphate and guanosine-5'-monophosphate due to presence of interfering phosphatase activity in the enzyme preparations. That required further purification of obtained proteins. For this purpose, the precipitate containing inclusion bodies was washed with 1 M urea solution containing 2% Triton X-100. Purity of obtained protein preparations was determined using SDS-PAGE (Figure 2). The results of conducted purification steps are summarized in Table 1.

There was an approximate 25-fold increase of enzymatic activity in the protein preparations after purification procedure. It might be caused by presence of proteins with interfering phosphatase activity in the initial cell lysate. In the purified protein preparations no phosphatase activity was detected. Purification grade of washed inclusion bodies estimated by program TotalLab TL120 was 31% for DGC^{R158A} and 50% for DGC^{W}. Thus, from 1 L of culture liquid we obtained 80 mg DGC^{R158A} and 410 mg DGC^{W} with specific activities 60.25 and 8.8 U/g of target protein, respectively. The specific activity of mutant DGC is 7-fold higher than that of native protein.

The obtained proteins were used for small-scale c-di-GMP synthesis. Reaction mixture containing 50 mM Tris-HCl (pH 8.0), 250 mM NaCl, 5 mM $MgCl_2$, 1 mM GTP and 2.5 μM DGC^{R158A} or 9.5 μM DGC^{W} was incubated at 55°C. Reaction progress was monitored using high-performance liquid chromatography (Figure 3).

Table 1. Results of DGC^{R158A} and DGC^{W} purification

Fraction	DGC^{R158A}					DGC^{W}				
	Total activity, U	Total protein, g	Specific activity, U/g protein	Relative purification factor	Reco-very, %	Total activity, U	Total protein, g	Specific activity, U/g protein	Relative purification factor	Reco-very, %
Crude cell lysate	2	1.27	1.57	1	100	1.4	1.9	0.74	1	100
Precipitate of the cell lysate	2.3	0.55	4.23	2.7	115	1.9	1.032	1.84	2.5	135
Precipitate of the cell lysate after washing with 1 M urea + 2% Triton X-100	4.8	0.26	18.68	12	240	3.6	0.825	4.4	5.9	259

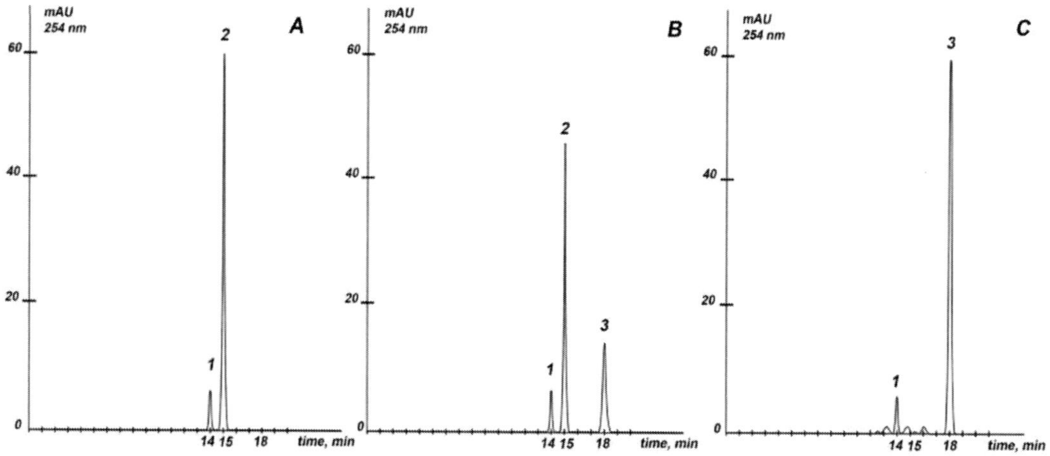

Figure 3. HPLC analysis of aliquots from the reaction mixture before (A) and after incubation with DGC for 20 min (B) and 60 min (C).

It was shown that under described conditions approximately 100% of 1 mM GTP was converted into c-di-GMP during 1 h by 2.5 μM DGC^{R158A} and 95% of the substrate at the same concentration was transformed into the end product in 5 h by 9.5 μM DGC^{W}.

As reported earlier, under the same conditions the reaction was completed in 1 h by 10 μM fragment of mutated *T. maritima* DGC [8]. The latter enzyme level is 4-fold higher than the amount of the mutant full-length DGC of *T. maritima* used in the current work.

CONCLUSION

This work resulted in the construction of strains *E. coli* pTmadgc-w7 and *E. coli* pTmadgc producing full-length recombinant DGC of *T. maritima* with native and mutant allosteric sites. The obtained strains have several-fold higher DGC producing capacity than those described in literature. It has been shown that target enzymes are produced substantially in the form of inclusion bodies that facilitate the protein purification process and subsequent isolation of the enzyme from the reaction mixture. 7-fold increase in activity of the *T. maritima* DGC was triggered by mutation introduced into allosteric site. Moreover, full-length mutant enzyme DGC^{R158A} is more active than the truncated protein described previously by Rao et al. [8] and it allows the synthesis of c-di-GMP with lower enzyme amounts to be accomplished, probably due to the presence of a hydrophobic domain.

ACKNOWLEDGMENTS

The work was supported by the grant No. 3.07 from the Belarus State Research Program «Fundamental Basics of Biotechnologies».

REFERENCES

[1] U. Romling, M. Gomelsky, M. Y. Galperin: *Mol. Microbiol.,* 57 (3), 629 (2005).

[2] U. Jenal, J. MALONE: *Annu. Rev. Genet.,* 40, 385 (2006).

[3] D. K. R. Karaolis, M. H. Rashid, R. Chythanya, W. Luo, M. Hyodo, Y. Hayakawa: Antimicrob. *Agents Chemother.*, 49 (3), 1029 (2005).

[4] E. Brouillette, M. HYoDO, Y. Hayakawa, D. K. Karaolis, F. MALOUIN: *Antimicrob. Agents Chemother.*, 49 (8), 3109 (2005).

[5] D. K. R. Karaolis, T. K. Means, D. Yang, M. Takahashi, T. Yoshimura, E. MUrAILLE, D. Philpott, J. T. Schroeder, M. Hyodo, Y. Hayakawa, B. G. Talbot, E. Brouillette, F. Malouin*: J. Immunol.,* 178 (4), 2171 (2007).

[6] A. D. Ogunniyi, J. C. Paton, A. C. Kirby, J. A. Mccullers, J. Cook, M. Hyodo, Y. Hayakawa, D. K. R. Karaolis: *Vaccine*, 26 (36), 4676 (2008).

[7] D. K. R. Karaolis, K. Cheng, M. Lipsky, A. Elnabawi, J. Catalano, M. Hyodo: *Biochem. Biophys. Res. Commun.,* 329 (1), 40 (2005).

[8] F. RAO, S. Pasunooti, Y. NG, W. Zhuo, L. Lim, A. W. Liu, Z. H. Liang: *Anal. Biochem.,* 389 (2), 138 (2009).

[9] I. Kiburu, A. Shurer, L. Yan, H. O. Sintim: Mol. Biosyst., 4 (6), 518 (2008).

[10] C. Chan, R. Paul, D. Samoray, N. C. Amiot, B. Giese, U. Jenal, T. *Schirmer: Proc. Natl. Acad. Sci.* USA, 101 (49), 17084 (2004).

[11] V. Spehr, R. Warrass, K. Hocherl, T. ILG: Appl. Biochem. Biotechnol., 165 (3-4), 761 (2011).

[12] F. Zahringer, C. Massa, T. Schirmer: Appl. Biochem. Biotechnol., 163 (1), 71 (2011).

[13] M. J. Mcpherson, S. G. Moller: PCR. *The basics from background to bench.* BIOS, New York, 2000. 276 p.

[14] F. W. Studier: *Protein Expr. Purif.,* 41 (1), 207 (2005).
[15] R. M. C. Dawson, D. C. Eleiott, W. H. Eleiott, K. M. *Jones: Data for biochemical research.* Mir, Moscow, 1991. 544 p. (in Russian).
[16] R. Rudolph, H. Lilie: *Faseb J.,* 10 (1), 49 (1996).
[17] J. Nahalka, V. Patoprsty: *Org. Biomol. Chem.*, 7 (9), 1778 (2009).
[18] E. Garcia-Fruitos, E. Vazquez, C. Diez-GIL, J. L. Corchero, J. Seras-Franzoso, I. Ratera, J. Veciana, A. Villaverde: T*rends Biotechnol.,* 30 (2), 65 (2012).

In: Quantitative Chemistry, Biochemistry and Biology
Editors: G.E. Zaikov, O.V. Stoyanov et al.
ISBN: 978-1-62948-332-0

Chapter 10

The Segmental and Translational Movements of Polystyrene into Solutions: Investigation of Kinetic Parameters

Yuriy Medvedevskikh and Oksana Khavunko*
Physical Chemistry of Combustible Minerals Department
Institute of Physical–Organic Chemistry & Coal Chemistry
named after L. M. Lytvynenko
National Academy of Sciences of Ukraine, Lviv, Ukraine

Abstract

An analysis of the characteristic times of the polystyrene segmental motion has been done, which were estimated based on the elastic and frictional viscosity components of the polystyrenes solutions. It was estimated the characteristic times of translational motion and also the diffusion coefficients of polystyrene in diluted and concentrated solutions.

Keywords: Viscosity of solution, characteristic time of the segmental motion, coefficient of diffusion

1. Introduction

Kinetics of the chemical processes proceeding with the formation of polymers (*namely*, polymerization, polycondensation) or with their participation (polymeranalogical transformations) in the most cases actually depends on the dynamic properties of the polymeric chains, *that is* on the characteristic times of their translational movement (t_t^*) and the rotary movement (t_r^*). Since the monomeric links connected into a chain, the all of these

* e–mail: hav.ok@yandex.ru.

types of the movements are realized purely in accordance with the reputation mechanism, i.e. via the segmental movement with the characteristic time τ_s.

2. EXPERIMENTAL

In order to obtain the statistically important experimental data we studied the gradient dependence of the viscosity for the diluted and concentration solution of the polystyrene in toluene [1,2]. The experiments have been carried out with the use of rotary viscometer "Rheotest 2.1" by VEB MLW production with the working cylinder containing two rotation surfaces.

Thus, experimentally measuring the effective viscosity of polymeric solution as the function of the angular rate ω (rps) of the working cylinder, we have found the frictional η_f and elastic η_e components of the viscosity and parameter b applying the optimization method with the use of the Origin 5.0 program [1,2].

3. RESULTS AND DISCUSSION

The characteristic times of the segmental motion τ were obtained based on the experimental values of η_e and b. That is why let's analyze and generalize the obtained experimental data [**1**, **2**] of the characteristic times of the segmental movement τ_s of the chains of polystyrene into solutions, estimated based on the elastic component of the viscosity η_e. Besides, let's add to this analysis also the characteristic times of the segmental movement, estimated based on the coefficient of the frictional component of the viscosity of diluted (η_{sm}) [**1**] and concentrated (η°_{pz}) [**2**] solutions. The values τ_s will be used in the sequel for the estimation of the characteristic time of the translation movement t^*_t and of the coefficient of the diffusion D of the polystyrene chains into solutions.

Accordingly to the experimental data [1, 2] the temperature dependence τ_s, estimated based on the elastic component of the viscosity η_e, is described by the equations:

into diluted solution (temperature range 20 – 35 °C)

$$\ln\tau_s = -44.07 + 6660/T, \quad (1)$$

into concentrated solution (temperature range 25 – 40 °C)

$$\ln\tau_s = -42.23 + 5950/T, \quad (2)$$

Let's write also the temperature dependencies of the coefficients of a frictional component of the viscosity:

into diluted solution

$$\ln\eta_{sm} = -29.04 + 7300/T, \quad (3)$$

into concentrated solution

$$\ln \eta^{\circ}_{pz} = -16.67 + 4800/T \, . \qquad (4)$$

Characteristic time of the segmental movement let's describe by the expression [**3**]:

$$\ln \tau_s = \ln 2\frac{h}{kT} - \frac{\Delta S_s}{R} + \frac{E_s}{RT}, \qquad (5)$$

where $\ln 2\frac{h}{kT}$ = -28.78 at T = 303 *K*. Using of these values and comparing (5) and (1), (2), we will obtain the numerical estimations for the activation entropy of the segmental movement $\Delta S_s / R$, which represented in Table, in which also the activation energies E_s of the segmental movement and the value τ_s at T = 303 K are represented too. It can be seen from the Table, that the numerical values both τ_s, and the thermodynamic characteristics ($\Delta S_s / R$ and E_s) of the segmental movement into diluted and concentrated solutions are differed only within the limits of the experimental error of their estimations. Let's assume, that the coefficients of the frictional component of the viscosity of polymeric chains are described by as same general expression [**3**], as the coefficients of the viscosity of low–molecular solution. At that time it can be written:

$$\eta_{sm} = 3\frac{RT}{V}\tau_{sm}, \qquad (6)$$

$$\eta^{\circ}_{pz} = 3\frac{RT}{V}\tau_{pz}, \qquad (7)$$

where *V* is the partial–molar volume of a monomeric link of the chain; τ_{sm} and τ_{pz} per sense, the characteristic times of the segmental movement of free polymeric chain into diluted solution and overlapping one with others polymeric chains into the concentrated solution taking into account the all possible gearing effects, correspondingly.

Table. Characteristic parameters of the segmental and translational movements of the polystyrene into solutions

System	E_s, $\frac{kJ}{mole}$	$\frac{\Delta S_s}{R}$	E_{pz}, $\frac{kJ}{mole}$	$\frac{\Delta S_{pz}}{R}$	T = 303 K			
					τ_s, c	τ_{pz}, c	t_t^*, c	D, м^2/c
Diluted solutions	55.3	15.3	—	—	$2.0 \cdot 10^{-10}$	—	$1.3 \cdot 10^{-6}$	$1.4 \cdot 10^{-12}$
Concentrated solution	49.4	13.5	39.9	6.0	$2.0 \cdot 10^{-10}$	$6.0 \cdot 10^{-9}$	$2.9 \cdot 10^{-4}$	$1.0 \cdot 10^{-13}$

Since the partial–molar volume *V* of the monomeric link of the polystyrene is unknown, then for the next calculations it can be assumed without a great error to be equal to the molar volume of the monomeric link into the melt: $V = \rho/M_0$, where $\rho = 1.05 \cdot 10^6$ g/m^3 is a density of the polystyrene melt; M_0 = 104.15 g/mole is the molar mass of styrene. Let us write (6) and (7) in general form:

$$\ln \eta = \ln 3RT\frac{\rho}{M_0} + \ln \tau . \tag{9}$$

At this, $\ln 3RT\frac{\rho}{M_0} = 18.15$ at T = 303 K. Taking into account of this value and comparing the (9) and (3) and (4), we will obtain the temperature dependences τ_{sm} and τ_{pz}:

for diluted solution

$$\ln \tau_{sm} = -47.15 + 7300/T , \tag{10}$$

for concentrated solution

$$\ln \tau_{pz} = -34.82 + 4800/T , \tag{11}$$

in accordance of which the τ_{pz} and τ_{sm} have been calculated at T = 303 K. Taking into account a general equation (5) it has been found also the value of the activation entropy $\Delta S_{pz}/R$ (see Table). Comparing the parameters of the equation (1) for τ_s and (10) for τ_{sm}, it can be seen, that the difference between them is adequately kept within the error limits of their estimation. The values of τ_s and τ_{sm} at T = 303 K, equal to $2.5 \cdot 10^{-10}$ s and $1.0 \cdot 10^{-10}$ s correspondingly prove of this fact. Thus, it can be assumed that $\tau_s \equiv \tau_{sm}$, and that is why the coefficient of the frictional component of the viscosity η_{sm} of the polymeric chains can be described by as same general expression (6) as for the coefficient of the low–molecular solution. The values of τ_{pz} calculated accordingly to the expression (11) for concentrated solution at T = 303 K (see Table), are essentially differed from τ_s: $\tau_{pz} > \tau_s$, approximately on two orders. An analysis of the parameters of the equations (2) and (11) showed that the difference between τ_s and τ_{pz} is caused by two factors, which abhorrent the one of the other: by insignificant decreasing of the activation energy ($E_{pz} < E_s$), that should be decreased the τ_{pz}, and by a sharp decreasing of the activation entropy $(\Delta S_{pz} < \Delta S_s)$, that increases of τ_{pz}.

The coefficient of the frictional component of viscosity η_{pz} in concentrated solutions caused by the movement of the overlapping between themselves and polymeric chains relatively the one of the other and characterizes the efficiency of the all possible gearings. However, the mechanism of this movement is also reputational that is realized via the segmental movement. Correspondingly, between the times τ_s and τ_{pz} some relationship should exist. Let's assume the thermodynamical approach for the determination of this relationship is one among many possibilities.

Let's determine the notion «gearing» as the thermodynamical state of a monomeric link of the chain, at which its segmental movement is frozen. This means that under the relative movement of the intertwining between themselves and polymeric chains the reputational mechanism of the transfer at the expense of the segmental movement takes place, but under condition that the part of the monomeric links of a chain is frozen.

Let the ΔG_z° is a standard free energy of the monomeric link transfer from a free state into the frozen one. Then the probability of the frozen states formation or their part should be proportional to the value $\exp\{-\Delta G_z^\circ/RT\}$. That is why, if the k_s is a constant rate of the free segmental transfer, and k_{pz} is the rate constant of the frozen segmental transfer, then between themselves the relationship should be existing:

$$k_{pz} = k_s \exp\{-\Delta G_z^\circ/RT\}. \tag{12}$$

Then k_{pz}, additionally to k_s, has a free activation energy equal to the standard free defrosting energy of the frozen state.

Since $k_s = \tau_s^{-1}$, $k_{pz} = \tau_{pz}^{-1}$, we obtained

$$\tau_{pz} = \tau_s \exp\{\Delta G_z^\circ/RT\}. \tag{13}$$

By assigning

$$\Delta G_z^\circ = \Delta H_z^\circ - T\Delta S_z^\circ, \tag{14}$$

and taking into account the experimentally determined ratios $\tau_{pz} > \tau_s$, $E_{pz} < E_s$ and $\Delta S_{pz} < \Delta S_s$, we conclude, that in (14) $\Delta G_z^\circ > 0$, $\Delta H_z^\circ < 0$ and $\Delta S_z^\circ < 0$, and besides the entropy factor $T\Delta S_z^\circ$ should be more upon the absolute value than the enthalpy factor ΔH_z°. These ratios per the physical sense are sufficiently probable. A contact of the links under the gearing can activate a weak exothermal effect $(\Delta H_z^\circ < 0)$ at the expense of the intermolecular forces of interaction, and the frosting of the segmental movement activates a sharp decrease of the entropy of monomeric link $\Delta S_z^\circ < 0$, but at this $|T\Delta S_z^\circ| > |\Delta H_z^\circ|$. Let's rewrite the (13) with taking into account of (14) in a form

$$\ln \tau_{pz} = \ln \tau_s + \frac{\Delta H_z^\circ}{RT} - \frac{\Delta S_z^\circ}{R}. \tag{15}$$

Comparing the expressions (2) and (11) and taking into account (15) we obtained: $\Delta G_z^\circ = 9.0$ kJ/mole, $\Delta H_z^\circ = -9.6$ kJ/mole, $\Delta S_z^\circ / R = -7.4$. Next, let's use the obtained numerical values of the characteristic times of the segmental movement τ_s for the estimation of dynamical properties of the polystyrene chains that is their characteristic time of the translational movement t_t^* and coefficient of diffusion D into solutions. Accordingly to [**4**], the values t_t^* and D are determined by the expressions:

into diluted solutions

$$t_t^* = N^{8/5}\tau_s, \text{ and } D = \frac{a^2}{2\tau_s} N^{-3/5}.$$

into concentrated solutions

$$t_t^* = N^{3.4}\left(\frac{\rho}{\rho_0}\right)^{2.5} \tau_s, \text{ and } D = \frac{a^2}{2\tau_s} \Big/ \left(\frac{\rho}{\rho_0}\right)^{2.5} N^{2.4}.$$

a is a length of a link of the polymeric chain, and N is their number; ρ is the concentration of a polymer into solution, g/m^3.

It was assumed for the calculations $a = 1.86 \cdot 10^{-10}$ m, N = 10^3 and $\rho = 0.5 \cdot 10^6$ g/m^3 for concentrated solution (see Table). As we can see, the characteristic time of the translational movement t_t^* of the polystyrene chains is on 4 and 6 orders higher than the characteristic time of their segmental movement; this is explained by a strong dependence of t_t^* on the length of a chain. The coefficients of diffusion weakly depend on the length of a chain, that is why their values into solutions is on 2–3 order less, than the coefficients of diffusion of low–molecular substances, which are characterized by the order 10^{-9} m^2/ s.

CONCLUSION

Thus, the carried out analysis shows that the studies of the viscosity of polymeric solutions permits sufficient accurately in estimating the characteristic times of the segmental and translational movements, on the basis of which the coefficients of diffusion of polymeric chains into solutions can be calculated.

REFERENCES

[1] Medvedevskikh Y. G. Frictional and elastic components of the viscosity of polysterene–toluene diluted solutions/ Medvedevskikh Y. G., Khavunko O. Yu. // *«Chemistry and Chemical Technology»* – 2011. – vol. 5. – № 3 – p. p. 291 – 302.

[2] Medvedevskikh Yu. G. Frictional and Elastic Components of the Viscosity of Concentrated Solutions «Polystyrene in Toluene»/ Medvedevskikh Yu. G., Khavunko O. Yu., Bazylyak L. I., Zaikov G. E. // Collected Papers: «Kinetics, Catalysis and Mechanism of Chemical Reactions: From Pure to Applied Science» / Series: «Chemistry Research and Applications» – Nova Science Publisher Edn. – ISBN: 978–1–61470–712–7. 2012. – vol. 2: *«Tomorrow and Perspectives»*. – Chapter 6. – P. 59–81.

[3] Medvedevskikh Yu. G. Phenomenological coefficients of viscosity of the low–molecular liquids and solutions // Collected «Papers of the Scientific Society named after Shevchenko. Series: *«Chemistry & Biochemistry»*. – 2011. – v. XXVIII. – p. p. 70–83 *(in Ukrainian).*

[4] Medvedevskikh Yu. G. Diffusion Coefficient of Macromolecules into Solutions and Melts / Yu. G. Medvedevskikh // Conformation of Macromolecules Thermodynamic and Kinetic Demonstrations – N. Y.: *Nova Sci. Publishing,* 2007 – P. 107 - 123.

In: Quantitative Chemistry, Biochemistry and Biology
Editors: G.E. Zaikov, O.V. Stoyanov et al.
ISBN: 978-1-62948-332-0

Chapter 11

INTERACTION OF HYDROXY ORGANOCYCLOTETRASILOXANES WITH PROPYL BUTYRATE SIDE GROUPS AND POLYMER ELECTROLYTE MEMBRANES ON THEIR BASIS: RING OPENING POLYMERIZATION REACTIONS

***T. Tatrishvili*[1,2], *E. Markarashvili*[1,2], *I. Esartia*[1,2], *J. Aneli*[2], *G.E. Zaikov*[3] *and O. Mukbaniani*[1,2*]**

[1]Iv. Javakhishvili Tbilisi State University, Tbilisi, Georgia
[2] Iv. Javakhishvili Tbilisi State University,
Institute of Macromolecular Chemistry and Polymeric Materials, Tbilisi, Georgia
[3]Institute of Chemical Physics, Russian Academy of Sciences, Moscow, Russia

ABSTRACT

Ring opening polymerization reactions of organocyclotetrasiloxanes with propyl butyrate, hydroxyl and ethyltriethoxysilane side groups in solution and in melt condition, in the temperature range 60-120^{0}C, in the presence of nucleophilic powder-like anhydrous potassium hydroxide, alkali metal fluoride as well as quaternary ammonium hydroxide have been investigated. Catalyst activity in ring opening polymerization reactions, of organocyclotetrasiloxanes, has been determined.

The synthesized polymers were characterized by FTIR, ^{1}H, ^{13}C and ^{29}Si NMR spectroscopy, GPC and wide angle X ray investigations have been carried out.

Electro physical properties (electric conductance and voltamogrammes) of the membranes were investigated.

Keywords: Organocyclotetrasiloxane, ring opening polymerization, NMR spectroscopy, gel-permeation chromatography, wide-angle X-ray analyzes, electric conductance

* E-mail: omarimu@yahoo.com.

AIMS AND BACKGROUND

It was widely recognized that solid polymer electrolytes could be potentially applied in many solid electrochemical devices such as high-energy density batteries, electrochromic windows and light-emitting devices [1-5]. The polymer electrolytes to be used in the devices have to satisfy several requirements, including high ambient ionic conductivity, fast charge transfer at the electrode interface, electrochemical stability, mechanical strength and processability.

After the pioneering work discovering ionic conductivity in alkali metal salt complexes of poly-*For correspondence (ethylene oxide) (PEO) by P.V. Wright in 1973, polymer electrolytes were proposed for batteries by M.B. Armand in 1978. But unfortunately, this material has low ion migration that mainly takes place in the amorphous part of the polymer and the crystalline domains will impede ion mobility both by blocking the paths of ions and by reducing the overall polymer flexibility.

In order to improve ionic conductivity, the various approaches and strategies have been employed. The main requirements for high conductivity are ability to solvate ions and a low glass transition temperature affording the facile ion transport. The promotion of the anion-cation dissociation is desirable, as it should lead to enhancement of ionic conductivity via an increase in the free ion concentration.

In the aspect of low T_g, polysiloxane has been studied extensively. It has the highly flexible backbone with the barrier to the bond rotation being only 0.8 kJ mol^{-1} as well as very low glass-transition energies (T_g) and high free volumes. Polysiloxane with PEO side chains was synthesized by Hall et al. for the first time [6] and Smid and co-workers [7]. Later, similar systems were investigated by Spindler and Shriver [8]. However, the chemical stabilities were not good enough due to the hydrolysies of Si-O-C branched side chains [9]. Smid and co-workers enhanced their chemical stability by connecting the PEO branches via a Si-C bond [3].

EXPERIMENTAL

THF and toluene were dried over sodium benzophenone ketyl and distilled in an atmosphere of dry nitrogen before use.

Initial compounds 2.4.6-tris(propylbutyrate)-8-hydroxy-2.4.6.8-tetramethylcyclo-tetrasiloxane (I), 2.4-bis(propylbutyrate)-6-ethyltriethoxysilane-8-hydroxy-2.4.6.8-tetra-methylcyclotetrasiloxane (II) are prepeared by the hydrosilylation reaction of 2.4.6.8-tetrahy-dro-2.4.6.8-tetramethylcyclotetrasiloxane (D_4^H) with allyl butyrate and vinyltriethoxysilane [10].

Characterization

FTIR spectra were recorded on a Nicolet Nexus 470 machine with MCTB detector. ^{1}H, ^{13}C NMR and ^{29}Si spectra were recorded on a 500 MHz Brukker NMR spectrometer, using $CDCl_3$ as the solvent and an internal standard. Differential scanning calorimetric investigation (DSC) was performed on a Perkin Elmer DSC-7 apparatus. Thermal transitions including glass transition temperatures T_g were taken as the maxima of the peaks. The heating and cooling scanning rates were 10 K/min. All DSC analyses were performed under a nitrogen atmosphere.

Gel-permeation chromatographic investigation was carried out with the use of Waters Model 6000A chromatograph with an R 401 differential refractometer detector. The column set comprised 10^3 and 10^4 Å Ultrastyragel columns. Sample concentration was approximately 3% by weight in toluene and typical injection volume for the siloxane was 5 μL flow rate - 1,0 ml/min. standardization of the GPC was accomplished by the use of styrene or polydimethylsiloxane standards with the known molecular weight.

Wide-angle X-ray analyses were performed on a DRON-2 (Burevestnik, Saint Petersburg, Russia) instrument. A-CuK_α was measured without a filter; the angular velocity of the motor was $\omega \approx 2\ ^0$/min.

Polymerization of 2.4.6-tris(propylbutyrate)-8-hydroxy-2.4.6.8-tetramethylcyclotetrasiloxane in the presence of potassium hydroxide. 1.2500 g (1.953 mmol) compound and the solvent dry toluene were placed into 100 ml flask under the dried condition and was polymerized with 0.01 wt.% of anhydrous powder-like potassium hydroxide at 50^0C for 48 hours under nitrogen. After 30 ml chloroform was added, the product is washed by water. The crude product was stirred with $MgSO_4$ for 6 hours, filtered and evaporated. The product is precipitated with THF and hexane (30/50 v/v), the viscous liquid product dried in a vacuum and 1.1 g (90%) polymer has been obtained. The hydrosilylation reactions in the presence of other catalysts and various temperatures as well as in melt condition were carried out according to the above-mentioned method.

On the basis of synthesed polymer IV and lithium trifluoromethylsulfonate (triflate) or lithium bis(trifluoromethylsulfonyl)imide thin (0.2-0.3 mm) membranes were obtained.

General Procedure for Preparation of Cross-Linked Polymer Electrolytes. The 0.75 g of the base oligomer IV was dissolved in 4 mL of dry THF and thoroughly mixed for half an hour before the addition of the catalytic amount of acid (one drop of 0.1 N HCl solution in ethyl alcohol) to initiate the cross-linking process. After stirring for another 3 h the required amount of lithium triflate from the previously prepared stock solution in THF was added to the mixture and further stirred for 1 h. The mixture was then poured onto a Teflon mould with a diameter of 4 cm and the solvent was allowed to evaporate slowly overnight. Finally, the membrane was dried in an oven at 70°C for 3 days and at 100°C for 1 h. Homogeneous and transparent films with average thickness of 200 μm were obtained in this way. These films were insoluble in all solvents, only swollen in THF.

Ac Impedance Measurements. The total ionic conductivity of samples was determined by locating an electrolyte disk between two 10 mm diameter brass electrodes. The electrode/electrolyte assembly was secured in a suitable constant volume support which allowed extremely reproducible measurements of conductivity to be obtained between repeated heating–cooling cycles. The cell support was located in an oven and the sample temperature was measured using a thermocouple positioned close to the electrolyte disk. The bulk conductivities of electrolytes were obtained during a heating cycle using the impedance technique (Impedance meter BM 507 –TESLA for frequencies 50 Hz-500 kHz) over a temperature range between 20 and 100°C.

Measurements of the electric conductance of the samples were performed on the disk shape membranes with diameter 10 mm by two electrodes (made from brass plates) method with using impedance device in the temperature range 25 – 90°C at constant rate of it change both at increasing and decreasing of the temperature. The membranes were sandwiched between two cylindrical brass electrodes in a PTFE cell. This cell was used also at voltammeter measurements operating at 20 mV /s scan rate in the potential range of 1 to 5 V. Experiments on voltammeter analysis were performed at room temperature.

RESULTS AND DISCUSSION

In this presented work we report the synthesis of new polysiloxane having propylbutyrate as a electro donor side group, ethyltriethoxysilane and hydroxyl group as a cross-linkers.

In ring opening polymerization (ROP) as an initial compounds 2.4.6-tris(propylbutyrate)-8-hydroxy-2.4.6.8-tetramethylcyclotetrasiloxane (I), 2.4-bis(propylbutyrate)-6-ethyltriethoxysilane-8-hydroxy-2.4.6.8-tetramethylcyclotetrasiloxane (II) with following structure have been used:

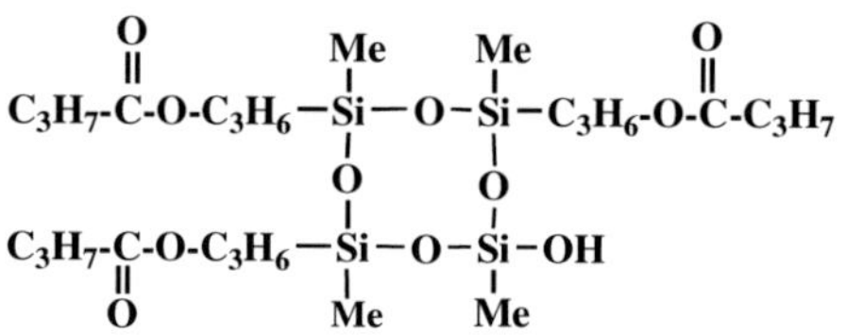

I

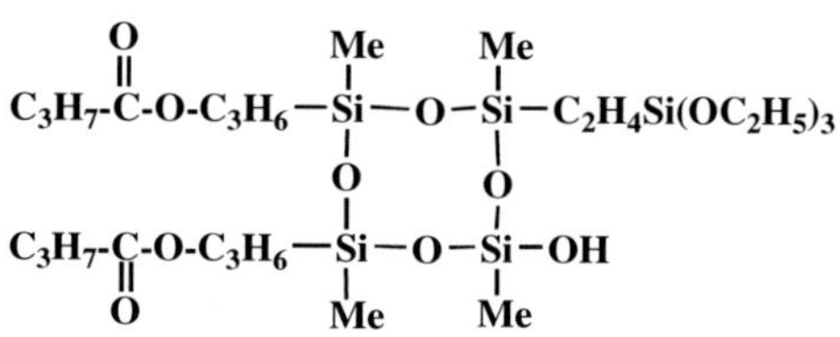

II

It is known that ionic fluorides show high catalytic activity, similar to typical nucleophils to catalysts in reactions of siliconorganic compounds: reduction [11], dehydrocondensation [12], and hydrolysis [13]. It was shown also, that ionic fluorides catalyze polymerization reactions of organocyclosiloxanes [14, 15]. It is shown, that polymerization eight members' cycles proceeds very slowly [16].

Polymerization reactions in the presence of catalysts (0.01 - 0.005% of total mass) alkali fluorides (KF, LiF and CaF_2) and tetramethyl ammonium hydroxide have been carried out in inert atmosphere, in melt condition in temperature range 50-120^{0}C and in toluene solution at 50-110^{0}C. By gas-liquid chromatography (GLC) it was established that alkali fluorides in this condition does not promote obtaining of polymers and only 25-30% of initial compounds take place in ring opening polymerization in case of KF. So the yield of polymers is very low.

Studying of catalytic activity of various ionic fluorides in relation of polymerization of organocyclotetrasiloxanes has shown that activity, decreases in a line: CaF_2<LiF<KF.

So this type of catalysts is not convenient for polymerization of these types of organocyclotetrasiloxanes with such substituted donor type side groups. It must be denote that during polymerization reaction of compound II in the presence of fluorides one can observe obtaining of 3D structure, which can be explained by side parallel reactions of hydroxyl and ethoxy groups. Polymerization reaction of compounds I and II also have been studied in inert atmosphere, in melt condition and in toluene solution in temperature range 60-70^{0}C in the presence of powder-like anhydrous potassium hydroxide (0.05 - 0.01% of total mass). The reaction has been carried out without terminated agent (hexamethyldisiloxane). It was established that ring-opening polymerization in the presence of potassium hydroxide proceeds during 48-64 h. The optimal condition of polymerization reaction has been determined and it was established that it's better to carry out the polymerization reactions in solution in temperature range 50-60^{0}C.

On the Figure 1, one can see the dependence of viscosity on the time during ring opening polymerization of compound I at various temperatures. With the rise of temperature the viscosity of obtained polymers decreases.

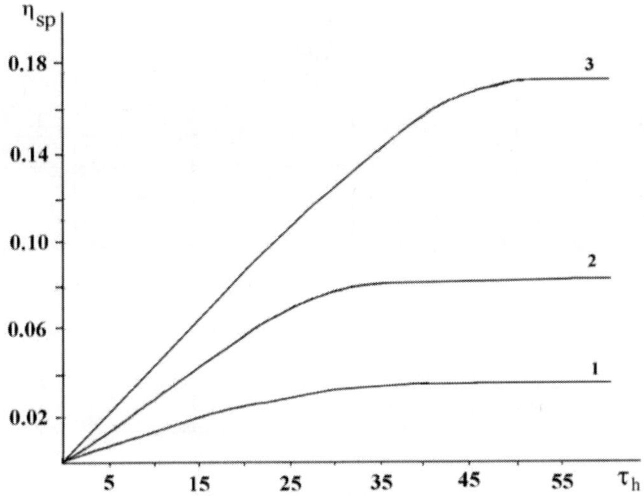

Figure 1. Dependence of viscosity on the time, during ring opening polymerization of compound I at various temperatures. Where the curve 1 corresponds to 110^{0}C, curve 2 -80^{0}C and 3 – 60^{0}C.

Polymerization reaction proceeds according to the following scheme:

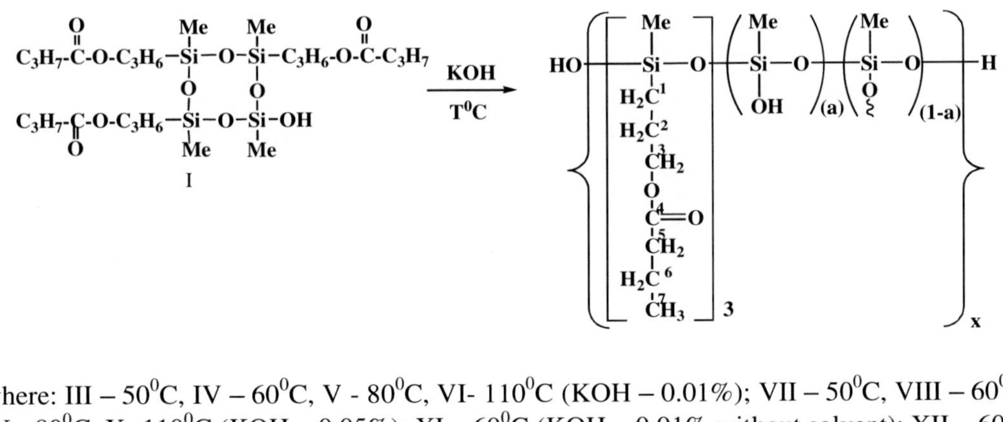

where: III – 50^{0}C, IV – 60^{0}C, V - 80^{0}C, VI- 110^{0}C (KOH – 0.01%); VII – 50^{0}C, VIII – 60^{0}C, IX - 80^{0}C, X- 110^{0}C (KOH – 0.05%); XI – 60^{0}C (KOH – 0.01% without solvent); XII – 60^{0}C (KOH – 0.01% without solvent).

The synthesized polymers are vitreous viscous products, which are well soluble in organic solvents with the specific viscosity $\eta_{sp} \approx 0.035 - 0.2$. Structures and compositions of the polymers were determined by elemental and functional analyses, FTIR, ^{1}H, ^{13}C and ^{29}Si NMR spectral data. Some physical-chemical properties of obtained polymers are presented in Table 1.

Table 1. Some physical-chemical properties of obtained polymers

№	Polymer structure	Catalyst concentration	Reaction T^0C	Duration of ROP hour	η_{sp}
III	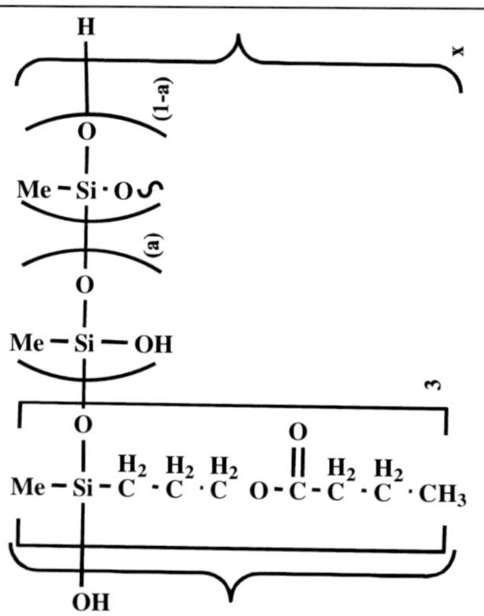	0.01	50	48	0.16
IV		0.01	60	55	0.18
V		0.01	80	55	0.08
VI		0.01	110	52	0.038
VII		0.05	50	55	0.045
VIII		0.05	60	52	0.038
IX		0.05	80	52	0.35
X		0.05	110	52	0.025
*XI		0.01	60	56	0.0248
*XII		0.01	70	55	0.082

* in melt condition.

In the FTIR spectra of obtained product one can see the absorption bands at 1080 and 1188 cm^{-1}, characteristic for asymmetric valence oscillation of ≡Si-O-Si≡ and ≡CO-O-C≡ bonds.

Also one can observe absorption bands at 1265 and 2800-3100 cm^{-1} region, characteristic for ≡Si-C≡ bonds and for valence oscillation of ≡C-H bonds. In the spectra one can see absorption bands for hydroxyl bonds in the range 3100-3600 cm^{-1}.

^{29}Si NMR spectra of obtained polymer indicate the occurrence of a branching process in cyclosiloxane molecules (Figure 2).

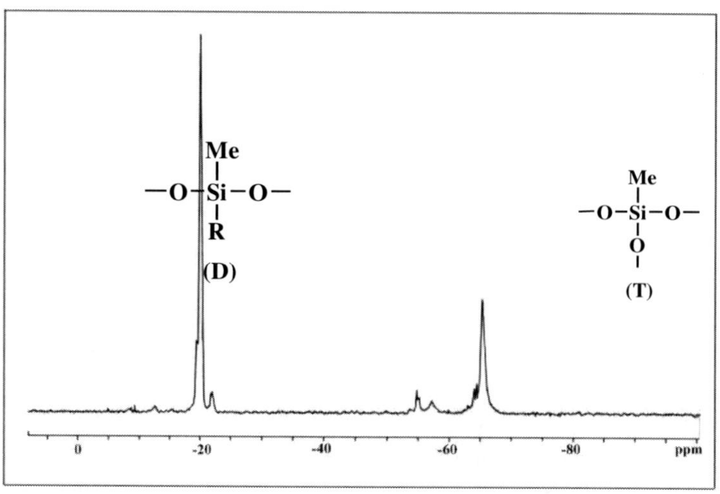

Figure 2. ^{29}Si NMR spectra of polymer III.

The presence of the signal at δ=-22 ppm corresponds to the presence of RR'SiO (D) units and the signal at -66 ppm correponds to T units.

This indicate the occurence of branchig process among the cyclosiloxane molecules, on the basis of ≡Si-OH bonds. The same side reactions take place during hydrosilylation reaction of allyl carbamates with D_4^H [17].

In the ^1H NMR spectra of compound III one can see multiplet signals with center of chemical shifts at δ=0.9; 1.6; 2.2 and 3.9 ppm characteristic for the methylene protons in C^7H_3, C^6H_2 and C^2H_2 (this signals overlap), C^5H_2 and C^3H_2 accordingly Figure 3.

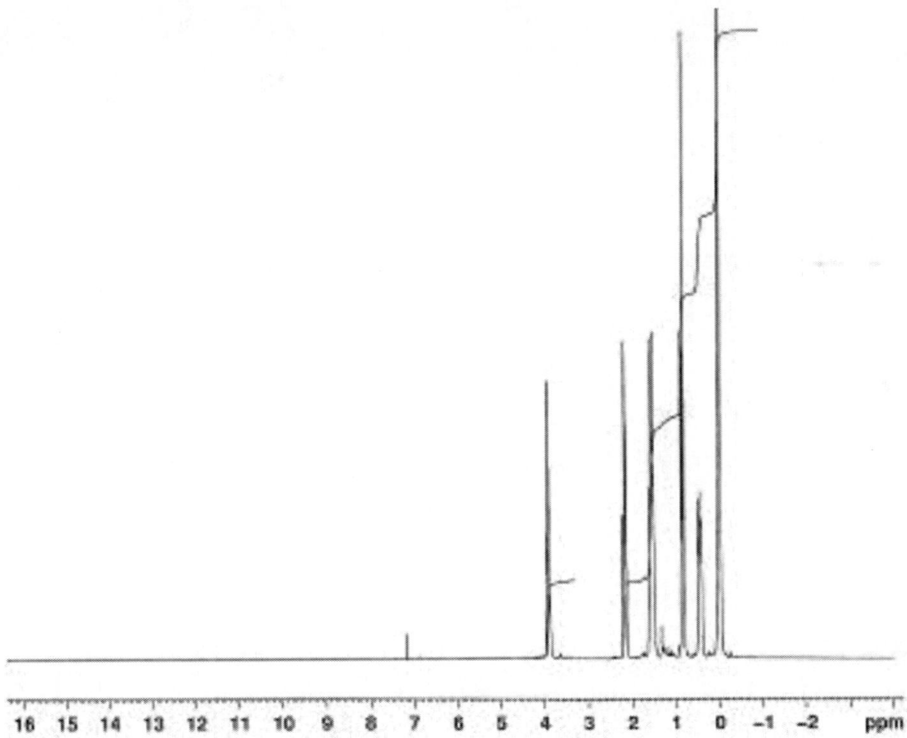

Figure 3. ^1H NMR spectra of polymer III.

In the ^{13}C-NMR spectra of polymer III one can observed signals characteristic for ≡Si-CH_3 groups with chemical shift of δ≈-0.2 ppm and signals, for carbon atoms with chemical shift δ≈14.6 ppm 19.27, 23.12, 37.02 and 67.17 ppm characteristic for C^1, C^2, C^7, C^6, C^5, C^3 carbon atoms and at 174.56 ppm characteristic for carbon in carbonyl group (Figure 4).

For synthesized polymer III gel permeation chromatographic investigation was studied. Figure 5 and 6 shows the molecular weight distribution of polymers III and IV. The polymer has tetra-modal and monomodal molecular weight distribution accordingly. For polymers the average molecular weights were determined: $\overline{M}_n$≈1.978- 32.822x10^3, $\overline{M}_\omega$=12,7-42.592x10^4 and polydispersity is equal to D=1.3-6.34.

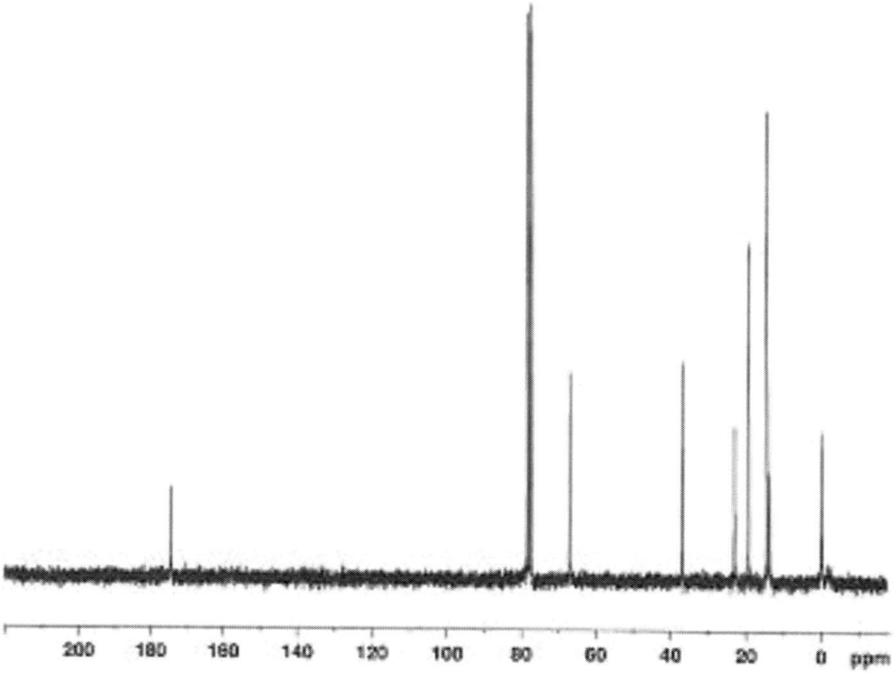

Figure 4. ^{13}C NMR spectra of polymer III.

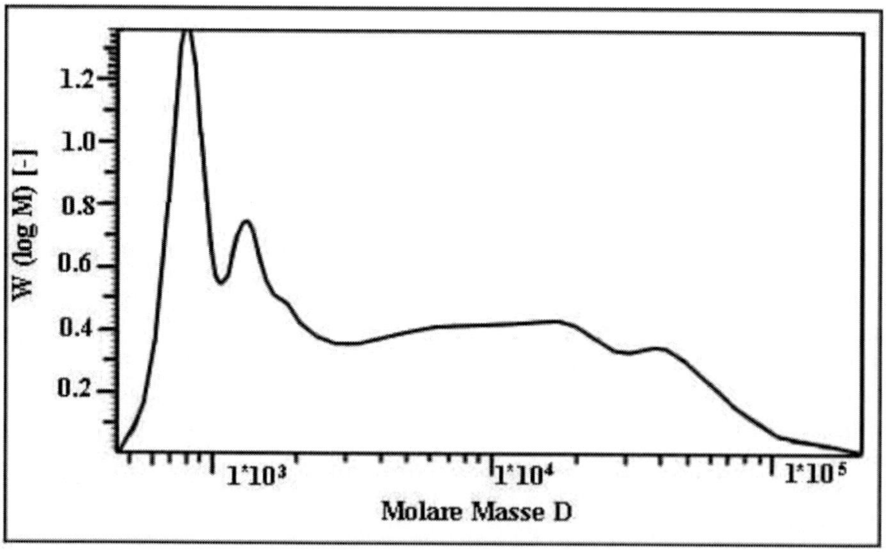

Figure 5. GPC of polymer III.

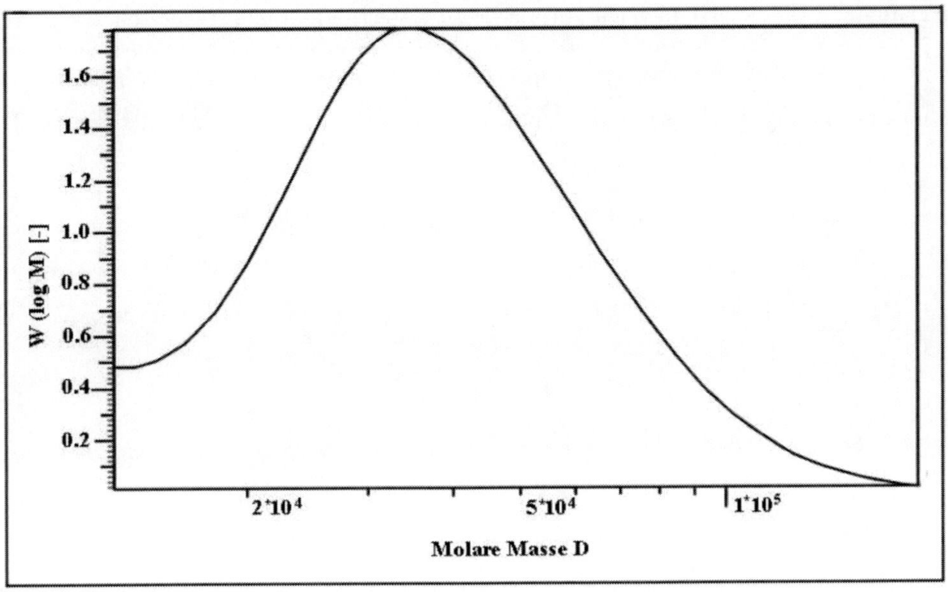

Figure 6. GPC of polymer IV.

Table 2. GPC data of polymers III and IV

№	Detector	Mn(g/mol)	Mw(g/mol)	D(Mw/Mn)	Vp(ml)	Mp(g/mol)	Area(ml*V)
III	RI 101	1978.56	12716.60	6.43	29.92	800.19	0.03
IV	RI 101	32822.40	42592.10	1.30	23.63	34456.50	0.01

During the ring opening polymerization of compound II in the presence of potassium hydroxide 3D structures are obtained. The obtaining of insoluble products may be explained by condensation reactions hydroxy and ethoxy groups. In this condensation reactions may take parts also terminal hydroxyl groups. In general during ring oprening polymerization the following condensation reaction take place:

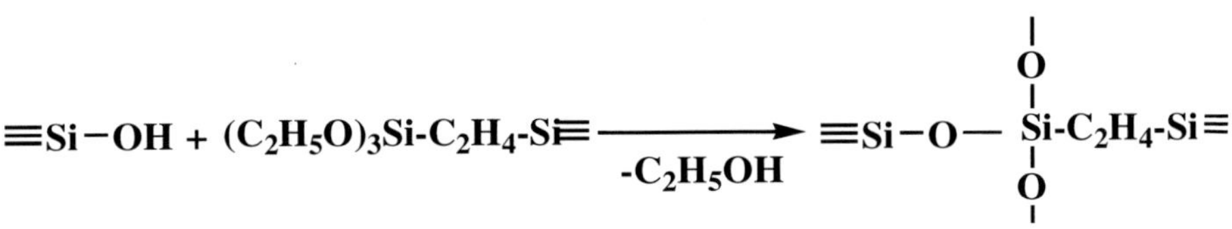

A wide angle X-ray investigation has been carried out and it was shown that polymers are represented as one phase amorphous systems with the value of interchain distances d_1=11.2 Å.

By DSC investigation it was shown, that the polymers III and IV are characterized by one glass transition temperature (T_g) in the range -82÷-93^{0}C. With an increase of molecular masses of polymers the T_g of polymers slightly increases.

Investigation of electro physical properties of the membranes made on the basis of obtained polymer IV with two types of trifluoromethylsulfonate (triflate) and lithium bis(trifluoromethylsulfonyl)imide shows that the dependence of the conductivity of these membranes essentially depends both on type and concentration of the used salts (Figure 7).

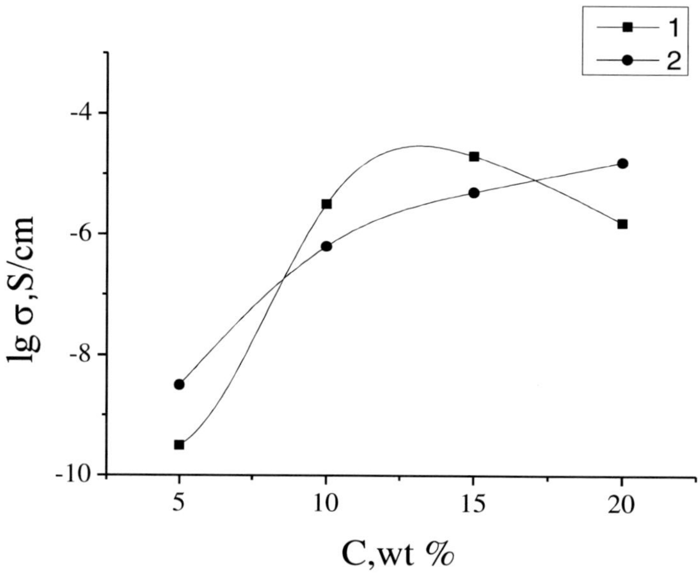

Figure 7. Dependence of the conductivity of the obtained membranes on the concentration of the Li-salts of types II (1) and I (2) at room temperature.

In accordance with the Figure 7 electrical conductivity of the membranes on the basis of the polymer P and salt I increases at increasing of concentration of this salt, although at relatively high values of this parameter the conductivity growth slows and the conductivity approaches the limiting value. Moreover In the case of the membranes containing the salt II with concentration 15 wt% is higher than that for analogs based on the same polymer and salt at containing of last 20 wt.%. This result looks at oneself as paradox. However in accordance with results from [18] at increasing of the salt concentration the probability of formation of the ion pairs enhances. In result of this phenomenon the mobility of these pairs decreases and consequently the general conductivity of PE decreases adequately. It must be noted that besides of this phenomenon no small reason may play the polymer microstructure, mobility of ions and interaction of type macromolecules – salt ions, which effectively influences on the ionization of the salt. Probably the ionization potential of the salt II is lower than that for salt I. However the noted processes operate on the charge transport in different ways. For example, the ionization of salt can flow readily at a low mobility of the corresponding ions. Thus, the end result will affect the ratio of these quantities. with that side, we can assume that in the membrane with 15% of the salt II ionization is more intense than in membranes with same amount of the salt I and is higher than the effect of mobility, which is from it side higher for the ions of salt I. However, as it was noted above, at arising of the salt II and consequently of ions the probability of formation of the ion pairs in the membranes with salt II increases, which decreases the conductivity of the membranes containing this salt at concentrations higher 15wt%.

On the Figures 8 and 9 there are presented the curves of temperature dependences of conductivity for membranes with salts of type I and II.

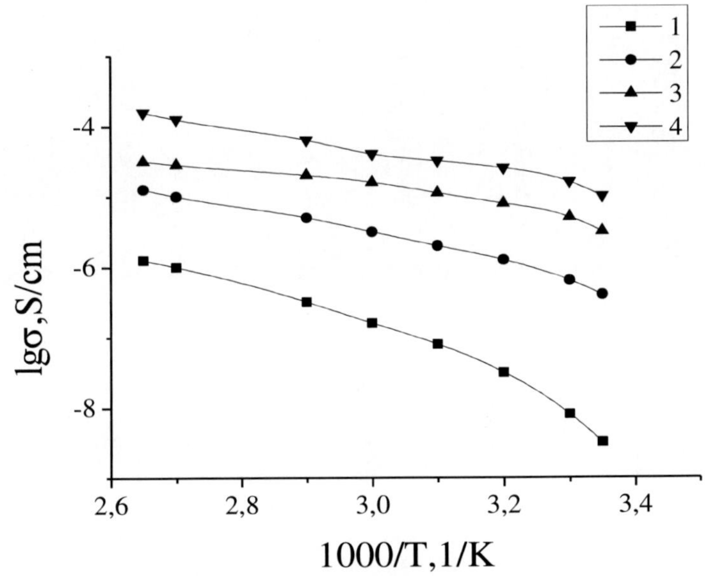

Figure 8. Arrhenius plots of conductivity of the membranes based on polymer IV containing 5 (1), 10 (2), 15 (3) and 20 wt (4) of the salt I.

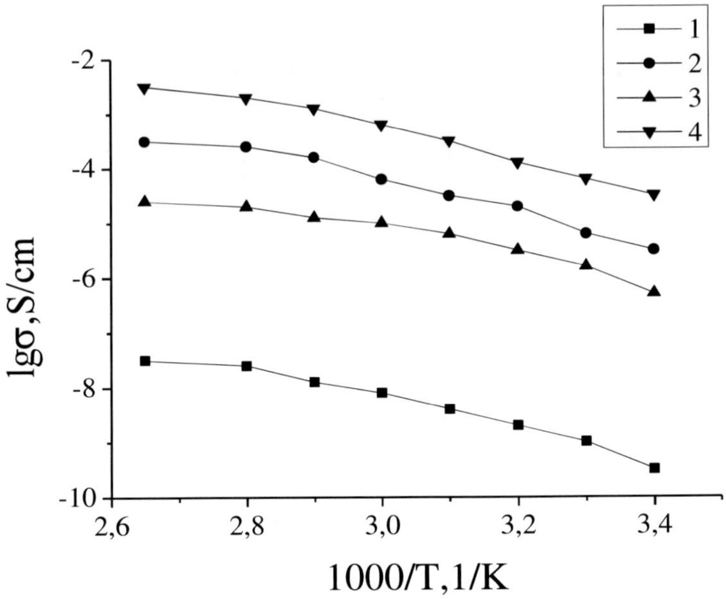

Figure 9. Arrhenius plots of conductivity of the membranes based on polymer IV containing 5 (1), 10 (2), 20 (3) and 15 wt (4) of the salt II.

The phenomena described above are practically repeated at more high temperatures of measure of polyelectrolyte's conductivity - the higher is conductivity of materials at room temperature the higher are the values of their conductivity at high temperatures because of corresponding enhancing of the ions mobility. At cyclic heat loading of the investigated polyelectrolyte's the behavior of the curves conductivity temperature are repeated with small hysteresis.

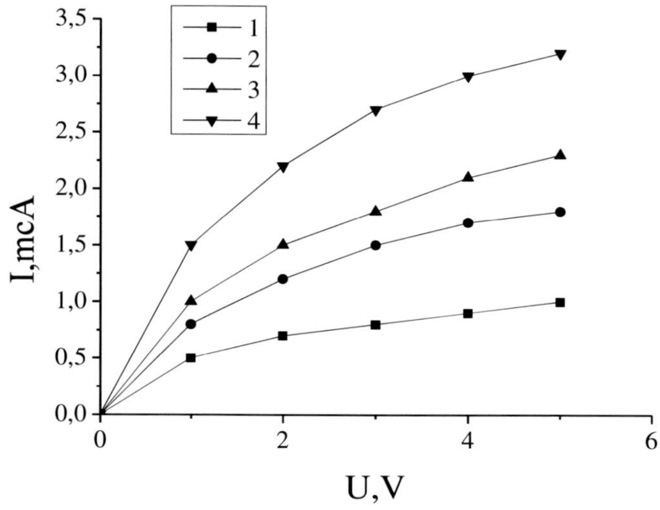

Figure 10. Voltamogrammes of the membranes based on polymer IV containing 5 (1), 10 (2), 15 (3) and 20 wt (4) of the salt I.

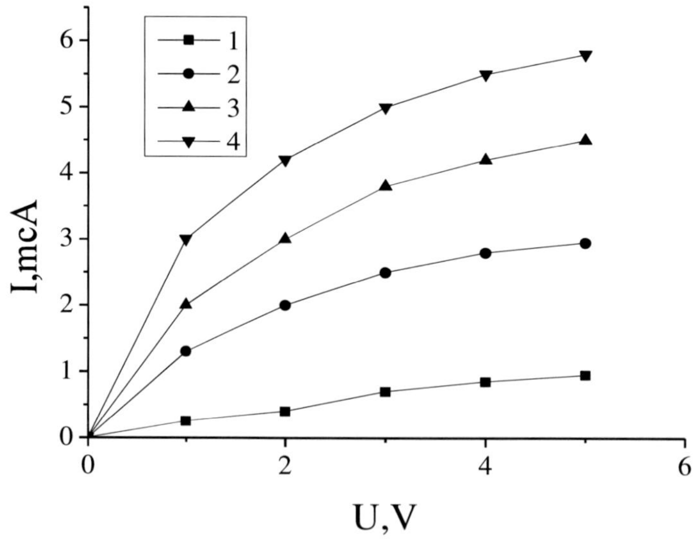

Figure 11. Voltamogrammes of the membranes based on polymer IV containing 5 (1), 10 (2), 20 (3) and 15 wt (4) of the salt II.

On the Figures 10 and 11 the voltamogrammes at room temperature of considered membranes were obtained. It is constructed the curves of current vs constant voltage in conditions of monotonic increasing of the voltage in the range 0 -5 V. It was expected that the values of the current through the membranes "run" curves inclines of which are the higher the higher is the initial conductivity (at room temperature) of membranes.

Conclusion

Ring opening polymerization reactions of some hydroxyorganocyclotetrasiloxanes in solution and in melt condition, in the temperature range 60-120°C, in the presence of nucleophilic powder-like anhydrous potassium hydroxide, alkali metal fluoride as well as quaternary ammonium hydroxide have been investigated and comb-type organosiloxane polymers with propylbutyrate, hydroxyl and ethyltriethoxysilane side groups has been obtained. The optimal condition of polymerization of methylorganocyclotetrasiloxanes has been determined. Via sol-gel processes of doped with lithium trifluoromethylsulfonate (triflate) or lithium bis(trifluoromethylsulfonyl)imide oligomer systems solid polymer electrolyte membranes have been obtained.

The values of the specific volumetric electric conductivity of the investigated membranes essentially depend both on type and concentration of the Li-salts. The concentration dependence of conductivity of the polyelectrolyte membranes has not ordinary character – these dependences are characterized with maximums at definite significant of the salt concentration, higher of which the membranes conductivity decreases. This phenomenon is explained by formation of so called ion pairs, replacing in polymer matrix as one charge carrier, the mobility of which is lower than that for separate charge carriers and consequently the integral conductivity for membranes containing relatively low concentration of Li salts separate ions may be higher than in case of the membranes containing more concentration of another Li imide salt, ions of which creates the pairs.

Acknowledgments

This work has been fulfilled by financial support of Science & Technology center in Ukraine (Grant #5055).

References

[1] Mukbaniani O., Tatrishvili T., Marqarashvili E., Esartia E. Hydrosilylation reaction of tetramethylcyclotetrasiloxane with allyl butyrate and vinyltriethoxysilane. *Georgian Chemical Journal*, 2(11), 153-155, (2011).

[2] Armand M.B. Polymer Electrolyte. *Annu. Rev. Mater. Sci.*, 16, 245-261 (1986).

[3] Fish D., Khan I.M., Wu E., and Smid J. Polymer electrolyte complexes of $LiClO_4$ and comb polymers of siloxane with oligooxyethylene side chains. *British Polymer Journal*, 20(3), 281-288 (1988).

[4] Ratner M.A., Shriver D.F. Ion transport in solvent-free polymers. *Chem. Rev.*, 88, 109-124 (1988).

[5] Watanabe M., Ogata N. Ionic conductivity of polymer electrolytes and future applications. Br. *Polym. J.,* 20, 181-192 (1988).

[6] Papke B.L., Ratner M.A., Shriver D.F. Conformation and ion-transport models for the structure and ionic conductivity in complexes of polyethers with alkali metal salts. *J. Electrochem. Sci.*, 129(8), 1694-1701 (1982).

[7] Oh B., Vissers D., West R., Tsukamoto H., Amine K. New interpenetrating network type poly(siloxane-g-ethylene oxide) polymer electrolyte for lithium battery. *J. Power Sources*, 119-121, 442-447 (2003).

[8] Fish D., Khan I.M., Smid J. Conductivity of solid complexes of lithium perchlorate with poly[ω-methoxyhexa(oxyethylene)ethoxymethylsiloxane]. *Makromol.Chem.Rapid* Comm., 7, 115-129 (1986).

[9] Spindler R., Shriver D.F. Investigations of a siloxane-based polymer electrolyte employing carbon-13, silicon-29, lithium-7, and sodium-23 solid-state NMR spectroscopy. *J. Am. Chem. Soc.,* 10, 3036-3043 (1988).

[10] Zhu Z., Einset A.G., Yang C.Y., Chen W.X., Wnek G.E. Synthesis of Polysiloxanes Bearing Cyclic Carbonate Side Chains. Dielectric Properties and Ionic Conductivities of Lithium Triflate Complexes. *Macromolecul*es, 27, 4076-4079 (1994).

[11] O. Mukbaniani, T. Tatrishvili, E. Marqarashvili, E. Esartia. Hydrosilylation reaction of tetramethylcyclotetrasiloxane with allyl butyrate and vinyltriethoxysilane. *Georgia Chemical Journ* (in presss)

[12] Tartakovskaya L.M., Makarov I.G., Kazakova B.M., Kopilov B.M., Zhdanov A.A. Reduction of nitrobenzole by organohydrosilanes in the presence of tetraalkylammonium fluoride. Izv. *AN USSR, Ser.Khim.*, № 12, 2818 (1980).

[13] Andrianov K.A., Tartakovskaya L.M., Kopilov B.M., Semencova Zh.P. About reaction of diphenylmethylsilane with alcohols in the presence of tetramethylamoniumfluoride. *Zhurn. Obshch. Khimii*., 45(1), 112-120 (1975).

[14] Corey E.J., Venkateswarlu A. Protection of hydroxyl groups as tert-butyldimethylsilyl derivatives. *J. Amer. Chem. Soc.*, 94(17), 6190-6191 (1972).

[15] Andrianov K.A., Tartakovskaya L.M., Kopilov B.M. Method of synthesis of polyorganosiloxanes. Author's Certificate. 485931 (USSR*).- Published in Bulletin of Inventions*. № 16, 306 (1980).

[16] Andrianov K.A., Zhdanov A.A., Zavin B.G., Evdokimov A.M. Polymerization of organocyclosiloxanes in the presence of amines. *Dokl. AN USSR* 12, 557-560 (1971).

[17] Tartakovskaia L.M., Kopilov B.M., Zhdanov A.A. Polymerization of organocyclosiloxanes in the presence of Fluorides. *Visokomol. Soed.*, 26B (3), 234-238 (1984).

[18] Yu M.J., Teissie D., Boileau S. Hydrosilylation of allyl carbamates by poly(methylhydrosiloxane) and related side-reactions. *Jorn. of Polymer Sci.: Part A; Polymer chemistry*, 31, 2373-2381 (1993).

[19] Zhang Z.C., Sherlock D., West R., Lyons L.J., Amine K. Cross-linked network polymer electrolytes based on a polysiloxane backbone with oligo(oxyethylene) side chains: synthesis and conductivity. *Macromolecules*, 36, 9176-9180 (2003).

In: Quantitative Chemistry, Biochemistry and Biology
Editors: G.E. Zaikov, O.V. Stoyanov et al.
ISBN: 978-1-62948-332-0

Chapter 12

UV-Protective Properties of Nanocomposites Films Based on Low Density Polyethylene

***A. A. Olkhov*[1*], *S. V. Romanov*[2], *A. A. Ischenko*[1] and *G. E. Zaikov*[3]**

[1]Moscow M.V. Lomonosov State University of Fine Chemical Technology, Moscow, Russia
[2]"VNIPIenergoprom," Moscow, Russia
[3]N.M. Emanuel Institute of Biochemical physics Russian Academy of Sciences Moscow, Russia

Abstract

High-strength polyethylene films containing 0.5-1.0 wt. % of nanocrystalline silicon (nc-Si) were synthesized. Samples of nc-Si with an average core diameter of 7-10 nm were produced by the plasmochemical method and the laser-induced decomposition of monosilane. Spectral studies revealed almost complete (up to ~95%) absorption of UV radiation in 200-400 nm spectral region by 85 micron thick film if the nc-Si content approaches to 1.0 wt. %. The density function of particle size in the starting powders and polymer films containing immobilized silicon nanocrystallites were obtained using the modeling a complete profile of X-ray diffraction patterns, assuming spherical grains and the lognormal distribution. The results of X-ray analysis have shown that the crystallite size distribution function remains almost unchanged and the crystallinity of the original polymer increases to about 10% with the implantation of the initial nc-Si samples in the polymer matrix.

Keywords: Polyethylene, nanocrystalline silicon, UV-protective film, polymer nanocomposites, spectroscopy, X-ray diffraction analysis

[*] E-mail: aolkhov72@yandex.ru.

INTRODUCTION

In recent years, considerable efforts have been devoted for research on new functional nanocomposite materials with unique properties that are lacking in their traditional analogues. Control of these properties is an important fundamental problem. The use of nanocrystals as one of the elements of a polymer composite opens up new possibilities for targeted modification of its optical properties because of a strong dependence of the electronic structure of nanocrystals on their sizes and geometric shapes. An increase in the number of nanocrystals in the bulk of composites is expected to enhance long-range correlation effects on their properties. Among the known nanocrystals, nanocrystalline silicon (nc-Si) attracts high attention due to its extraordinary optoelectronic properties and manifestation of quantum size effects. Therefore, it is widely used for designing new generation functional materials for nanoelectronics and information technologies. The use of nc-Si in polymer composites calls for a knowledge of the processes of its interaction with polymeric media. Solid nanoparticles can be combined into aggregates (clusters), and, when the percolation threshold is achieved, a continuous cluster is formed.

An orderly arrangement of interacting nanocrystals in a long-range potential minimum leads to formation of periodic structures. Because of the well-developed interface, an important role in such systems belongs to adsorption processes, which are determined by the structure of the nanocrystal surface. In a polymer medium, nanocrystals are surrounded by an adsorption layer consisting of polymer, which may change the electronic properties of the nanocrystals. The structure of the adsorption layer has an effect on the processes of self-organization of solid-phase particles, as well as on the size, shape, and optical properties of resulting aggregates. According to data obtained for metallic [1] and semiconducting [2] clusters, aggregation and adsorption in three-phase systems with nanocrystals have an effect on the optical properties of the whole system. In this context, it is important to reveal the structural features of systems containing nanocrystals, characterizing aggregation and adsorption processes in these systems, which will make it possible to establish a correlation between the structural and the optical properties of functional nanocomposite systems.

Silicon nanoclusters embedded in various transparent media are a new, interesting object for physicochemical investigation. For example, for particles smaller than 4 nm in size, quantum size effects become significant. It makes possible to control the luminescence and absorption characteristics of materials based on such particles using of these effects [3, 4]. For nanoparticles about 10 nm in size or larger (containing ~10^4 Si atoms), the absorption characteristics in the UV and visible ranges are determined in many respects by properties typical of massive crystalline or amorphous silicon samples. These characteristics depend on a number of factors: the presence of structural defects and impurities, the phase state, etc. [5, 6]. For effective practical application and creation on a basis nc-Si the new polymeric materials possessing useful properties: sun-protection films [7] and the coverings [8] photoluminescent and electroluminescent composites [9, 10], stable to light dyes [11], embedding of these nanosized particles in polymeric matrixes becomes an important synthetic problem.

The method of manufacturing silicon nanoparticles in the form of a powder by plasma chemical deposition, which was used in this study, makes possible to vary the chemical composition of their surface layers. As a result, another possibility of controlling their spectral characteristics arises, which is absent in conventional methods of manufacture of nanocrystalline silicon in solid matrices (for example, in α–SiO_2) by implantation of charged silicon particles [5] or radio frequency deposition of silicon [2]. Polymer composites based on silicon nanopowder are a new object for comprehensive spectral investigation. At the same time, detailed spectral analysis has been performed for silicon nanopowder prepared by laser induced decomposition of gaseous SiH_4 (see, for example, [6, 12]). It is of interest to consider the possibility of designing new effective UV protectors based on polymer containing silicon nanoparticles [13]. An advantage of this nanocomposite in comparison with other known UV protectors is its environmental safety, i.e., ability to hinder the formation of biologically harmful compounds during UV-induced degradation of components of commercial materials. In addition, changing the size distribution of nanoparticles and their concentration in a polymer and correspondingly modifying the state of their surface, one can deliberately change the spectral characteristics of nanocomposite as a whole. In this case, it is necessary to minimize the transmission in the wavelength range below 400 nm (which determines the properties of UV-protectors [13]) by changing the characteristics of the silicon powder.

Objects of Research

In this study, the possibilities of using polymers containing silicon nanoparticles as effective UV protectors are considered. First, the structure of nc-Si obtained under different conditions and its aggregates, their adsorption and optical properties was studied in order to find ways of control the UV spectral characteristics of multiphase polymer composites containing nanocrystalline silicon. Also, the purpose of this work was to investigate the effect of the concentration of silicon nanoparticles embedded in polymer matrix and the methods of preparation of these nanoparticles on the spectral characteristics of such nanocomposites. On the basis of the data obtained, recommendations for designing UV protectors based on these nanocomposites were formulated.

nc-Si consists of core–shell nanoparticles in which the core is crystalline silicon coated with a shell formed in the course of passivation of nc-Si with oxygen and/or nitrogen. nc-Si samples were synthesized by an original procedure in an argon plasma in a closed gas loop. To do this, we used a plasma vaporizer/condenser operating in a low-frequency arc discharge. A special consideration was given to the formation of a nanocrystalline core of specified size. The initial reagent was a silicon powder, which was fed into a reactor with a gas flow from a dosing pump. In the reactor, the powder vaporized at 7000 − 10000 °C. At the outlet of the high-temperature plasma zone, the resulting gas–vapor mixture was sharply cooled by gas jets, which resulted in condensation of silicon vapor to form an aerosol. The synthesis of nc-Si in a low-frequency arc discharge was described in detail in [3].

The microstructure of nc-Si was studied by transmissionelectron microscopy (TEM) on a Philips NED microscope. X-ray powder diffraction analysis was carried out on a Shimadzu Lab XRD-6000 diffractometer. The degree of crystallinity of nc-Si was calculated from the integrated intensity of the most characteristic peak at $2\theta = 28°$. Low-temperature adsorption isotherms at 77.3 K were measured with a Gravimat-4303 automated vacuum adsorption apparatus. FTIR spectra were recorded on in the region of 400-5000 cm^{-1} with resolution of about 1 cm^{-1}.

Three samples of nc-Si powders with specific surfaces of 55, 60, and 110 m^2/g were studied. The *D* values for these samples calculated by Eq. (2) are 1.71, 1.85, and 1.95, respectively; i.e., they are lower than the limiting values for rough objects. The corresponding *D* values calculated by Eq. (3) are 2.57, 2.62, and 2.65, respectively. Hence, the adsorption of nitrogen on nc-Si at 77.3 K is determined by capillary forces acting at the liquid–gas interface. Thus, in argon plasma with addition of oxygen or nitrogen, ultra disperse silicon particles are formed, which consist of a crystalline core coated with a silicon oxide or oxynitride shell. This shell prevents the degradation or uncontrollable transformation of the electronic properties of nc-Si upon its integration into polymer media. Solid structural elements (threads or nanowires) are structurally similar, which stimulates self-organization leading to fractal clusters. The surface fractal dimension of the clusters determined from the nitrogen adsorption isotherm at 77.3 K is a structurally sensitive parameter, which characterizes both the structure of clusters and the morphology of particles and aggregates of nanocrystalline silicon.

As the origin materials for preparation film nanocomposites served polyethylene of low density (LDPE) marks 10803-020 and ultradisperse crystal silicon. Silicon powders have been received by a method plazmochemical recondensation of coarse-crystalline silicon in nanocrystalline powder. Synthesis nc-Si was carried out in argon plasma in the closed gas cycle in the plasma evaporator the condenser working in the arc low-frequency category. After particle synthesis nc-Si were exposed microcapsulating at which on their surfaces the protective cover from SiO_2, protecting a powder from atmospheric influence and doing it steady was created at storage. In the given work powders of silicon from two parties were used: nc-Si-36 with a specific surface of particles ~36 m^2/g and nc-Si-97 with a specific surface ~97 m^2/g.

Preliminary mixture of polyethylene with a powder nc-Si firms "Brabender" (Germany) carried out by means of closed hummer chambers at temperature 135±5°C, within 10 minutes and speed of rotation of a rotor of 100 $minutes^{-1}$. Two compositions LDPE + nc-Si have been prepared: (1) composition PE + 0.5 % ncSi-97 on a basis ncSi-97, containing 0.5 weights silicon %; (2) composition PE + 1 % ncSi-36 on a basis ncSi-36, containing 1.0 weights silicon %.

Formation of films by thickness 85±5 micron was spent on semiindustrial extrusion unit ARP-20-150 (Russia) for producing the sleeve film. The temperature was 120-190°C on zones extruder and extrusion die. The speed of auger was 120 $minutes^{-1}$. Technological parameters of the nanocomposites choose, proceeding from conditions of thermostability and the characteristic viscosity recommended for processing polymer melting.

Experimental Methods

Mechanical properties and an optical transparency of polymer films, their phase structure and crystallinity, and also communication of mechanical and optical properties with a microstructure of polyethylene and granulometric structure of modifying powders nc-Si were observed.

Physicomechanical properties of films at a stretching (extrusion) measured in a direction by means of universal tensile machine EZ-40 (Germany) in accordance with Russian State

Standard GOST-14236-71. Tests are spent on rectangular samples in width of 10 mm, and a working site of 50 mm. The speed of movement of a clip was 240 mm/minutes. The 5 parallel samples were tested.

Optical transparency of films was estimated on absorption spectra. Spectra of absorption of the obtained films were measured on spectrophotometer SF-104 (Russia) in a range of wavelengths 200 – 800 nanometers. Samples of films of polyethylene and composite films PE + 0.5 % ncSi-36 and PE + 1 % ncSi-36 in the size 3x3 cm were investigated. The special holder was used for maintenance uniform a film tension.

X-ray diffraction analysis by wide-angle scattering of monochromatic X-rays data was applied for research phase structure of materials, degree of crystallinity of a polymeric matrix, the size of single-crystal blocks in powders nc-Si and in a polymeric matrix, and also functions of density of distribution of the size crystalline particles in initial powders nc-Si

X-ray diffraction measurements were observed on Guinier diffractometer: chamber G670 Huber [14] with bent Ge (111) monochromator of a primary beam which are cutting out line $K\alpha_1$ (length of wave $\lambda = 1.5405981$ Å) characteristic radiation of x-ray tube with the copper anode. The diffraction picture in a range of corners 2θ from 3° to 100 ° was registered by the plate with optical memory (IP-detector) of the camera bent on a circle. Measurements were spent on original powders nc-Si-36 and nc-Si-97, on the pure film LDPE further marked as PE, and on composite films PE + 0.5 % ncSi-97 and PE + 1.0 % ncSi-36. For elimination of tool distortions effect diffractogram standard SRM660a NIST from the crystal powder LaB_6 certificated for these purposes by Institute of standards of the USA was measured. Further it was used as diffractometer tool function.

Samples of initial powders ncSi-36 and ncSi-97 for x-ray diffraction measurements were prepared by drawing of a thin layer of a powder on a substrate from a special film in the thickness 6 microns (MYLAR, Chemplex Industries Inc., Cat. No: 250, Lot No: 011671). Film samples LDPE and its composites were established in the diffractometer holder without any substrate, but for minimization of structure effect two layers of a film focused by directions extrusion perpendicular each other were used.

Phase analysis and granulometric analysis was performed with the interpretation of the X-ray diffraction data. For these purposes the two different full-crest analysis methods [15, 16] were applied: (1) method of approximation of a profile diffractogram using analytical functions, polynoms and splines with diffractogram decomposition on making parts; (2) method of diffractogram modeling on the basis of physical principles of scattering of X-rays. The package of computer programs WinXPOW was applied to approximation and profile decomposition diffractogram ver. 2.02 (Stoe, Germany) [17], and diffractogram modeling at the analysis of distribution of particles in the sizes was spent by means of program PM2K (version 2009) [18].

RESULTS AND DISCUSSION

Results of mechanical tests of the prepared materials are presented to tab. 1 from which it is visible that additives of particles nc-Si have improved mechanical characteristics of polyethylene.

Table 1. Mechanical characteristics of nanocomposite films based of LDPE and nc-Si

Sample	Tensile strength, kg/cm^2	Relative elongation-at-break, %
PE	100 ± 12	200 – 450
PE + 1% ncSi-36	122 ± 12	250 – 390
PE + 0.5% ncSi-97	118 ± 12	380 – 500

The results presented in the table show that additives of powders of silicon raise mechanical characteristics of films, and the effect of improvement of mechanical properties is more expressed in case of composite PE + 0.5 % ncSi-97 at which in comparison with pure polyethylene relative elongation-at-break has essentially grown.

Transmittance spectra of the investigated films are shown on Figure 1.

It is visible that additives of powders nc-Si reduce a transparency of films in all investigated range of wavelengths, but especially strong decrease transmittance (almost in 20 times) is observed in a range of lengths of waves of 220-400 nanometers, i.e. in UV areas.

The wide-angle scattering of X-rays data were used for the observing phase structure of materials and their component. Measured x-ray diffractograms of initial powders ncSi-36 and ncSi-97 on intensity and Bragg peaks position completely corresponded to a phase of pure crystal silicon (a cubic elementary cell of type of diamond – spatial group $Fd\bar{3}m$, cell parameter a_{Si} = 0.5435 nanometers).

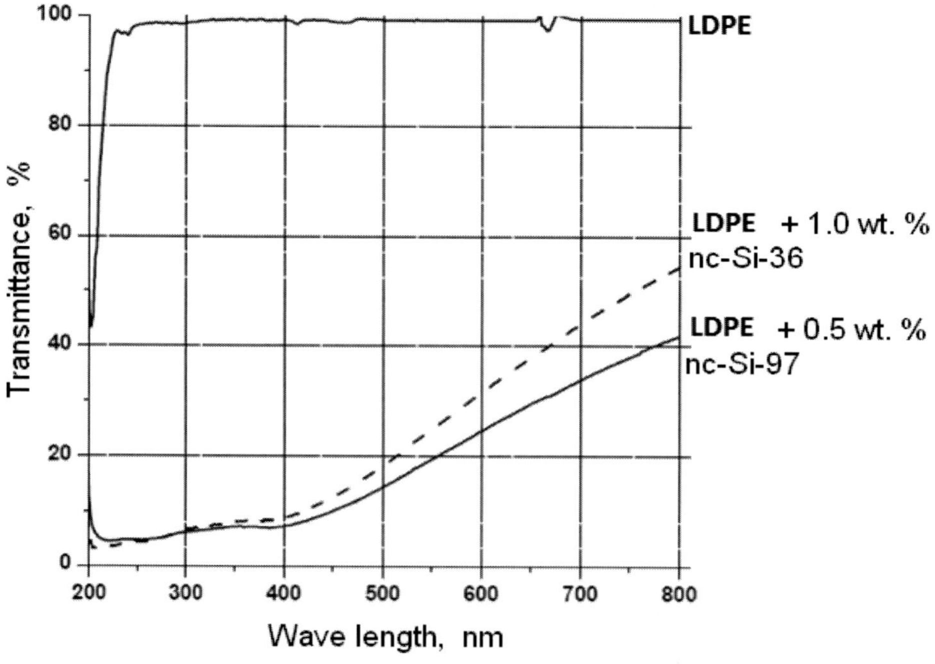

Figure 1. Transmittance spectra of the investigated films LDPE and nanocomposite films PE + 0.5 % ncSi-97 and PE + 1.0 % ncSi-36.

For the present research granulometric structure of initial powders nc-Si is of interest. The density function of particle size in a powder was restored on an X-ray diffractogram a powder by means of computer program PM2K [18] in which the method [19] modellings of a full profile diffractogram based on the theory of physical processes of diffraction of X-rays is realized. Modeling was spent in the assumption of the spherical form of crystalline particles and logarithmically normal distributions of their sizes. Deformation effects from flat and linear defects of a crystal lattice were considered. Received function of density of distribution of the size crystalline particles for initial powders nc-Si are represented graphically on Figure 2, in the signature to which statistical parameters of the found distributions are resulted. These distributions are characterized by such important parameters, as *Mo(d)* – position of maximum (a distribution mode); $<d>_V$ – average size of crystalline particles based on volume of the sample (the average arithmetic size) and *Me(d)* – the median of distribution defining the size *d*, specifying that particles with diameters less than this size make half of volume of a powder.

The results represented in Figure 2, show that initial powders nc-Si in the structure have particles with the sizes less than 10 nanometers which especially effectively absorb UV radiation. The both powders modes of density function of particle size are very close, but median of density function of particle size of a powder ncSi-36 it is essential more than at a powder ncSi-97. It suggests that the number of crystalline particles with diameters is less 10 nanometers in unit of volume of a powder ncSi-36 much less, than in unit of volume of a powder ncSi-97. As a part of a powder ncSi-36 it is a lot of particles with a diameter more than 100 nanometers and even there are particles more largely 300 nanometers whereas the sizes of particles in a powder ncSi-97 don't exceed 150 nanometers and the basic part of crystalline particles has diameter less than 100 nanometers.

The phase structure of the obtained films was estimated on wide-angle scattering diffractogram only qualitatively. Complexity of diffraction pictures of scattering and structure don't poses the quantitative phase analysis of polymeric films [20]. At the phase analysis of polymers often it is necessary to be content with the comparative qualitative analysis which allows watching evolution of structure depending on certain parameters of technology of production. Measured wide-angle X-rays scattering diffractograms of investigated films are shown on Figure 3. Diffractogramms have a typical form for polymers. As a rule, polymers are the two-phase systems consisting of an amorphous phase and areas with distant order, conditionally named crystals. Their diffractograms represent [20] superposition of intensity of scattering by the amorphous phase which is looking like wide halo on the small-angle area (in this case in area 2 θ between 10 ° and 30°), and intensity Bragg peaks scattering by a crystal phase.

Data on Figure 3 is presented in a scale of relative intensities (intensity of the highest peak is accepted equal 100 %). For convenience of consideration curves are represented with displacement on an axis of ordinates. The scattering plots without displacement represented a complete overlapping of diffractogram profiles of composite films with diffractogram of a pure LDPE film, except peaks of crystal silicon which weren't present on PE diffractogram. It testifies that additives of powders nc-Si practically haven't changed crystal structure of polymer.

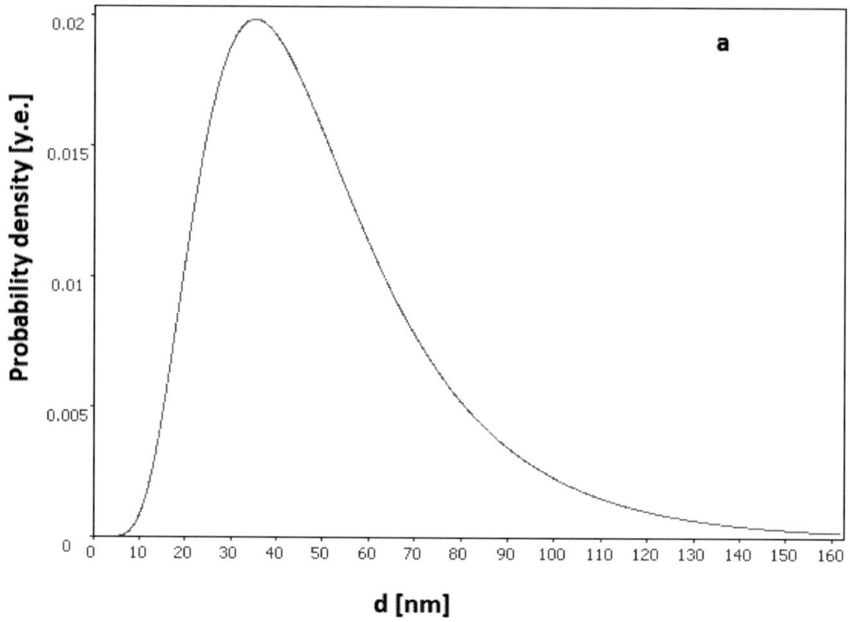

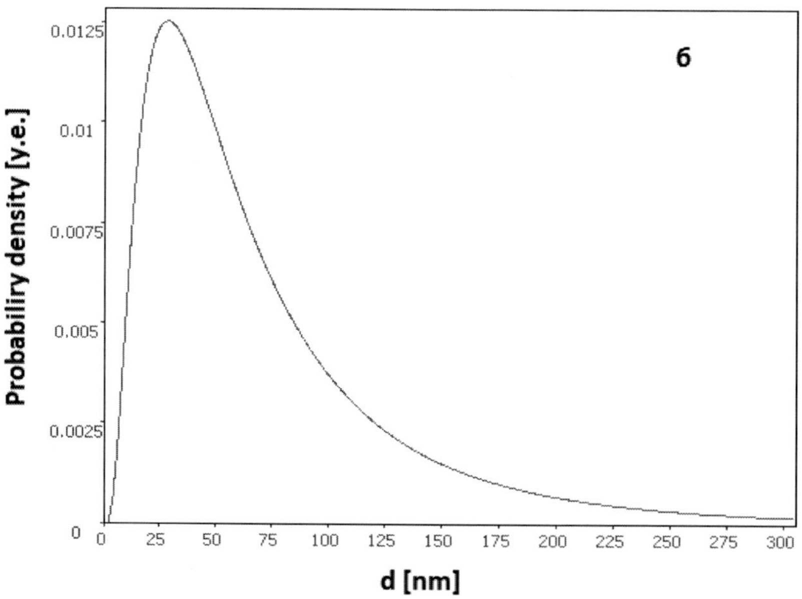

(a) – ncSi-97	Mo(d) = 35 nm	Me(d) = 45 nm	$<d>_V$ = 51 nm;
(б) – ncSi-36	Mo(d) = 30 nm	Me(d) = 54 nm	$<d>_V$ = 76 nm.

Figure 2. Density function of particle size in powders ncSi, received from x-ray diffractogram by means of program PM2K:

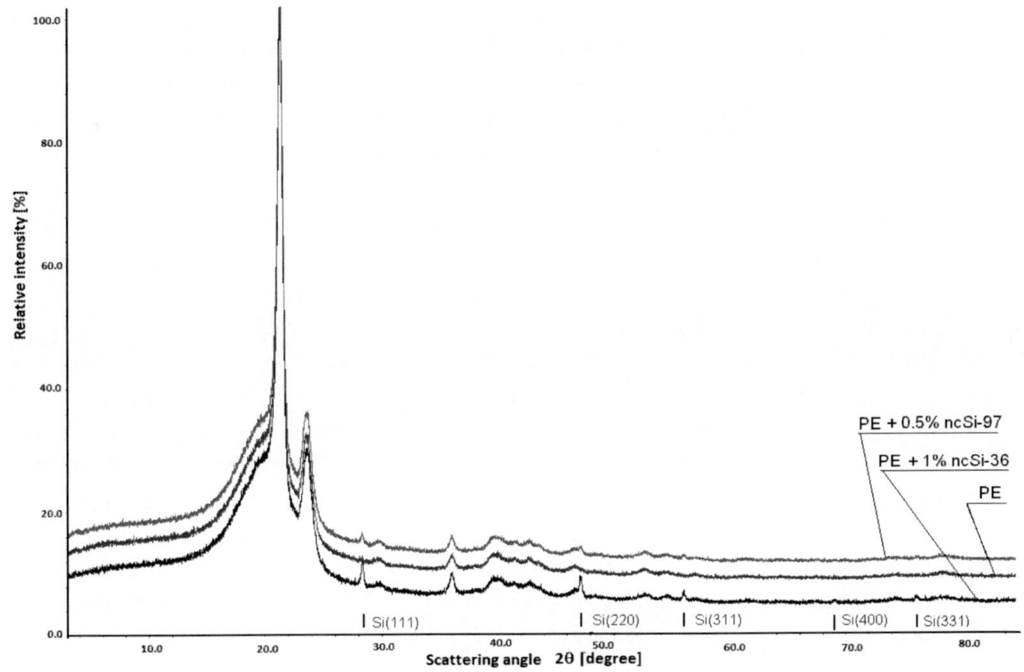

Figure 3. Diffractograms of the investigated composite films in comparison with diffractogram of pure polyethylene. Below vertical strokes specify reference positions of diffraction lines of silicon with their interference indexes (hkl).

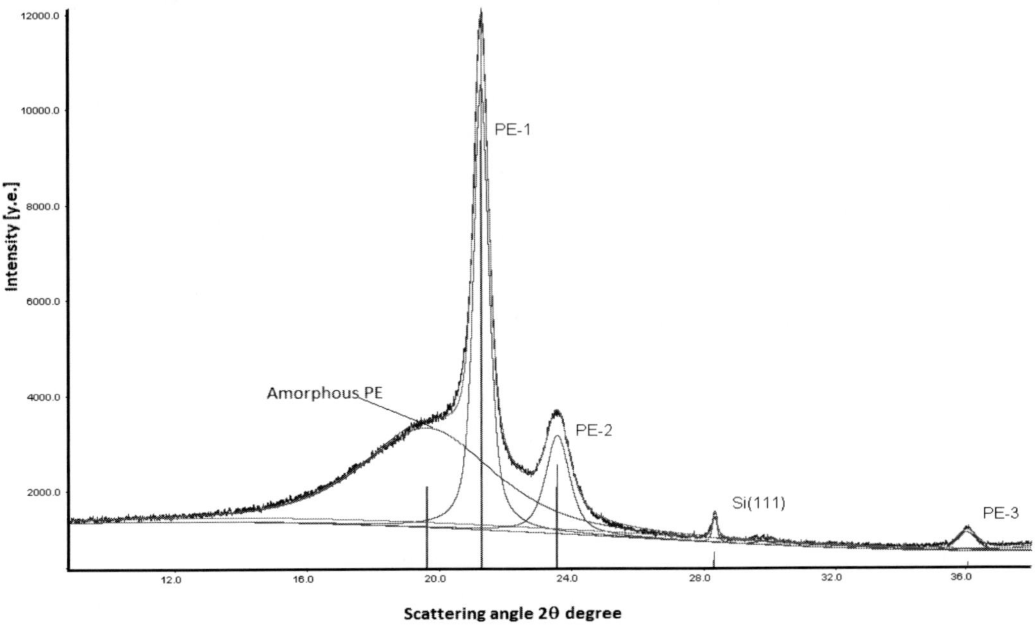

Figure 4. Diffractogram decomposition on separate peaks and a background by means of approximation of a full profile by analytical functions on an example of the data from sample PE+1%ncSi-36 (Figure 3). PE-n designate Bragg peaks of crystal polyethylene with serial numbers n from left to right. Si (111) – Bragg silicon peak ncSi-36. Vertical strokes specify positions of maxima of peaks.

The peaks of crystal silicon are distinguishable on diffractograms of films with silicon (the reference positions with Miller's corresponding indexes are pointed below). Heights of the peaks of silicon with the same name (i.e. peaks with identical indexes) on diffractograms of the composite films PE + 0.5 % ncSi-97 and PE + 1.0 % ncSi-36 differ approximately twice that corresponds to a parity of mass concentration Si set at their manufacturing.

Degree of crystallinity of polymer films (a volume fraction of the crystal ordered areas in a material) in this research was defined by diffractograms Figure 3 for a series of samples only semiquantitative (more/less). The essence of the method of crystallinity definition consists in analytical division of a diffractogram profile on the Bragg peaks from crystal areas and diffusion peak of an amorphous phase [20], as is shown in Figure 4.

Peaks profiles of including peak of an amorphous phase, were approximated by function pseudo-Foigt, a background 4 order polynoms of Chebysheva. The nonlinear method of the least squares minimized a difference between intensity of points experimental and approximating curves. The width and height of approximating functions, positions of their maxima and the integrated areas, and also background parameters were thus specified. The relation of integrated intensity of a scattering profile by an amorphous phase to full integrated intensity of scattering by all phases except for particles of crystal silicon gives a share of amorphy of the sample, and crystallinity degree turns out as a difference between unit and an amorphy fraction.

It was supposed that one technology of film obtaining allowed an identical structure. It proved to be true by coincidence relative intensities of all peaks on diffractograms Figure 3, and samples consist only crystal and amorphous phases of the same chemical compound. Therefore received values of degree of crystallinity should reflect correctly a tendency of its change at modification polyethylene by powders nc-Si though because of a structure of films they can quantitatively differ considerably from the valid concentration of crystal areas in the given material. The found values of degree of crystallinity are represented in table 2.

Table 2. Characteristics of the ordered (crystal) areas in polyethylene and its composites with nc-Si

PE			PE + 1% ncSi-36			PE + 0.5% ncSi-97		
Crystallinity	46%		47,5%			48%		
2θ [°]	d [E]	ε	2θ [°]	d [E]	ε	2θ [°]	d [E]	ε
21.274	276	8.9	21.285	229	7.7	21.282	220	7.9
23.566	151	12.8	23.582	128	11.2	23.567	123	11.6
36.038	191	6.8	36.035	165	5.8	36.038	162	5.8
Average values	206	9.5×10^{-3}		174	8.2×10^{-3}		168	8.4×10^{-3}

One more important characteristic of crystallinity of polymer is the size d of the ordered areas in it. For definition of the size of crystalline particles and their maximum deformation ε in X-ray diffraction analysis [21] Bragg peaks width on half of maximum intensity (Bragg lines half-width) is often used. In the given research the sizes of crystalline particles in a polyethylene matrix calculated on three well expressed diffractogram peaks Figure 3. The peaks of polyethylene located at corners 2 θ approximately equal 21.28 °, 23.57 ° and 36.03 °

(peaks PE-1, PE-2 and PE-3 on Figure 4 see) were used. The ordered areas size *d* and the maximum relative deformation ε of their lattice were calculated by the joint decision of the equations of Sherrera and Wilson [21] with use of half-width of the peaks defined as a result of approximation by analytical functions, and taking into account experimentally measured diffractometer tool function. Calculations were spent by means of program $WinX^{POW}$ *size/strain*. Received *d* and ε, and also their average values for investigated films are presented in table 2. The updated positions of maxima of diffraction peaks used at calculations are specified in the table.

The offered technology allowed the obtaining of films LDPE and composite films LDPE + 1 % ncSi-36 and LDPE + 0.5 % ncSi-97 an identical thickness (85 microns). Thus concentration of modifying additives ncSi in composite films corresponded to the set structure that is confirmed by the X-ray phase analysis.

By direct measurements it is established that additives of powders ncSi have reduced a polyethylene transparency in all investigated range of lengths of waves, but especially strong transmittance decrease (almost in 20 times) is observed in a range of lengths of waves of 220-400 nanometers, i.e. in UV areas. Especially strongly effect of suppression UV of radiation is expressed in LDPE film + 0.5 % ncSi-97 though concentration of an additive of silicon in this material is less. It is possible to explain this fact to that according to experimentally received function of density of distribution of the size the quantity of particles with the sizes is less 10 nanometers on volume/weight unit in a powder ncSi-97 more than in a powder ncSi-36.

Direct measurements define mechanical characteristics of the received films – durability at a stretching and relative lengthening at disrupture (Tab. 1). The received results show that additives of powders of silicon raise durability of films approximately on 20 % in comparison with pure polyethylene. Composite films in comparison with pure polyethylene also have higher lengthening at disrupture, especially this improvement is expressed in case of composite PE + 0.5 % ncSi-97. Observable improvement of mechanical properties correlates with degree of crystallinity of films and the average sizes of crystal blocks in them (Tab. 2). By results of the X-ray analysis the highest crystallinity at LDPE film + 0.5 % ncSi-97, and at it the smallest size the crystal ordered areas that should promote durability and plasticity increase.

This work is supported by grants RFBR № 10-02-92000 and RFBR № 11-02-00868 also by grants FCP "Scientific and scientific and pedagogical shots of innovative Russia", contract № 2353 from 17.11.09 and contract № 2352 from 13.11.09.

REFERENCES

[1] S.V. Karpov, V.V.Slabko, *Optical and Photophysical Properties of Fractally Structured Metal Sols, Novosibirsk*: Sib. Otd. Ross. Akad. Nauk, 2003.

[2] A.E. Varfolomeev, A.V. Volkov, D.Yu. Godovskii, et al., *Pis'ma Zh. Eksp. Teor. Fiz.*, 62, 344 (1995).

[3] C. Delerue, G. Allan, and M. Lannoo, *J. Lumin.* 80, 65 (1999).

[4] R.K. Soni, L.F. Fonseca, O. Resto, et al., *J. Lumin.* 83–84, 187 (1999).

[5] I.S. Altman, D. Lee, J.D. Chung, et al., *Phys. Rev. B: Condens. Matter Mater. Phys.* 63, 161402 (2001).

[6] S. Knief and W. von Niessen, Phys. Rev. B: Condens. *Matter Mater. Phys.* 59, 12940 (1999).

[7] A.A. Olkhov, M.A. Goldschtrakh, A.A. Ischenko *RU Patent* № 2009145013 (2009).

[8] V.N. Bagratashvili, I.A. Tutorskii, A.I. Belogorokhov et al. // Reports of Academy of Sciences. *Physical Chemistry.* v. 405, 360 (2005).

[9] Kumar V. (editor) *Nanosilicon.* Elsevier Ltd.– xiii + 368 p (2008).

[10] Nanostructured *Materials. Processing, Properties, and Applications.* / Edited by Carl C. Koch. NY: William Andrew Publishing. 752 (2009).

[11] A.A. Ischenko, S.G. Dorofeev, N.N. Kononov, et al. *RU Patent* №2009146715 (2009)

[12] G.P. Kuzmin, M.E. Karasev, E.M. Khokhlov, et al., *Laser Phys.* 10, 939 (2000)

[13] J. Beckman, A.A. Ischenko, *RU Patent No.* 2 227 015 (2003)

[14] K. Stehl The Huber G670 imaging-plate Guinier camera tested on beamline I711 at the MAX II synchrotron // *J. Appl. Cryst.* (2000). Vol. 33, p. 394-396.

[15] G.V. Fetisov The X-ray phase analysis. Chapter 11, p. 153-184. // *Analytical chemistry and physical and chemical methods of the analysis.* T. 2. / Red. A.A. Ischenko. M.: ITc Academy, 2010, – 416 p.

[16] P. Scardi and M. Leoni Line profile analysis: pattern modelling versus profile fitting. // *J. Appl. Cryst.* (2006). V. 39, 24-31.

[17] WINX^{POW} Version 1.06. // STOE & CIE GmbH Darmstadt/Germany – 1999.

[18] M. Leoni, T. Confente and P. Scardi // PM2K: a flexible program implementing Whole Powder Pattern Modelling // *Z. Kristallogr. Suppl.* (2006). V. 23. P. 249–254.

[19] P. Scardi *Recent advancements in whole powder pattern modeling* // Z. Kristallogr. Suppl. 2008. V. 27. P. 101-111

[20] N. Strbeck X*-ray scattering of soft matter.* Springer-Verlag Berlin Heidelberg. (2007). – xx + 238 p.

[21] V.I. Iveronova, U.P. *Revkevich The theory of scattering of X-rays.* M.: MGU. 1978. – 278 p.

In: Quantitative Chemistry, Biochemistry and Biology
Editors: G.E. Zaikov, O.V. Stoyanov et al.
ISBN: 978-1-62948-332-0

Chapter 13

Biodegradable Compositions of Starch with Synthetic Polymers and Natural Polysaccharides: Production, Morphology and Properties

S. Z. Rogovina, K. V. Aleksanyan*[*] *and E. V. Prut
Semenov Institute of Chemical Physics, Russian Academy of Sciences,
Moscow, Russia

Abstract

Binary compositions of starch with low density polyethylene and their ternary compositions with the addition of poly(ethylene oxide) as well as polysaccharides chitin or chitosan as a third component were obtained under action of shear deformations in a rotor disperser at different initial ratios of the components. It was shown that the composition of the powder fractions is identical to the original blend composition. Using X-ray diffraction analysis and differential scanning calorimetry method the structure of the initial and forming powder blends was investigated, the mechanical characteristics of the prepared films were measured and the features of thermo- and photooxidative degradation were studied. It was established that the addition of third component promotes an increase in the biodegradability of compositions and, consequently, results in the expansion of the possible areas of application of the related materials.

Keywords: Starch, biodegradability, low density polyethylene, chitin, chitosan, mixing

Because of constantly increasing production of polymers leading to environmental pollution, the problem of their utilization is becoming more and more urgent. This problem can be solved by creation of biodegradable polymer materials which can be decomposed under the environmental conditions to give harmless products for nature (H_2O, CO_2, biomass,

[*] E-mail: christine.alexanyan@gmail.com.

and other natural compounds). Despite the fact that the cost of biodegradable polymers is higher than the cost of traditional synthetic polymers, they intensively lock in the consumer markets. The rise in application of the biodegradable materials is primarily connected with irreplaceable consumption of the world reserves of oil products representing main raw material for production of synthetic polymers. With increase of the production capacities of biodegradable polymer compositions and improvement of their production technology, their cost will gradually decrease and approach the cost of traditional synthetic polymers.

One of the most promising ways for producing partially or completely biodegradable polymer materials is based on the blending of synthetic and natural polymers. This concept of treating the enhancement of biodegradability of a polymer material via addition of a biodegradable component was proposed by Griffin in the early 1970s [1, 2]. Among such compositions, the blends based on polyolefins and renewable natural polysaccharides characterized by easy biodegradability are of special interest [3-6]. The methods of production of the blends based on polysaccharides with synthetic polymers are widely covered in the literature [7-9]. The properties of such compositions (thermal, sorption, strength) are generally determined by both bond type between the components and their compatibility and character of forming morphology. Starch is a most popular polysaccharide, which is added to synthetic polymers, therefore, the blends based on it are the most extensively studied [10-12]. The production of packaging packets for stores was first launched in Europe in 1974 on the basis of starch–PE blends. Similar materials are also popular in production of biomedical items [13]. Starch has advantages such as low cost, wide availability, and total compostability without toxic residues, so the blends on its basis can be used as the substitutes to petroleum-based plastics [14].

Starch is widely used as a filler for production of reinforced plastics [15, 16]. Even small amount of starch (6-30%) mixed with synthetic polymer increases the biodegradability of the final products [17]. However because of differences in chemical structure and hydrophilic characteristics of these two components, the polymer blends have low compatibility that leads to deterioration of mechanical properties. This problem can be solved by chemical modification of one of the components with the aim to introduce functional groups capable to react with the second component [18]. Thermoplastic blends of starch with synthetic polymers are obtained with the use of starch plasticized by water, glycerol, etc. [19]. In this case the components are usually mixed in an extruder at ~ 150 °C providing well gelatinization of starch [20].

Since there are, as a rule, no common solvents for natural polysaccharides and synthetic polymers, the production of their blends by solid-phase mixing is of great interest. Besides, due to the absence of solvents the solid-phase mixing has an advantage as high ecological safe method.

Among various methods of polymer mixing in solid phase (e.g., centrifugal and ball mills) the extrusion mixing is quite promising one. This method, where the principle of joint action of temperature, pressure and shear deformations on material is realized, is widely studied by the Russian scientist [21]. This method is based on destruction of the initial components under action of uniform compression and shear deformations. The physical principle underlying the method proposed is in the fact that energy stored in material under action of pressure and shear deformations is realized in formation of new surface, small shearing force is sufficient. As a result of such action, the powders of different dispersion are formed. It should be noted that the grinding process is not a gradual breakage of the material to smaller particles, but sharp process in a restricted space that points to spontaneous destruction and allows one to assume the branched mechanism of breakdown site development. Use of this method allows one to obtain uniform homogeneous powder blends in the absence of common solvents.

The aim of this work is to produce binary biodegradable blends of starch with low density polyethylene (LDPE) (T_m = 108°C, M_n = 23000, MFI = 2 g/10 min (190°C, 2.16 kg)), their ternary blends with poly(ethylene oxide) (PEO) of various molecular weight (low-molecular PEO_{lm}, $M = 3.5 \times 10^4$ and high-molecular PEO_{hm}, $M = 5 \div 6 \times 10^6$) and natural polysaccharide chitin (content of basic compound 87.2%, degree of deacetylation 0.045) and its deacetylated derivative chitosan (degree of deacetylation 0.87 and $M = 4.4 \times 10^5$) that significantly expands the possible fields of application of such compositions. Structure, biodegradability, and physicomechanical properties of above listed blends were studied in detail.

Starch was blended with various polymers at 150°C via single pass of the components through a rotor disperser, in which material was subjected to joint action of temperature, pressure, and shear deformations. To create intensive shear deformations the rotor disperser was equipped with a grinding head designed as a cam element rotating inside a channeled cylinder and a cooling jacket. Whatever the blend composition was, homogeneous fine-dispersed powders were formed after passing of polymers through the rotor disperser.

The systems of the following composition were obtained: starch–LDPE binary compositions with starch content from 20 to 50 wt.%; starch–LDPE–PEO ternary compositions containing from 20 to 40 wt.% of starch, and starch–LDPE–chitin/chitosan ternary systems with 30 wt.% of starch. The obtained blends were fractionated with the use of a sieve set. Figure 1 shows the comparative histograms of binary and ternary compositions containing 30 wt.% of starch. As is seen from the given data, for starch–LDPE binary composition the basic fraction (about 60%) is composed of particles 0.315-0.63 mm in size. It should be noted that for binary composition the content of fractions with a large particle size increases with the rise of the starch content in the blend (Table 1). Essentially the same fraction distribution is observed for starch–LDPE–PEO_{lm} ternary composition. In this case two fractions with the particle size from 0.09 to 0.63 mm are basic, the fractions with particle size from 0.8 mm appear. Somewhat different distribution was obtained for compositions based on two polysaccharides (e.g., starch and chitin). In this case the basic fraction (above 70%) is composed of particles 0.09-0.315 mm in size. At the same time, new fraction containing small particles 0.071-0.09 mm in size appears. Thus, although the total content of polysaccharides increases from 50 to 60%, the powder with particles of finer size is formed during passing through the rotor that is likely to be connected with co-grinding process in this case [21].

The fractional composition of starch–LDPE blend was determined by iodine reaction. For this purpose, starch was washed with hot water up until the disappearance of the blue color of water in the presence of iodine. The residual LDPE was dried, and the component ratios in fractions were calculated (Table 2). As is seen from the given data, the fractional composition is nearly consistent with the initial component ratio in the blend that testifies the homogeneous distribution of the blend components.

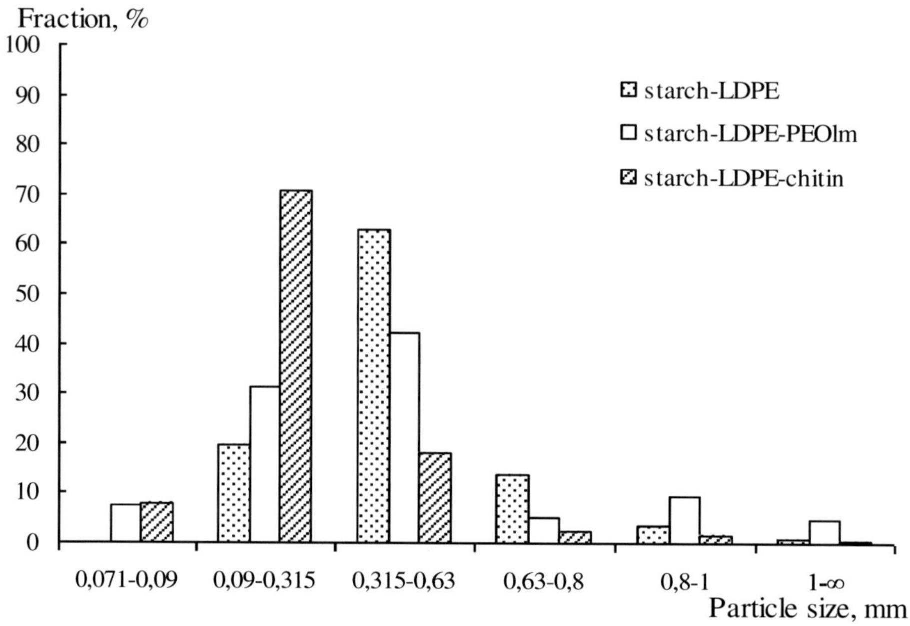

Figure 1. Histograms of particle size distribution of starch–LDPE (30:70 wt.%), starch–LDPE–PEO_{lm} (30:50:20 wt. %), and starch–LDPE–chitin (30:40:30 wt.%) blends.

Table 1. Fractional composition of starch–LDPE blend at different component ratios

Blend composition	Component ratio, %	Fraction with *d* (mm), %				
		0.071-0.09	0.09-0.315	0.315-0.63	0.63-0.8	0.8-1.0
Starch–LDPE	20 : 80	17.7	53.5	23.6	2.8	2.4
	30 : 70	0	19.6	62.8	13.8	3.8
	50 : 50	0	10.8	78.6	6.8	3.8

Table 2. LDPE content in fractions of starch–LDPE blend of different compositions determined by chemical method

Fraction, mm	Initial content of LDPE, wt.%		
	80	70	50
0.071-0.09	88	–	–
0.09-0.315	89	81	76
0.315-0.63	87	87	76
0.63-0.80	–	88	65

Structure of the binary compositions was investigated by the X-ray diffraction analysis. The powder samples were studied on a Rigaku RU-200 Rotaflex diffractometer (Japan) with a rotating copper anode in the transmission mode (40 kV, 140 mA, CuK_α radiation with λ = 0.1542 nm). Figure 2 shows the X-ray patterns of starch, LDPE, and their blend obtained under shear deformations. The X-ray pattern of starch exhibits a set of reflections at $2\theta = 5.7°$

and 16.9° characteristic for crystalline structure. In the X-ray pattern of starch–LDPE blend subjected to shear deformations, these reflections and characteristic peak of LDPE are weaker; i.e., the amorphization of the initial polymers takes place. Thus, the mixing of polymers under shear deformations leads to decrease of their crystallinity degree.

These results were confirmed by DSC data obtained during investigation of starch–LDPE compositions produced by mechanical mixing or under the action of shear deformations on a DSM-10m calorimeter. The weight of samples was in the range from 2 to 6 mg, the heating rate was 16 K/min. The melting enthalpies of individual LDPE and PEO, and binary and ternary blends were calculated per unit mass of LDPE and PEO, respectively, in blend.

It turned out that the position of peaks for LDPE on all three curves coincides, and peak areas differ between each other that is caused by both starch introduction and amorphization of blend obtained under shear deformations (Figure 3). As is seen from figure, the curve of starch–LDPE blend displays a single peak characteristic for LDPE. The position of this peak on the DSC curves for the blends corresponds to T_m of LDPE independently of the conditions of blend preparation. However the values of the melting enthalpies ΔH of the blends prepared via mechanical mixing or under shear deformations differ from the values of LDPE ΔH. In this case the value of melting enthalpy ΔH for blend prepared in the rotor disperser is lower than ΔH for blend obtained by mechanical mixing (Table 3). So, for starch–LDPE (30 : 70 wt.%) blend prepared under action of shear deformations the melting enthalpy equals 43 J/g, while for the blend obtained by mechanical mixing ΔH = 53 J/g, i.e., the crystallinity degree is lower for the blend obtained in the rotor disperser. Consequently, one may state that LDPE amorphization occurs during production of blends under the action of shear deformations.

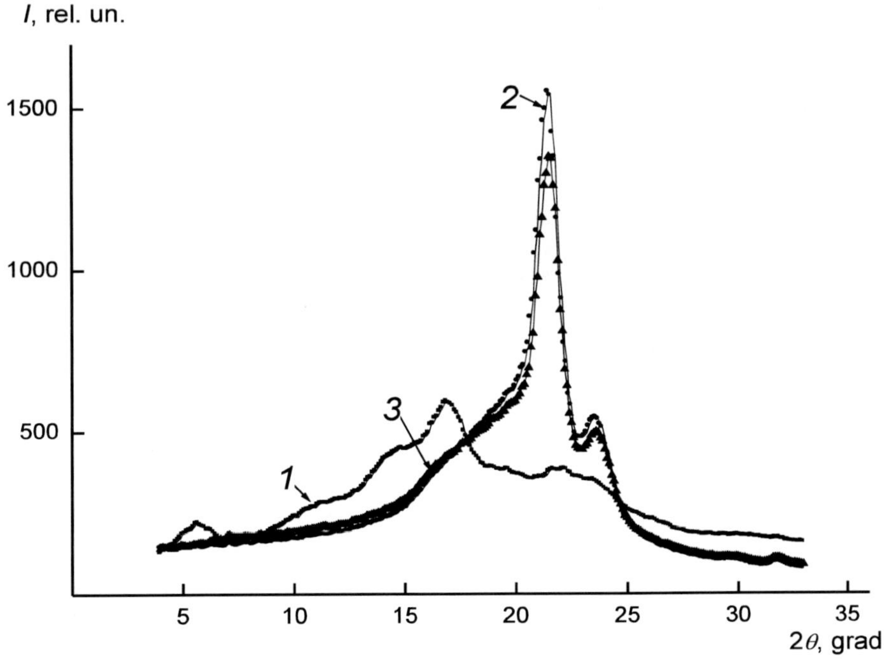

Figure 2. X-ray patterns of starch (*1*), LDPE (*2*), and starch–LDPE blend (30:70 wt. %) obtained under action of shear deformations (*3*).

Table 3. Basic parameters of melting (melting point T_m, temperature range of melting Δ_m, melting enthalpy ΔH) of LDPE and its blends with starch obtained under shear deformations (*1*) and by mechanical mixing (*2*)

System	T_m, °C	ΔT_m, °C	ΔH, mJ	ΔH, J/g
LDPE	109	92-126 34	155	52
Starch–LDPE (30 : 70 wt. %) (*1*)	108.5	93-124 31	107	43
Starch–LDPE (30 : 70 wt. %) (2)	108	93-125 32	128	53

Figure 3 shows the melting curves of LDPE, starch–LDPE (30 : 70 wt.%) binary composition, and starch–LDPE–PEO_{lm} (30 : 50 : 20 wt.%) ternary composition. As is seen from the presented data, the melting curve of the binary blend displays a single peak (characteristic for LDPE) and the melting curve of the ternary blend displays two peaks (characteristic for LDPE and PEO). Change in the melting enthalpies of LDPE and PEO, in addition to the enthalpy change due to amorphization, which is constant for these blends, depends also on the blend composition. Based on this fact, the content of the blend components can be calculated. As is seen from Table 4 presenting the data for the ternary blends based on PEO_{lm}, the calculated values are close to the initial ratio of components. Thus, DSC can be used as an indirect method to calculate starch content in the ternary blends with two synthetic polymers.

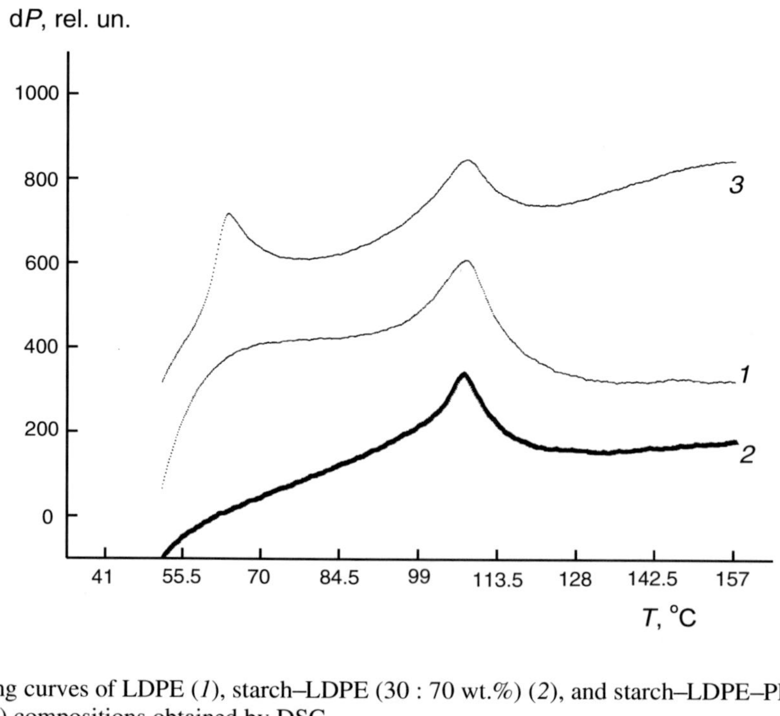

Figure 3. Melting curves of LDPE (*1*), starch–LDPE (30 : 70 wt.%) (2), and starch–LDPE–PEO_{lm} (30 : 50 : 20 wt%) (*3*) compositions obtained by DSC.

Table 4. Composition of starch–LDPE–PEO_{lm} blend determined by DSC

System	Component ratio, wt.%	ΔH (PEO_{lm}), J/g	ΔH (LDPE), J/g	Determined content of components, %		
				PEO_{lm}	LDPE	Starch
Starch–LDPE–PEO_{lm}	20 : 60 : 20	23	31	17.8	55.8	26.4
	30 : 50 : 20	29	29	22.3	51.5	26.2
	40 : 40 : 20	31	28	23.5	49.0	27.5

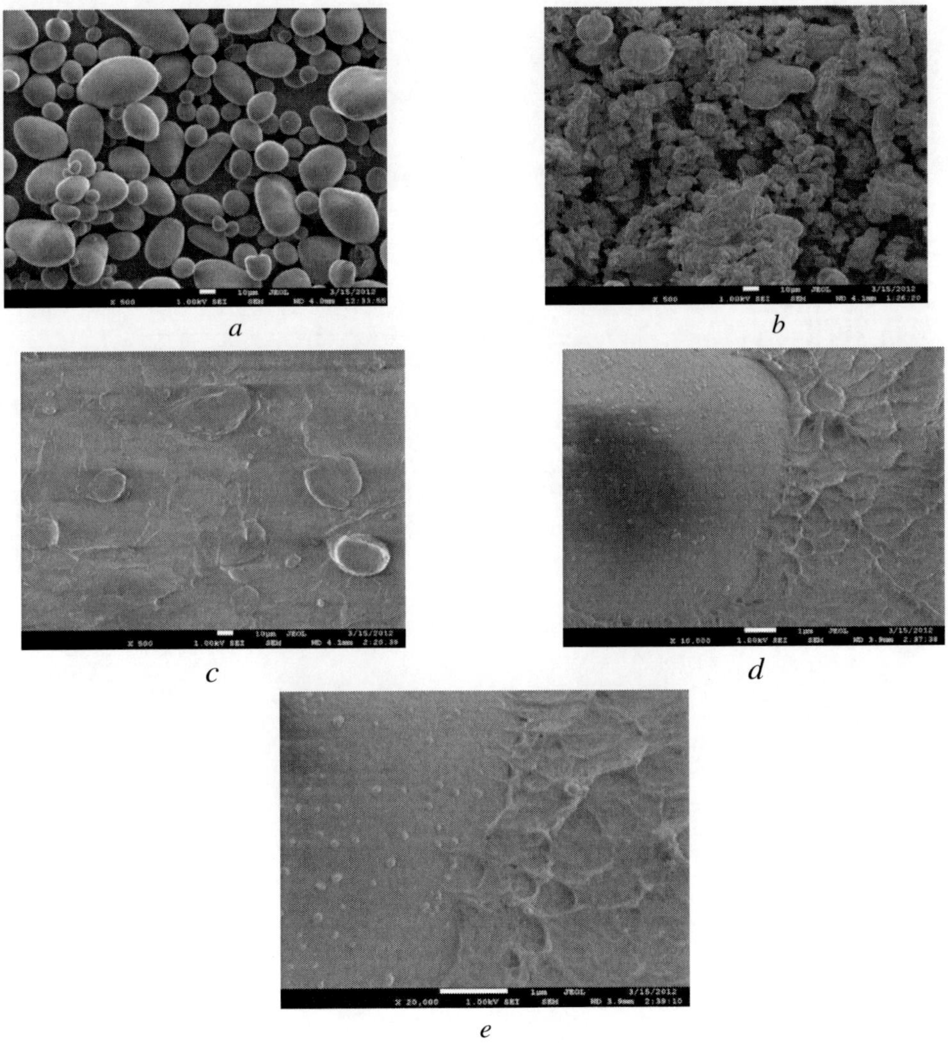

Figure 4. Electron micrographs of the initial starch (*a*), powder starch–LDPE (30:70 wt.%) blend (*b*) and cross-section of the film obtained from starch–LDPE (30 : 70 wt.%) blend (*c-e*) at various magnifications: *a*, *b*, *c* – 500; *d* – 10000; *e* – 20000.

The samples of initial powder starch, starch–LDPE (30 : 70 wt.%) blend, and cross-section of the film from starch–LDPE blend obtained after freezing in a liquid nitrogen were investigated with the use of scanning electron microscopy (SEM). The study was performed on SEM JEOL JSM-7001F at various magnifications. Electron micrographs show that the starch particles have ideal shape, and their size varies from 12 to 60 μm (Figure 4*a*). From consideration of electron micrographs of the powder starch–LDPE blend it follows that starch particles are uniformly distributed in LDPE matrix (Figure 4*b*). Figures 4*c-e* illustrate LDPE bands to starch particles in the cross-section of the film from starch–LDPE blend that testifies some interfacial interaction.

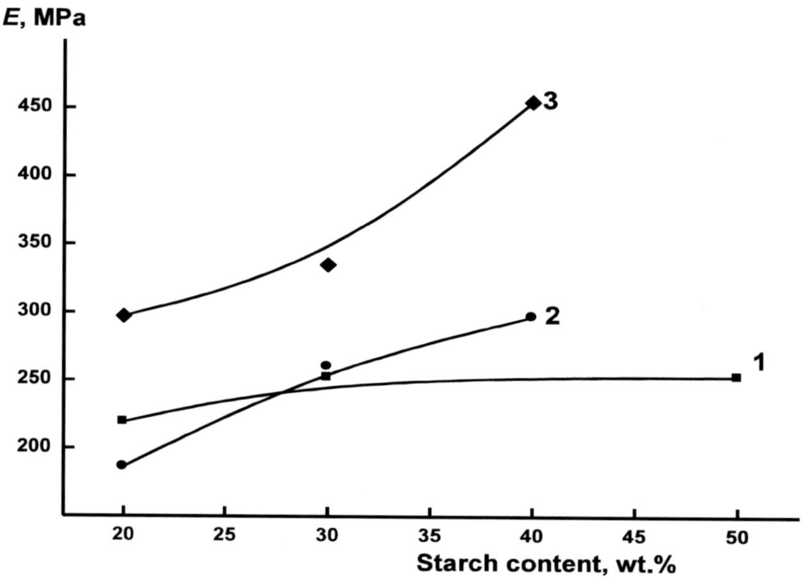

a

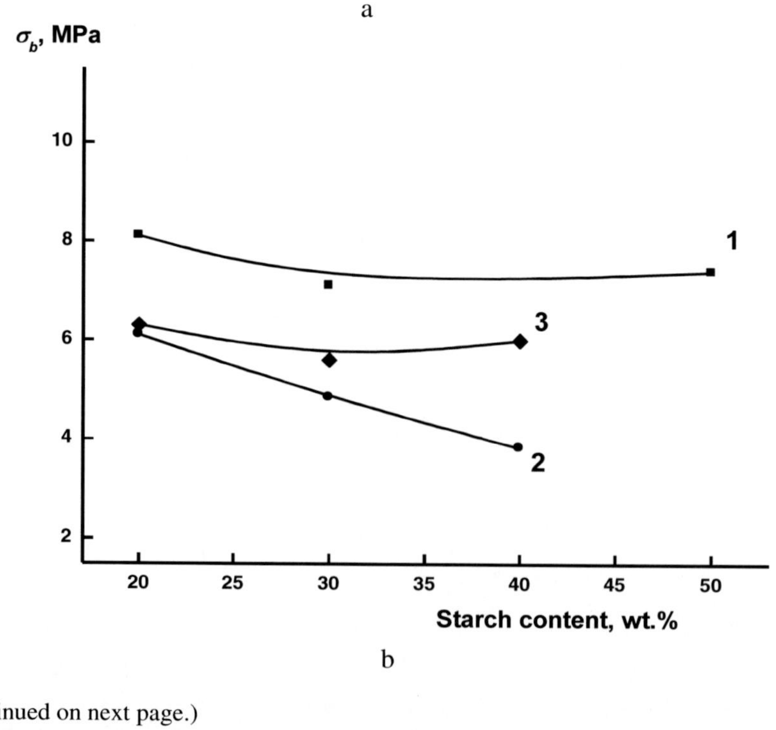

b

Figure 5. (Continued on next page.)

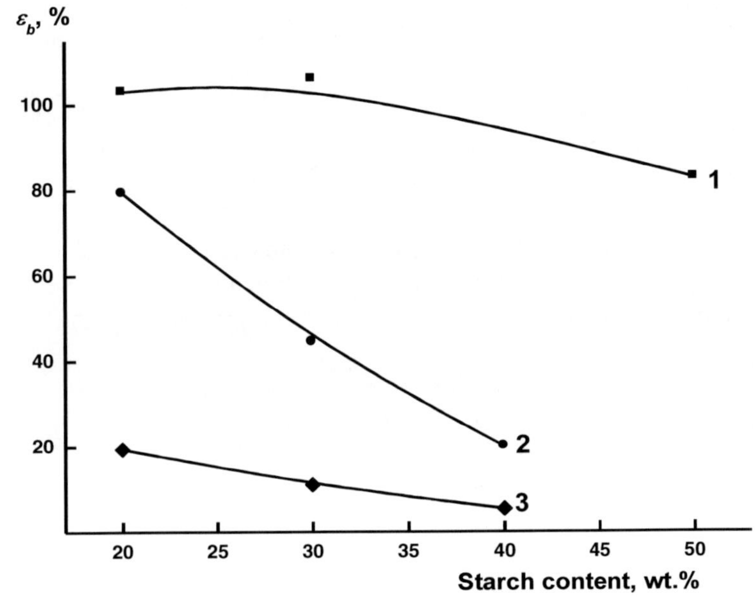

c

Figure 5. Dependence of elastic modulus E (*a*), ultimate tensile strength σ_b (*b*), and elongation at break ε_b (*c*) on starch content for starch–LDPE (*1*), starch–LDPE–PEO_{lm} (*2*), and starch–LDPE–PEO_{hm} (*3*) compositions.

Mechanical tests were performed on films obtained from binary and ternary powder compositions by pressing. Figure 5 shows the comparative dependences of elastic modulus E, ultimate tensile strength σ_b, and elongation at break ε_b of starch–LDPE binary composition and ternary compositions of starch with LDPE containing low-molecular and high-molecular PEO on starch content. It is seen from this figure that as starch content increases the value of elastic modulus E rises for all systems. Maximal values were obtained for the compositions containing high-molecular PEO. The character of curves of E dependence on starch content of the binary and ternary compositions with low-molecular PEO containing up to 30 wt.% of starch remains almost unchanged. With further increase of starch content the value of elastic modulus for ternary blends sharply rises.

The value of ultimate tensile strength σ_b of starch–LDPE and starch–LDPE–PEO_{hm} compositions changes insignificantly with increase of starch content, but it is obvious that the highest values are observed for the binary compositions. Ultimate tensile strength of starch–LDPE–PEO_{hm} composition drops monotonically with increase of starch content, while value of elongation at break ε_b decreases for all compositions. If for starch – LDPE (curve *1*) and starch–LDPE–PEO_{hm} (curve *3*) the drop of ε_b is not so substantial (on 20 and 14%, respectively), so for starch–LDPE–PEO_{lm} (curve *2*) ternary composition the value of elongation at break sharply decreases with increase of starch content (on 60%). Absolute values of ultimate tensile strength σ_b and elongation at break ε_b of binary compositions are higher than for the ternary compositions based on two synthetic polymers. Such character is apparently connected with different morphology of the compositions obtained.

Table 5 shows the dependences of mechanical parameters for binary and ternary compositions based on two synthetic polymers and two polysaccharides with the same starch content. From data presented it follows that elastic modulus E of starch–LDPE–chitin/chitosan composition is significantly higher than for the rest systems, that is caused by increased rigidity of chitin and chitosan in comparison with other polymers. Similar dependence is observed for the ultimate tensile strength. As one would expect, the values of elongation at break ε_b of the ternary systems based on two polysaccharides is significantly lower than of ternary compositions based on two synthetic polymers and starch–LDPE binary compositions.

Thus, a change in the mechanical characteristics of binary blends based on starch with LDPE and ternary blends with introduction of PEO or chitin/chitosan depends on both blend composition and component nature.

Table 5. Dependence of elastic modulus E, ultimate tensile strength σ_b, and elongation at break ε_b on blend composition

Composition, wt.%	E, MPa	σb, MPa	εb, %
Starch–LDPE (30 : 70)	290 ± 10	7.1 ± 0.2	105 ± 5
Starch–LDPE–PEOlm (30 : 50 : 20)	260 ± 10	4.85 ± 0.1	4.44 ± 0.3
Starch–LDPE–chitin (30 : 40 : 30)	968 ± 54	10.5 ± 20	2.1 ± 0.2
Starch–LDPE–chitosan (30 : 40 : 30)	796 ± 8	11.5 ± 0.3	3.05 ± 0.1

Thermal stability is one of the main characteristics of polymer compositions determining possible areas of their application. Therefore, investigation of the dependence of blend thermal stability on its composition is of both scientific and practical interest. Thermogravimetric analysis of the individual polymers and their blends was carried out on a synchronic thermal analyzer STA 449 F3 Jupiter (NETSCH, Germany) over the temperature range of 30–560°C with the rate of temperature change of 10 K/min in an argon atmosphere (rate of gas consumption was 10 mL/min). Figure 6 shows TGA curves of individual polymers and their binary and ternary blends. As is seen from Figure 6*a*, introduction of LDPE leads to increase of the composition thermal stability. Apart from thermal stability, photooxidation is also an important factor determining polymer system stability. It is known that photodegradation of such synthetic polymers as polyethylene and polypropylene leading to formation of low-molecular fractions causes increase of polymer biodegradability [22, 23]. Comparison of TGA curves of starch–LDPE binary composition before and after photooxidation (ISO 4892-1) showed that the temperature of the degradation onset for the irradiated blend is lower than for the blend before photooxidation.

Figure 6*a* also shows TGA curve of starch–LDPE–PEO_{lm} (30 : 50 : 20 wt.%) ternary composition. It is visible that addition of PEO leads to significant decrease of the system thermal stability compared to the binary systems. Figure 6*b* illustrates TGA curves of LDPE, starch, chitin and TGA curve of the ternary blend based on two polysaccharides, for example, starch–LDPE–chitin (30 : 50 : 30 wt.%). The comparison of TGA curves of binary and ternary compositions based on starch showed that the degradation temperature onset is approximately the same. In the temperature range from 300 to 420°C the binary composition reduces in weight on 7-10%, while the ternary composition based on two polysaccharides reduces in weight on 30-34%.

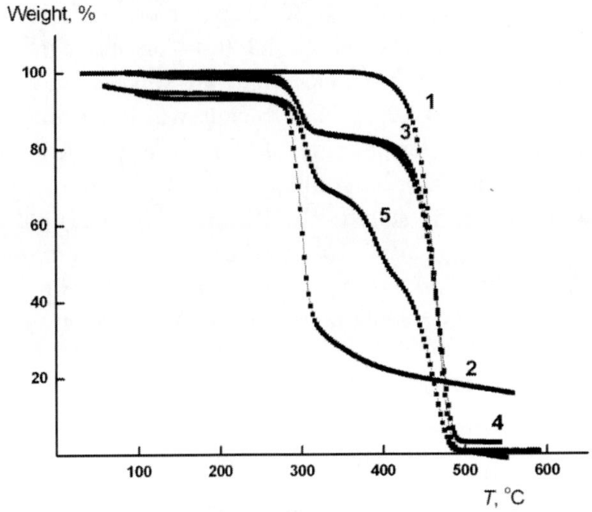

a

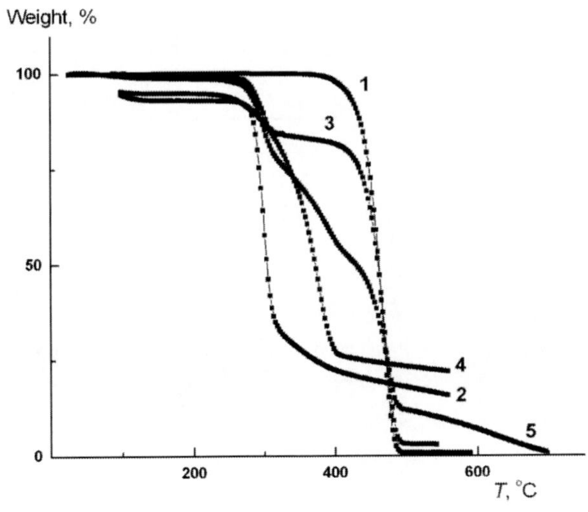

b

Figure 6. TGA curves of individual polymers and their blends: LDPE (*1*), starch (*2*), starch–LDPE (50:50 wt. %) before irradiation (*3*); *a* – starch–LDPE (50:50 wt. %) after irradiation (*4*), starch–LDPE–PEO_{Im} (40:40:20 wt. %) (*5*); *b* – chitin (*4*), starch–LDPE–chitin (30:40:30 wt. %) (*5*).

Data of thermogravimetric analysis allows one to state that thermal destruction of the compositions based on two synthetic polymers and two polysaccharides, where their total fraction is up to 60%, proceeds more intensively than degradation of the binary compositions based on starch with LDPE, where this factor does not exceed 50%.

The estimation methods of polymer composition biodegradability mainly based on analysis of waste products of the microorganisms are described in detail and introduced into standards of various countries [24-27]. They involves artificial infection of the samples by fungus test cultures (e.g., *Aspegilus*, *Penicillium*, etc.), which are more active in relation to a lot of classes of polymers [28].

Among various methods on investigation of the polymer composition biodegradability, generally accepted tests on fungus resistance and tests on biodegradation under natural conditions, according to which the samples placed into wet soil, were used.

The biodegradability of starch–LDPE binary blends was determined by the intensity of mold fungus growth on the samples. Various fungi from the All-Russia Collection of Microorganisms were used (Table 6). According to the tests on fungus resistance of the binary compositions based on starch, starch – LDPE (50 : 50 wt.%) blend shows maximum intensity of mold fungus growth, measured at 5 (Figure 7). The growth of fungi covering more than 90% of the surface is clearly seen by the naked eye. The films prepared from these blends have no fungus resistance and contain nutrients favoring fungi growth in the presence of mineral pollutants.

Table 6. Types and concentration of spores from All-Russia Collection of Microorganisms

Types of spores	Amount of spores	Concentration of spores, N/sm^3
Aspergillus niger van Tieghem	1550	7.8×10^6
A. terreus Thom	2050	1.0×10^7
A. oryzae (Ahlburg) Cohn	1200	6.0×10^6
Chaetomium globosum Kunze	200	1.0×10^6
Paecilliumces variotii Bainier	1650	8.3×10^6
Penicillium funiculosum	435	2.1×10^6
P. chrysogenum Thom	715	3.6×10^6
P. cyclopium Westling	465	2.3×10^6
Trichoderma viride Persoon: Fries	800	4.0×10^6

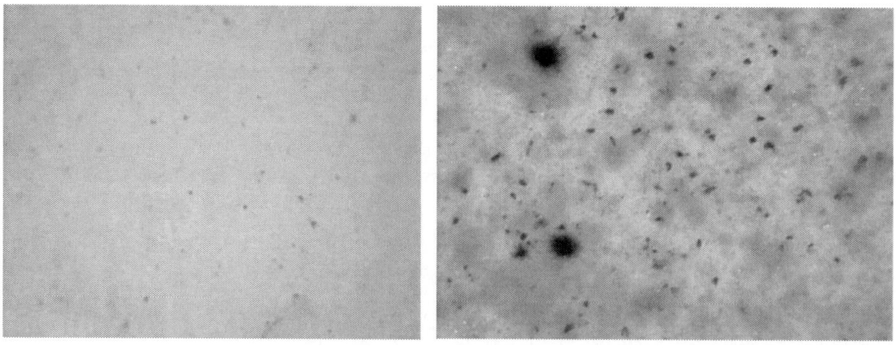

a b

Figure 7. Micrographs of the surface of starch–LDPE (50 : 50 wt.%) composition: *a* – original film, *b* – film infected with fungi spores for 28 days.

However, it turned out that this composition is not completely biodegradable. During holding the films in soil for several months they reduced insignificantly in weight, but in this case the films became more fragile with dark spots corresponding to the biodegradation areas.

The curves of weight loss for starch–LDPE–chitin/chitosan ternary compositions placed in soil are given in Figure 8. As illustrated in this figure, both blends reduce in weight up to 25%. It should be noted that the weight loss rate for all systems explored was the same during first 50 days, but on further holding in soil the rate of biodegradation for starch–LDPE–chitin composition was higher than for the similar system containing chitosan.

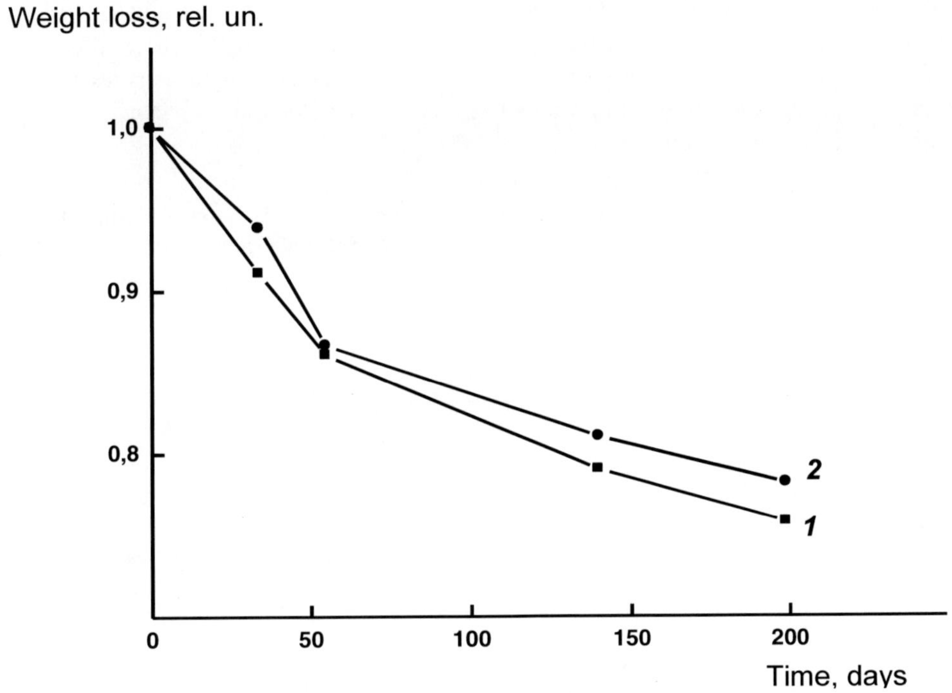

Figure 8. Weight loss curves of films from starch–LDPE–chitin (*1*) and starch–LDPE–chitosan (*2*) (30:40:30 wt.%) compositions.

Table 7 shows data characterizing the difference of the mechanical characteristics of ternary compositions based on starch before and after holding in soil. The data presented testify that the samples reduce not only in weight during the investigation, but also in strength.

Table 7. Change of mechanical characteristics of starch–LDPE–chitin/chitosan compositions after holding in soil

Composition (30 : 40 : 30 wt.%)	ΔE, MPa	$\Delta\sigma_b$, MPa	$\Delta\varepsilon_b$, %
Starch–LDPE–chitin	572 ± 32.5	5.5 ± 1.6	0.05 ± 0.2
Starch–LDPE–chitosan	571 ± 25.5	8.8 ± 0.3	1.05 ± 0.1

Biodegradable compositions based on starch with synthetic polymers and some other polysaccharides were obtained in the rotor disperser under conditions of shear deformations that allows one to obtain highly homogeneous powders. The results of X-ray analysis and DSC showed that the mixing of polymers under shear deformations leads to decrease of their crystallinity degree. Thermogravimetric analysis showed that the introduction of LDPE leads to increase of the binary system thermal stability, while the photooxidation and especially addition of the third component leads to significant decrease of the system thermal stability. The mechanical properties and biodegradability of these blends generally depend on the composition of blend. Both addition of PEO and other polysaccharide (chitin or chitosan) to starch–LDPE compositions leads to an increase of the material biodegradability with insignificant deterioration of the mechanical parameters. Thus, the solid-phase ecologically safe method of polymer mixing permits one to produce the biodegradable blends based on starch and natural/synthetic polymers with quite high mechanical characteristics that allows one to use them in considerably important fields as the production of packaging materials, domestic articles, medicine, etc.

We are grateful to S. M. Ozerskaya (Skryabin Institute of Biochemistry and Physiology of Microorganisms, Russian Academy of Sciences) for the help in microbiological experiments. The equipment of CKP MIPT and REC "Nanotechnology" of MIPT (SEM JEOL JSM-7001F) was used in this work.

REFERENCES

[1] Griffin G.J.L. Biodegradable fillers in thermoplastics. // *Adv.Chem. Ser.* –1974. – V. 134. – P. 159-170.

[2] Griffin G.J.L. *British Pat. Application.* 55195/73/ (1973).

[3] Vasnev V.A. Biodegradable polymers. // *Polym. Sci., Ser. B.* – 1997. – V. 39. – P. 474-485.

[4] Arvanitoyannis I. Totally and partially biodegradable polymer blends based on natural and synthetic macromolecules: preparation, physical properties, and potential as food packaging materials. // J. Macromol. S. – *Rev. Macromol. Chem. Phys.* – 1999. – V. 39, N 2. – P. 205-271.

[5] Albertson A.C. and Karlsson S. Chemistry and biochemistry of polymer biodegradation. // in *Chemistry and technology of biodegradable polymers.* Ed. by Griffin G.J.L. Glasgow: Blackie Academic and Professional, 1994. – P. 7-17.

[6] Bastioli C. // *Handbook of Biodegradable Polymers;* Shawbury, Shrewsbury, Shropshire: Rapra Technology ltd. – 2005.

[7] Huneault M.A. and Li H. Preparation and properties of extruded thermoplastic starch/polymer blends. // *J. Appl. Polym. Sci.* – 2012. – V. 126, N S1. – P. E96-E108.

[8] Xiong H.G., Tang S.W., Tang H.L., and Zou P. The structure and properties of a starch-based biodegradable film. // *Carbohydr. Polym.* – 2008. – V. 71, N 2. – P. 263-268.

[9] Kaseem M., Hamad K., and Deri F. Preparation and studying properties of polybutene-1/thermoplastic starch blends. // *J. Appl. Polym. Sci.* – 2012. – V. 124, N. 4. – P. 3092-3098.

[10] Aburto J., Thiebaud S., Alric I., Borredon E., Bikiaris D., Prinos J., and Panayiotou C. Properties of octanoated starch ad its blends with polyethylene. // *Carbohydr. Polym.* – 1996. – V. 34, N 1. – P. 101-112.

[11] Thakore I.M., Desai S., Sarawade D.D., and Davi S. Studies on biodegradability, morphology and thermo-mechanical properties LDPE/modified starch blends. // *Eur. Polym. J.* – 2001. – V. 37, N 1. – P. 151-160.

[12] Jang B.C., Huh S.Y., Jang J.G., and Bae Y.C. Mechanical properties and morphology of the modified HDPE/starch reactive blend. // *J. Appl. Polym. Sci.* – 2001. – V. 82, N 13. – P. 3313-3320.

[13] Sionkowska A. Current research on the blends of natural and synthetic polymers as new biomaterials: Review. // *Prog. Polym. Sci.* – 2011. – V. 36. – P. 1254-1276.

[14] Xie F., Halley P.J., and Avérous L. Rheology to understand and optimize processibility, structures and properties of starch polymeric materials // *Prog. Polym. Sci.* – 2012. – V. 37. – P. 595-623.

[15] Bagheri R. Effect of processing on the melt degradation of starch-filled polypropylene. // *Polym. Int.* – 1999. – V. 48, N 12. – P. 1257-1263.

[16] Evangelista R.L., Nikolov Z.L., Sung W., Jane J., and Gelina R.J. Effect of compounding and starch modification on properties of starch-filled low density polyethylene. // *Ind. Eng. Chem. Res.* – 1991. – V. 30, N 8. – P. 1841-1846.

[17] Röper H. and Koch H. The role of starch in biodegredable thermoplastic materials. // *Starch* – Stärke. – 1990. – V. 42. – P. 123-130.

[18] Pimpan V., Ratanajiajaroen P., and Laithumtaweekul L. Effect of chemical modification on the properties of cassava starch and medium density polyethylene blends. // *Abstracts of 9th European Symposium «Polymer Blends»*. Palermo, September 9-12, 2007. – P. 27.

[19] Russo M.A.L., O'Sullivan C., Rounsefell B., Halley P.J., Truss R., and Clarke W.P. The anaerobic degradability of thermoplastic starch: Polyvinyl alcohol blends: Potential biodegradable food packaging materials // *Bioresour. Technol.* – 2009. – V. 100. – P. 1705-1710.

[20] St.-Pierre N., Favis B.D., Ramsay J.A., and Verhoogt H. Processing and characterization of thermoplastic starch/polyethylene blends. // *Polymer.* – 1997. – V. 38, N 3. – P. 647-655.

[21] Prut EV. Creep instability and fragmentation of polymers (Review). // *Polym. Sci., Ser A.* – 1994. – V. 36, N 4. – P.601-607.

[22] Stranger-Johannessen M. Susceptibility of photodegradeed polyethylene to microbiological attack. // *J. Appl. Polym. Sci.: Appl. Polym. Symp.* – 1979. – V. 35. – P.415-421.

[23] Cornell J.H., Kaplan A.M., and Roger M.R. Biodegradability of photooxidized polyalkylenes. // *J. Appl. Polym. Sci.* – 1984. – V. 29, N 8. – P. 2581-2597.

[24] Calmon-Decriaud A., Bellon-Maurel V., and Silvestre F. Standard methods for testing the aerobic biodegradation of polymeric materials. Review and Perspectives. // *Adv. Polym. Sci.* – 1998. – V. 135. – P. 207-226.

[25] Vikman M., Itävaara M., and Poutanen K. Measurement of the biodegradation of starch-based materials by enzymatic methods and composting. // *J. Polym. Environ.* – 1995. – V. 3, N 1. – P. 23-29.

[26] Urstadt S., Augusta J., Müller R.-J., and Deckwer W.-D. Calculation of carbon balances for evaluation of the biodegradability of polymers. // *J. Polym. Environ.* – 1995. – V. 3, N 3. – P. 121-131.

[27] Kleeberg I., Hetz C., Kroppenstedt R.M., Müller R.-J., and Deckwer W.-D. Biodegradation of aliphatic-aromatic copolyesters by thermomonospora fusca and other thermophilic compost isolates. // *Appl. Environ. Microbiol.* – 1998. – V. 64. – P. 1731-1735.

[28] Ryabov S.V., Boiko V.V., Kobrina L.V., etc. Study and characterization of polyurethane composites filled with polysaccharides. // *Polym. Sci., Ser. A.* – 2006. – V. 48, N 8. – P. 841-853.

In: Quantitative Chemistry, Biochemistry and Biology
Editors: G.E. Zaikov, O.V. Stoyanov et al.
ISBN: 978-1-62948-332-0

Chapter 14

Kinetics and Mechanism of the Chemically Modificated Starches during Gelatinization

***A. I. Sergeev*[1,*], *N. G. Shilkina*[1], *L. A. Wasserman*[2,†] *and H. Staroszczyk*[3,‡]**

[1]Institute of Chemical Physics RAS, Moscow, Russia
[2]Institute of Biochemical Physics RAS, Moscow, Russia
[3]Gdansk University of Technology, Gdańsk, Poland

Abstract

Native potato starch was silicated by means of microwave-assisted reactions with varying doses of reagents. Resulting products were characterized by DSC and HMR-relaxation methods. It was observed that the hydratation (AUW) of the silicate starches during gelatinization decreased when compared to that of the native potato starch. The amount of unfrozen water (UW) for ungelitinizated silicate starches are less than in control samples. At the same time, the less mobile water population W_a in the experiment is three times more than in the control. These changes, just as the gelatinization temperature decrease, can be a consequence of starch distructure under microwave irradiation and chemical modification.

Keywords: Potato starch, silicated, microwave, DSC, HMR-relaxation, unfrozen water

Introduction

In recent years there has been an increased interest in starch chemical modification reactions influenced by microwave irradiation [1-4]. The microwave irradiation using in the solid state etherification of starch with inorganic acids, there salts, anhydrides and oxides

* E-mail: nismpa@mail.ru
† E-mail: lwasserma@mail.ru
‡ E-mail: hanna.staroszczyk@pg.gda.pl.

gave fast and uniform heating, better selectivity, less by-products, increasing the reaction rate [1]. Such modification produced anionic starches as a negative charge on them. The latter are essential for intermolecular interactions with other macromolecules which lead, for instance, to polysaccharide-protein complexes. The resulting binary systems are known as biodegradable materials. Recently natural biopolymers have been filled with layered silicates to improve their functional properties. They might be considered as environmental friendly components. The detailed synthesis of starch silicates described in [3, 4]. The physical and structural properties of starch strongly depend on molecular interaction with water. The objective of this study was the investigation of the starch chemical modification effects on the state of water and water mobility during gelatinization.

Different techniques have been used for the study of water-starch interaction. Differencial scanning calorimetry (DSC) has been extensively used for investigation gelatinization, glass and melting transition [5, 6]. X-ray diffraction method has been used to crystallization monitoring [7], small angle neutron scattering (SANS) - of probe water inside starch granules during gelatinization [8]. In wide spectrum of nuclear magnetic resonance (NMR) methods, NMR-relaxation is undoubtedly the most suitable technique for studying the dynamic state of water in such heterogeneous systems as starch [9-13].

A lot of scientific methods which were used in the recent works [3,4] gave a detailed functional properties of starch silicates. It is possible that the investigation of molecular dynamic of water interaction with such starches, the quality and quantity unfrozen water will give an additional understanding of nature of these properties.

Materials and Methods

The investigated starches were:

1. Native potato starch (PS) (Łomża, Poland)
2. Conventionally heated PS (2 hours at 100^0C) (PSC)
3. Conventionally heated the solid state mixture PS+ silicic acid (1:1 molar ratio) (PSCSi)
4. Microwave irradiated PS (30 min 450W) (PSW)
5. Microwave irradiated the solid state mixture PS+ H_2SiO_3 (1:1 molar ratio) (PSWSi)

The investigated samples were accomplished essentially as described by [4].

DSC

The differential scanning calorimeter DSC 204 F1 Phoenix (Netzsch, Germany) with Proteus Software on Windows was used to determine the frozen water (FW), the unfrozen water (UW), the additional unfrozen water (AUW) and the temperature of gelatinization (the peak melting) T_p in the starch-water systems. All samples were prepared at 1:1 ratio of starch to distilled water contents and stored for 24 hours in sealed aluminum pans. After experiment the aluminum pans were pierced and dried out to the constant weight. The DSC experiment consisted of several sets:

1. Sample cooling to -40^{0}C
2. Scanning from -40^{0}C to +80^{0}C at heating rate 10^0/min (first scan)
3. Sample re-cooling to -40^{0}C
4. Rescanning from -40^{0}C to +80^{0}C at heating rate 10^0/min (second scan)

The unfrozen water (UW) was calculated from the following equation:

UW= M_{com}- FW, where

M_{com} – water content arising from sample drying (g. water/ g dry starch),

FW – the frozen water, which taken from the area under the ice melting endotherm and heat of fusion of ice (334.7 J/g)

The additional unfrozen water (AUW) arising from gelatinization was calculated as difference between FW of sequential scans:

$AUW = FW_{1\text{-st scan}} - FW_{2\text{-nd scan}}$

All measurements were done in triplicate.

NMR-Relaxation Method

NMR-relaxation measurements were performed on Bruker Minispec PS 120 spectrometer operating at a proton resonance frequency of 20 MHz.

The samples were placed in 5 mm tube with a tefton plug inserted to prevent evaporation. The probe was equipped with a home made heating control system (± 0.5 ^{0}C) and the measurements were taken at different temperatures ranging from 25 ^{0}C to 95 ^{0}C. The sample had a thermal equilibration for 10 min at each temperature before measurement.

Spin-spin relaxation times (T_2) of water protons were measured using the Care-Purcell-Meiboom-Gill (CPMG) pulse sequence with 90^0-180^0 pulse spacing (τ) – 0,02 ms, inter-pulse sequence delay – 10 sec, number of points on curve -150. All data points of relaxation curve were acquired after each 10-th echo as an average of 16 repetitions. Relaxation curves obtained from CPMG pulse sequence were analyzed by the MULTY T_2 program (Bruker), using a multiexponential model. All measurements were done in triplicate.

DISCUSSION

In a starch-water mixture, the water of hydration can be found near the surface of biopolymers. The water molecular can interact with macromolecules of polymers and could remain unfrozen at subzero temperature. The extent of hydration depends on several factors such as the distance of water molecules from the surface of polymers, the charge of the polymers, the density of hydrogen bonds. The structural function of UW is mainly driven by the state of polymer chains. If hydration water should migrate to form ice, rearrangement of the polymer structure would be necessary to adjust to the loss of those water molecules. But this situation is difficult to realize because of the considerable energy required to change the polymer conformation at subzero temperature.

Therefore sufficient amount of UW is retained in the system. In work [9] the authors showed that the total water content influenced the gel structure and altered the magnitude and extent starch-water interactions. They found that maximum AUW for a water content of several starches corresponded to a semi-solid saturated suspension (about 1:1 weight ration) with a little or no additional supernatant water. In this condition the extra- granule water has been proposed to exist as thin layer between closely packed granules [14]. The gel structure of those concentrated starch systems has also been described as tightly packed system of swollen granules with the thin of amylose gel between the granules [15]. It is proposed that such a structure would be characterized by a short distance between hydration water and starch polymers and high local volume density of hydrogen bonds. As the initial water content increases AUW of the systems tended to decrease the indicating reduce starch-water interaction. All the starch types, despite the different amilose content, show a relatively similar trend of the change in AUW as a function of initial water content [9]. This was pointed out of important role of amilopectin in interaction between water and gelatinizated starches. Our results of DSC investigations of the starches during gelatinization present in Table 1.

AUW of native potato starch are close to those in the literature 0,38 g.H_2O/g. dry starch [16] and 0,43 g.H_2O/g. dry starch [9]. The amount of UW increases for all the samples after gelatinization. The convectional heating does not change AUW but microwave irradiation increases AUW almost twice. In work [4] it was shown that the changes in the granules structure induced by irradiation and etherification of starch weakened inter- and intra-molecular hydrogen bonds, facilitating the diffusion of water inside the granules/ But on the other hand the granules held together by cross linking became less water soluble . It has also provided granule agglomeration.

All these factors may be the reason of decreasing in AUW for PSWS and PSCSi in our experiment. The microwave irradiation of starch without silicic acid led to deterioration of granule structure whilst its silication inhibited the granularity in spite of granule swelling. The decrease of gelatinization temperature T_p (Table 1) is also the result of the crystal structure deterioration. Our results show that the change of AUW can reflect the destruction process in starch granules.

Table 1. The amount of unfrozen water after gelatinization (UW), Additional unfrozen water after gelatinization (AUW) and temperature of gelatinization (Tp) for different starch-water mixtures

	Unfrozen water after gelatinization g.H2O/ g. dry starch (UW)	Additional unfrozen water after gelatinization g.H2O/g. dry starch (AUW)	T_p Gelatinization temperature (peak) °C
PS	0,389±0,009	0,021±0,005	67,8±0,2
PSC	0,382±0,010	0,021±0,004	67,5±0,3
PSW	0,414±0,012	0,039±0,007	63,7±0,8
PSCSi	0,360±0,011	0,003±0,002	64,3±1,0
PSWSi	0,347±o,012	0,013±0,005	63,1±0,7

NMR Relaxation

Proton transverse relaxation curves deconvolution for ungelatinized starch-water mixture samples reveals two distinct water population. The population of less mobile W_a and the more mobile W_b water have spin-spin relaxation times T_{2a} and T_{2b} respectively, only W_b is dependent on the initial water content (data not shown). It's possible to suggest that the less mobile water fraction W_a occupies, the more constant environment, probably the interior of the granule and W_b- the exterior of the granule.

The same model of starch-water interaction has been reported [14]. It was based on the assumption that the water within different compartments is distinguishable in an NMR relaxation study only if the rate of diffusive exchange of water molecules between two distinct regions is low compared to their intrinsic relaxation rates. By investigating the temperature dependence of T_2 of water protons for saturated packed bed (1,1 g. water/g. dry starch) ungelatinized potato starch granules , three distinct water populations, assigned to extra-granular water, water in the amorphous growth rings and the water in the amorphous regions of the crystalline lamellae have been identified. T_2 of the first population is about 50 ms. The other two populations representing intra-granular water are in the fast diffusion exchange and have T_2 about 8 ms.

As the system is cooled down to 4°C the populations of intra-granular water enter slow, exchange regime and two T_2 appeared: the same water proton population for several starches discussed in work [9]. In [10] the authors reported about 4 populations for rice starch water mixture. But the population with the longest T_2 (1 sec) might be a result of water excess in the samples (3:1) g/H2O/g dry starch.

The identity of the experimental conditions in these works gives us the grounds to consider our assumptions of the water populations location fair.

Dynamic redistribution of extra- and intra-granule water and the change of relaxation times T_2 in starch-water samples during gelatinizationat are demonstrated on figures 1-2. The results can be divided into two groups:

1. *PS,PSC,PSW.* The heating causes structural disruption of starch granules, water interaction with biopolymers occurs easier and the W_a increases. After the partial gelatinization some of the structural barriers are destroyed leading to a faster exchange of water molecules between different regions, as shown by a reduction in less mobile water fraction W_a.

   It was reported in the work [11] that such changes can reflect the increased mobility of starch polymer chains because the exchangeable water protons could be strongly influenced by the chemical exchange with the hydroxyl protons in the biopolymer. They pointed out that the increase of inter-granular water (W_a) with temperature reflected the swelling of granules. Above the melting transition temperature (62°C) W_a decreased as amilose diffused out of the granule. The difference of W_a in melting point for different samples (W_a maximum for PS, PSC and PSW) can be explained by more destruction influence of microwave irradiation compared to convectional heating.

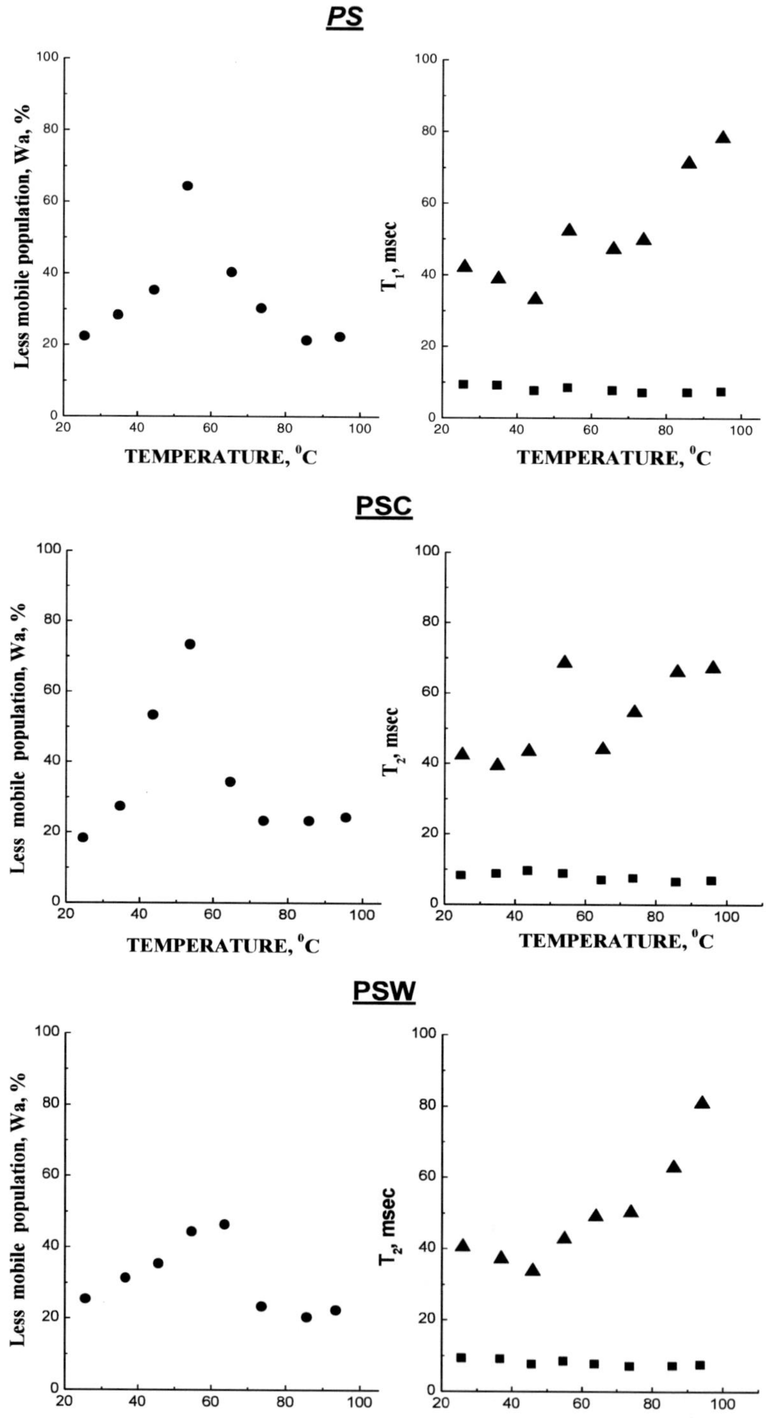

Figure 1. The changes less mobile proton population Wa (●) and spin-spin relaxation times T_{2a} (■) and T_{2b} (▲) in starches during gelatinization.

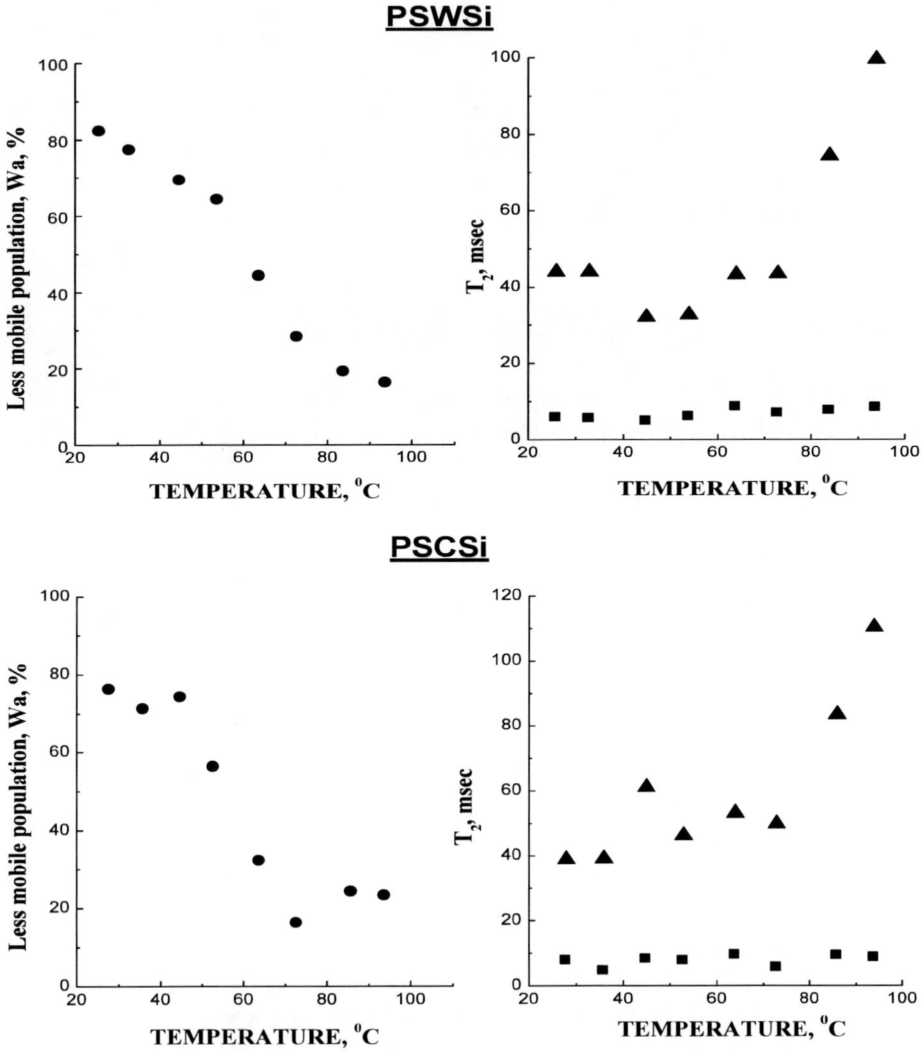

Figure 2. The changes less mobile proton population Wa (●) and spin-spin relaxation times T_{2a} (■) and T_{2b} (▲) in chemically modificated starches during gelatinization.

2. *PSCS,PSWSi* .W_a for the chemically modificated starch at t = 25°C are approximately four times more than for native potato. It is necessary to note that it contradicts with UW decrease for the same samples. Possibly it reflects the considerable reduction of extent of water interaction with biopolymers.

   The morphological proton changes of granules of silciated starch and DSC data which presented in [4] described the deterioration of granular structure. On the one hand the inter- and intra-molecular hydrogen bounds were weakened on the silication and granules could absorb water more easily, but on the other hand the granules held together by cross linking became less water soluble (Table 7,Figure 6 in [4]).

W_a of PSCSi and PSWSi monotonously decreases with heating. There is no considerable change in gelatinization area. Presumably it reflects the structural changes that chemical modificated starches have.

Spin-spin relaxation times T_{2a} and T_{2b} have the same behavior during the heating for all the samples. T_{2a} – approximately constant, T_{2b} – increases with the heating. This increase confirms our assumption of accessory this population to inter-granular water in which dipole-dipole mechanism of relaxation prevails.

Conclusion

The hydratation (AUW) of the silicate starches during gelatinization decreased in compared to that of native potato starch. Amount of unfrozen water (UW) for ungelitinizated silicate starches are less than in control samples. At the same time the less mobile water population W_a in experiment is three times more than in control. Possibly it reflects the reduction interaction energy of water with starch. These changes just as the gelatinization temperature decrease can be a consequence of starch distructure under microwave irradiation and chemical modification.

References

[1] Kappe C.O.& Dallinger D.(2009) Controlled microwave heating in modern organic synthesis: Highlights from 2004-2008 literature. *Molecular Diversity,* 13, 71-193.

[2] Staroszczyk H. (2009) Microwave – assisted boration of potato starch. *Polimery,* 54,31-41.

[3] Staroszczyk H (2009) Microwave – assisted silication of potato starch. *Carbohydrate Polymers* 77, 506-515.

[4] Staroszszyk H, Janas P. (2010) Microwave-assisted preparation of potato starch, silicated with silicic acid. *Carbohydrate Polymers,* 81, 599-606.

[5] J.W. Donovan (1979) Phase transition of the starch water system. *Biopolymers* , 18, 263-275.

[6] J.M.V. Blanshard (1987) Starch granule structure and function: A physicochemical approach in T.Gallard, Editor, *Starch: Properties and potential,* Wiley, Chictster, pp.16-54.

[7] Le Bail, H. Bizot, M Ollivon, G. Koller, C. Bourgaux and A. Buleon (1999) Monitoring the crystallization of amilose – lipid complexes during maize starch melting by synchrotron X-ray diffraction. *Biopolymers* 50, 99-110.

[8] P. Jenkins & A.M. Donald (1998) Gelatinization of starch: A combined SAXS/WAXS/DSC and SANS study. *Carbohydrate Research* 308, 133-147.

[9] K. Tananuwong, D.S. Reid (2004) DSC and NMR relaxation stages od starch-water interactions during gelatinization. *Carbohydrate Polymers* 58, 345-358.

[10] M. Ritita, R. Gianferri, R. Bucci and E. Brosio (2008) *Proton NMR relaxation study of swelling and gelatinization process in rice starch –water samples.*

[11] H.R. Tang, A. Brun, B. Hills (2001) A proton NMR relaxation study of gelatinization and acid hydrolysis of native potato starch. *Carbohydrate Polymers* 46, 7-18.

[12] S.G. Choi, W.L. Kerr (2003) Water mobility and textural properties of native and hydroxypropilated wheat starch gels. *Carbohydrate Polimers,* 51, 1-8.

[13] S.G. Choi, WL. Kerr (2003) 1H NMR Studies of molecular mobility in wheat starch. *Food Research International,* 36,341-348.

[14] H.R. Tang, J. Godward, B. Hills (2000) The distribution of water in native starch granules – a multinuclear NMR study. *Carbohydrate Polymers,* 43, 375-378.

[15] V. Garcia, V. Colonna, P. Bouchet, D.J. Gallant (1997). Structural changes of cassava starch granules after heating at intermediate water content. *Starch,* 49, 171-179.

[16] M. Wootton, A. Bamunuarachchi (1978) Water binding capacity of commercially produced and modified starches. *Starch,* 30, 306-309.

In: Quantitative Chemistry, Biochemistry and Biology
Editors: G.E. Zaikov, O.V. Stoyanov et al.
ISBN: 978-1-62948-332-0

Chapter 15

INVESTIGATION OF ETHYLENE COPOLYMERS SILANOL MODIFICATION

S. N. Rusanova[1], O. V. Stoyanov[1], A. B. Remizov[1], A. O. Yanaeva[1], V. K. Gerasimov[2], A. E. Chalykh[2] and G. E. Zaikov[3]

[1]Department of Plastics Technology of KSTU
[2]Laboratory of Structural and Morphological Studies of the Institute of Physical Chemistry and Electrochemistry named by Frumkin, Russia
[3]Polymer Division, N.M. Emanuel Institute of Biochemical Physics, Moscow, Russia

ABSTRACT

The chemical interaction of organosilicon compounds and copolymers of ethylene with vinylacetate and copolymers of ethylene with vinylacetate and maleic anhydride was studied by IR-spectroscopy translucent and FTIR. An enrichment of the surface layers of polymers by siloxane phase was found, that may be useful in the design of the chemical structure of adhesive materials for different purposes.

Keywords: IR-spectroscopy, chemical interaction, ethylene copolymers, organosilicon compounds

INTRODUCTION

It is possible to extend the sphere of application of ethylene copolymers by the modification of the original polymer or by the development of composite materials on their basis. Traditionally the compositions formation is effective in the presence of additives, interacting with the polymer during the processing. It results in the regulation of the material properties. One of the methods of chemical modification is the introduction of organosilicon compounds into polyolefins.

Earlier [1, 2], we studied the modification of ethylene copolymers with vinylacetate by ethyl silicates containing various amounts of ester groups. Similar studies were conducted by V.Bounor-Legaré et al [3], so it was interesting to study changes in the structure and properties of ethylene copolymers with vinylacetate containing other functional groups such as anhydride, to identify general trends and specific differences during the modification of ethylene copolymers, containing different reactive segments, by limiting alkoxysilane.

Subjects and Methods

Copolymers of ethylene with vinylacetate (SEVA), Evatane 2020 and Evatane 2805 (ATOFINA) and sevilen 11306-075 brand of JSC "Sevilen" (TU 6-05-1636-97); copolymer of ethylene with vinylacetate and maleic anhydride (SEVAMA) marks Orevac 9305 and Orevac 9707 (Arkema) were used as the objects of the study. The main characteristics of the polymers are listed in the Table 1.

Table 1. Main characteristics of the polymers

Characteristics	СЭВА 11306-075	Evatane 2020	Evatane 2805	Orevac 9305	Orevac 9307
1	2	3	4	5	6
Symbol	СЭВА 14	СЭВА 20	СЭВА 27	СЭВАМА 26	СЭВАМА 13
The content of vinyl acetate, %	14	19-21	27-29	26-30	12-14
Melt Flow Rate, g/10 min, 125°C	0,85	2,23	0,53	14,06	1,10
Density, kg/m^3	0,935	0,936	0,945	0,963	0,924
Melting temperature (max), °C	97	82	72	67	90
Tensile strength, MPa	18,15	14,28	17,40	6,72	18,88
Breaking elongation, %	650	660	830	800	670

Ethyl silicate ETS-32 (TU 6-02-895-78) being a mixture of tetraethoxysilane with geksaethoxydisiloxane with a small admixture of ethanol and oktaethoxytrisiloxane was used as a modifier. The silicon content in terms of silicon dioxide is 30-34%, tetraethoxysilane – 50-65%. Its density is 1.062 kg/m^3 and viscosity is 1.6 cP. Ethyl silicate is manufactured by Production Association "Chemprom" of Novocheboksarsk.

Reactive blending of polymers with ethyl silicate was carried out on the laboratory rolls at a rotational rolls speed of 12.5 m/min and at a friction of 1:1.2 during 10 minutes in the range of 100-120°C. The content of the modifier was varied in the range of 0-10 mass. %. The samples for investigations were prepared by the direct pressing in the restrictive frameworks. Pressing regime is under the temperature of 160°C and the unit pressure of 15 MPa; the time of preheating, injection boost time and the time of cooling is 1 min. for each 1 mm of the sample thickness. After the pressing all the compositions were subjected to aging at a room temperature for twenty-four hours.

Since the rolling of two-component systems as the formation of products of chemical interaction between the components and the simple mechanical mixtures is possible. That is why a purification of the modified polymers was carried out by fivefold reprecipitation under cold conditions by the ethanol from the solution in CCl_4. The samples for electron spectroscopy transmission were prepared by watering from solution in carbon tetrachloride on a substrate of KBr. The film samples with the thickness of 0.07-0.12 mm for FTIR were prepared by the direct pressing without restrictive frameworks on fluoroplastic plates.

IR spectra were registered by the infrared Fourier spectrometer «Spectrum BXII» of Perkin Company by transmission spectroscopy in the range of 450-4000 cm^{-1} and with the method of FTIR (frustrated total internal reflection) on the ZnS crystal in the range of 650-4400 cm^{-1} with the subsequent transformation by Kubelka-Munk. All spectra were normalized according to an internal standard and the intensity of the band of 720 cm^{-1} related to the deformation vibrations of CH_2 groups of the main chain not involved in the chemical reaction was assumed as the internal standard [1, 4]. The original spectra in the coordinates of the optical density - wave number were processed using the software package ACD/SpecManager (ACD/UV-IR Manager & UV-IR Processor. Version 6.0 for Microsoft Windows) to separate the individual components of the spectrum in the areas corresponding to strongly overlapping absorption bands. The contours shape during the spectra simulation is a mixed Gaussian-Lorentzian. After converting the spectra to eliminate the influence of the penetration depth of radiation and automatically determine the main peaks position a preliminary decomposition of the spectrum on these bands was conducted. The first derivative of the experimental contour and the deviation of the calculated spectrum from the experimental spectrum were analyzed. The most probable position of the characteristic peaks not recorded in the preliminary decomposition was determined by the deviation of the resulting deviation from the zero level and the position of the peaks on the graph of the derivative, taking into account the literature data. The addition of a decomposition component was carried out step by step with the expansion of the spectrum conduction, taking into account the added peak and the deviations of the calculated spectrum analysis.

The melt flow index (MFI) was determined by capillary viscometer IIRT-5M according to GOST 11645-73 under the temperature of 125°C and the load of 1.16 kg.

The intrinsic viscosity was determined by the standard method [5] in carbon tetrachloride solution at 25°C.

RESULTS AND DISCUSSION

The change of the macromolecules chemical structure as a result of polymer-analogous transformations can be analyzed by various spectroscopic methods. Qualitative differences between the IR spectra of modified and unmodified samples confirm this fact (Figures 1, 2). On spectra of polymers modified with ethyl silicates the bands related to silicone fragments in the areas of 1020-1090 cm^{-1}, 780-830 cm^{-1}, specific for the stretching vibrations of Si-O, Si-O-Si, Si-OC appears, as well as the band of 971 cm^{-1} appears, characterizing Si-O and Si-O-Si bonds in the cross-linked siloxane fragments. The difference in the nature of optical density changes is possibly due to the different structures formed during the reaction of organosilicon fragments.

It was established earlier [1, 2] that the introduction of tetraethoxysilane in the sevilen leads to a splitting in the IR spectra of characteristic band of 1240 cm^{-1} corresponding to the stretching vibrations of C-O bond in ester groups caused by the substitution of acetyl fragment of the copolymer for the remainder of organosilicon modifier. A similar splitting was observed for ethylene copolymers with vinylacetate and maleic anhydride (Figure 3). The vibrations of C-O bonds are in the strong interaction with the vibrations of C-C bonds due to the small differences in power coefficients and the proximity of the atoms masses forming the bond. Therefore, the contour of the C-O bands is characterized by the presence of satellites due to rotational isomerism with respect to σ-bonds [6, 7]. The characteristic band shift of the stretching vibrations of C-O bond in the direction of higher frequencies during the substitution of the acetate fragment by the silicone is due to the tension connection by the steric interactions of volume replacement groups [8].

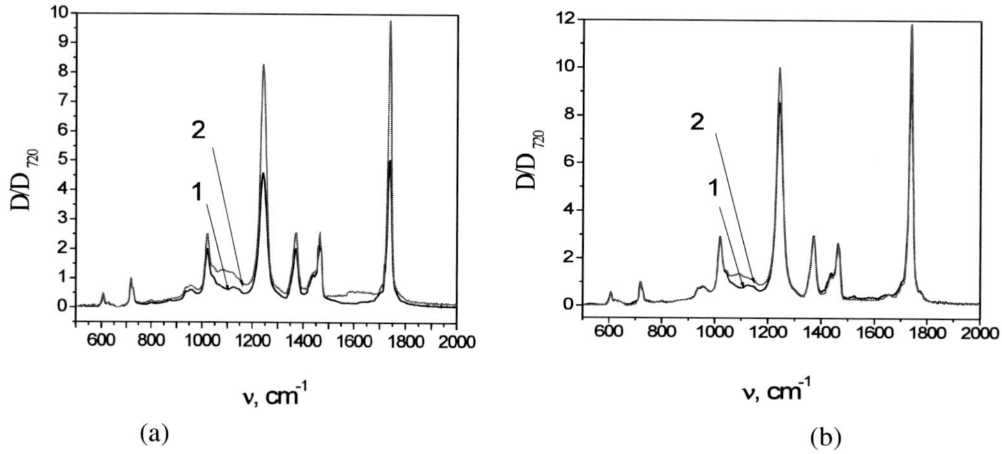

Figure 1. IR spectra of SEVA 27 (a) and SEVAMA 26 (b) initial (1) and modified by ethyl silicates (2).

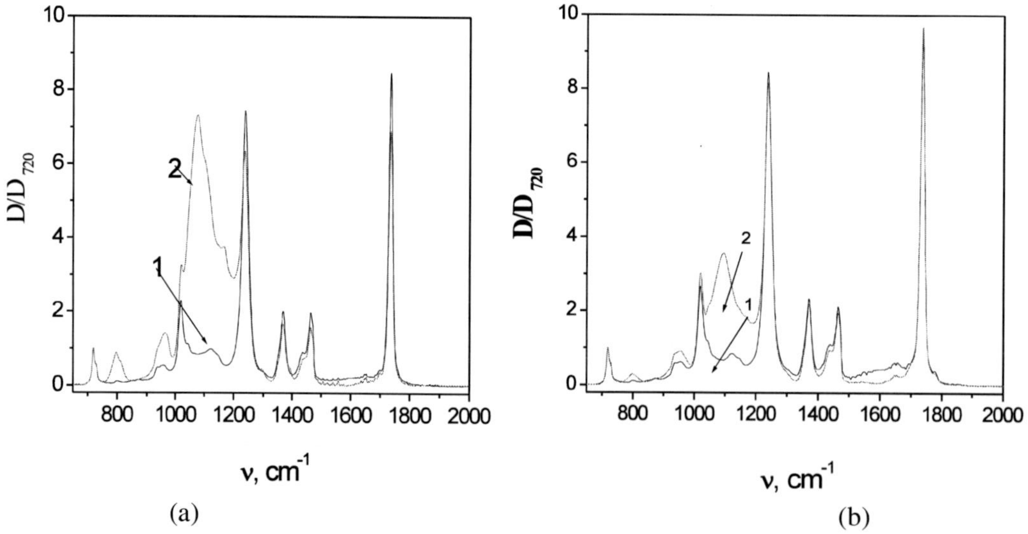

Figure 2. IR- FTIR spectra of SEVA 27 (a) and SEVAMA 26 (b) initial (1) and modified by ethyl silicates (2).

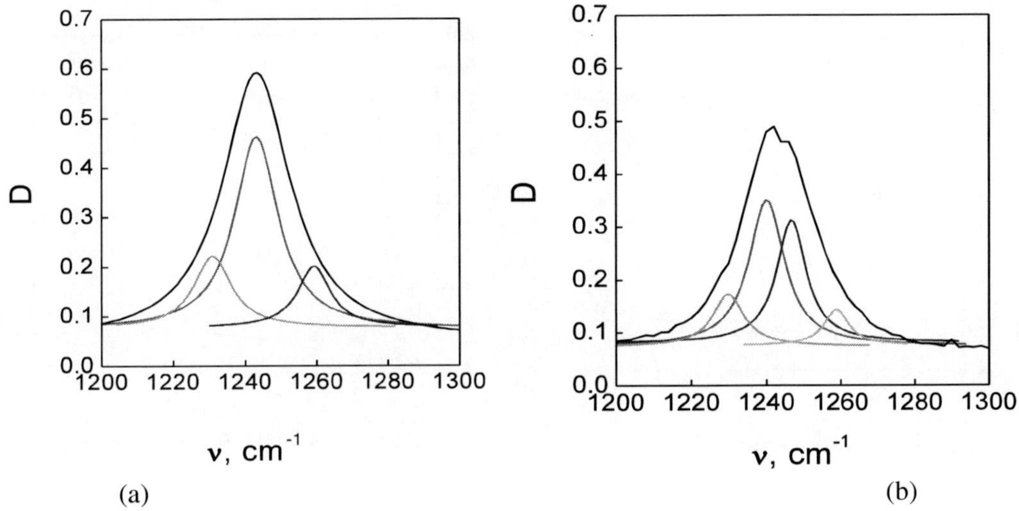

Figure 3. Computer decomposition of the infrared spectra in the region of stretching vibrations of C-O bond in ester groups of SEVAMA initial (a) and modified by ethyl silicates (b).

A decrease in the intensity of the characteristic band of 1462 cm^{-1} corresponding to the deformation vibrations of methyl groups in vinylacetate has been observed for SEVAMA as well as in the modification of SEVA (Figure 4), which confirms the participation of both copolymers SEVA and SEVAMA in the transesterification reaction of acetyl fragment.

The absence of characteristic bands doublets of 1790 cm^{-1} and 1850 cm^{-1} related to the stretching vibrations of C=O groups of maleic anhydride, both in the modified and unmodified terpolymer is of great interest, though in IR - spectra of films obtained from SEVAMA granules, it is presented. We can assume that it is due to the intense thermomechanical effects in the oxygen environment during the rolling, which results in the anhydride cycle opening with the formation of carboxylic acid, reacting further with alkoxy groups of the modifier.

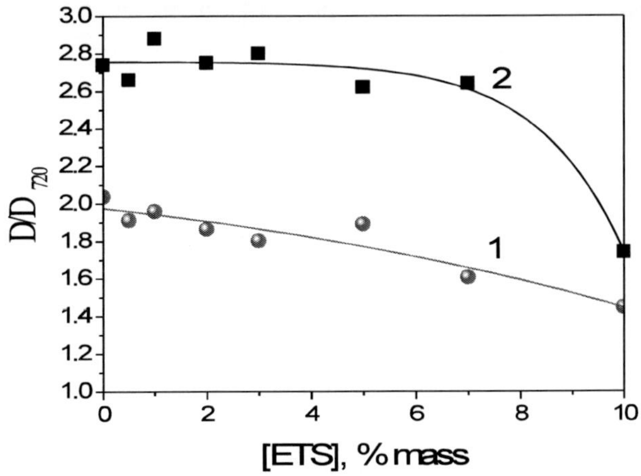

Figure 4. Relative optical density of the band of 1462 cm^{-1} in SEVA 27 (1) and SEVAMA 26 (2) from the content of ETS.

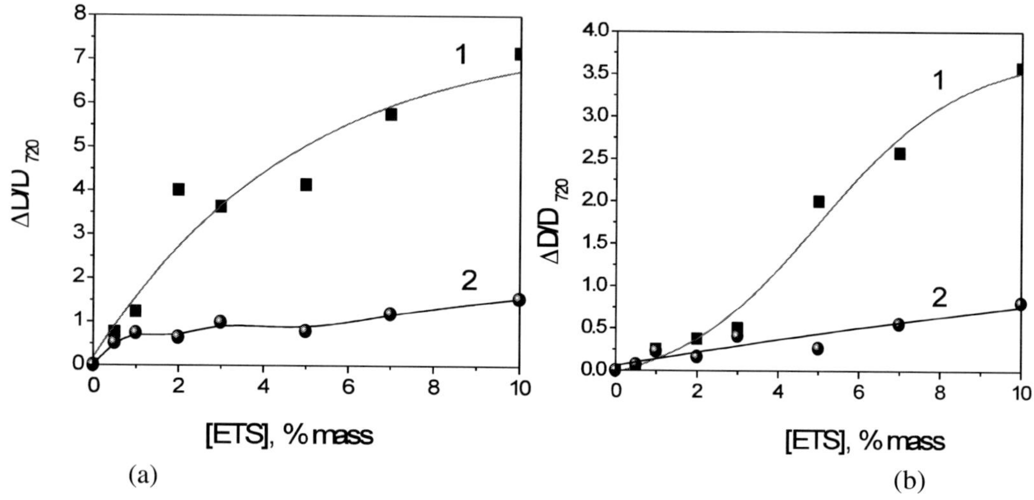

Figure 5. Relative optical density of the band of 1070 cm^{-1} in the FTIR spectra (1) and IR spectra (2) in SEVA 27 (a) and SEVAMA 26 (b) from the content of ETS.

In the spectra of organosilicon compounds the band of 1070 cm^{-1} corresponding to the stretching vibrations of Si-O-Si groups is indicated vividly. This band presents, respectively, in the spectra of the modified copolymers. The introduction of the modifier into the polymer, naturally, leads to an increase of the optical density of this band corresponding to the total content of Si-O-Si links, which is in the direct ratio to the amount of the injected additive (Figure 5). However, it was found that the surface layers of the polymer are enriched by siloxane phase, as evidenced by the differences in the growth of the characteristic band of 1070 cm^{-1} in the spectra obtained by FTIR (surface layer) and by the transmission infrared spectroscopy (integral). It was established earlier [9] that SEVA modified by ethyl silicate is a two-component heterophase system. Since polysiloxanes and polyolefins have different segmental mobility and large differences in free surface energy, it was suggested that the migration of grafted siloxane fragments into the subsurface layers of material is possible during the formation of the samples. The saturation of the polymer surface layer during the silanol modification may be of great interest in the development of adhesive materials.

Conclusion

In the course of studies, the effect of maleic anhydride contained in SEVAMA on the chemical interaction between polymer and ethyl silicate and the change of the properties of copolymers in the process of silanol modification haven't been detected, probably due to the small (not more than 2%) amounts of the comonomer. Previously, it had been found that the introduction of the ethyl silicate into the polymer composites leads to a change in its phase structure [9] and to an increase of the adhesive strength of the material to the metal [1]. Therefore, the data of the effect of ETS on the chemical structure of modified polymers, as well as the fact of the enrichment of the polymers surface layers by the siloxane phase, may be useful in designing of the chemical structure of adhesive materials for different purposes.

REFERENCES

[1] Русанова, С.Н. Дисс. … канд. техн. наук: 02.00.16 / Русанова Светлана Николаевна. - Казань: Казанский гос. технол. ун-т. - 2000. - 119с.

[2] Стоянов, О.В. Химическое строение сополимеров этилена с винилацетатом, модифицированным предельным алкоксисиланом, по данным ИК - спектроскопии / О.В.Стоянов, С.Н.Русанова, О.Г.Петухова, А.Б.Ремизов // Журн. прикл. химии. - 2001. - Т. 74. - № 7. - С. 1174-1177.

[3] Bounor-Legaré, V. New transesterification between ester and alkoxysilane groups: application to ethylene-co-vinyl acetate copolymer cross linking / V.Bounor-Legaré, I.Ferreira, A.Verbois, Ph.Cassagnau, A.Michel // Polymer: The International Journal for the Science and Technology of Polymers (including Polymer Communications). - 2002. - 43. - №23. - С. 6085-6092. - Англ.

[4] Богатырев, В.Л. Модифицирование полиэтилена гидролизованным метилдиэтоксисиланом / В.Л. Богатырев, Г.А. Максакова, Г.В. Виллевальд, В.А. Логвиненко // Изв. Сиб. Отд. АН СССР Сер. хим. наук. - 1986. - №2, вып.1. - С.104-105.

[5] Куренков, В.Ф. Практикум по физике и химии полимеров / В.Ф.Куренков, Л.А. Бударина, А.Е. Заикин - М.: Химия-КолосС. - 2008.

[6] Колебательная спектроскопия. Современные воззрения. Тенденции развития / Под ред. А. Барнса, У Орвилл Томаса. - М.: Мир. - 1981.

[7] Внутреннее вращение молекул / Под ред. Дж Орвилл Томаса. - М.: Мир. - 1987.

[8] Белопольская, Т.В. Влияние температуры на параметры компонент полосы 1050-1300 см$^{-1}$ в спектрах поглощения некоторых полимеров метиленового ряда и интерпретация ее структуры / Т.В.Белопольская // Высокомолек. соед. - 1972. - Т.14. - С.640-645.

[9] Чалых, А.Е. Формирование фазовой структуры силанольно-модифицированных сополимеров этилена с винилацетатом / А.Е. Чалых, В.К. Герасимов, С.Н. Русанова, О.В. Стоянов, О.Г. Петухова, Г.С. Кулагина, С.А. Писарев // Высокомолек. соед. А. - 2006. - Т.48. - № 10. - С.1801-1810.

In: Quantitative Chemistry, Biochemistry and Biology
Editors: G.E. Zaikov, O.V. Stoyanov et al.
ISBN: 978-1-62948-332-0

Chapter 16

DETERMINING THE SURFACE FREE ENERGY PARAMETERS

A. Starostina*, O. V. Stoyanov, N. V. Makhrova, G. E. Zaikov and R. Ya. Deberdeev

Kazan National Research Technological University, Kazan, Tatarstan, Russia
N. M. Emanuel Institute of Biochemical Physics, Moscow, Russia

ABSTRACT

According to acid–base theory, an important role in the formation of adhesion bonds are played by the interphase acid–base interactions between the adhesive and the substrate. Within this theory, the van Oss– Chaudhury–Good approach is most popular [1], in which characteristics of the acid and base interactions are nonadditive parameters γ^+ and γ^- of the surface free energy. Unfortunately, the values of γ^+ and γ^- of solids are impossible to determine if it is impossible to obtain the values of the corresponding parameters of test liquids from wettability data. The theory's authors calculated that the surfaces of most polymers (polystyrene, poly(vinyl alcohol), poly(methyl methacrylate), cellulose nitrate) are mainly basic (this fact is referred to in the scientific literature as the "basic disaster"), which prompted to introduce the term "monopole surfaces" [1]. The surface is called monopole if it is either only basic ($\gamma^- = 0$) or only acidic ($\gamma^+ = 0$). The simplicity of calculating the acid and base parameters of liquids and solids within the said hypothesis has interested many researchers, and this method is now being used quite widely (see, e.g., [2, 3]). Typically used acidic and basic monopole surfaces are those of poly(vinyl chloride) [2] and poly(methyl methacry-late) [3], respectively. However, the γ^+ and γ^- values of test liquids that are determined by this method are often inconsistent with their chemical nature. Therefore, the existence of monopole polymer surfaces requires detailed investigation. This is all the more so, as the Lewis and Bronsted theories say that there are no pure acids and pure bases and that any substance has both acid and base properties.

* chembio@sky.chph.ras.ru.

Our IR spectroscopic study of polymer samples confirmed a complex and ambiguous composition of poly(methyl methacrylate) and poly(vinyl chloride). The transmission and multiple attenuated internal reflection IR spectra in the region of C=O group vibrations show not only the major band at 1740–1725 cm^{-1}, which characterizes the absorption of the ester carbonyl group, but also a sextet of bands caused by high sensitivity of the carbonyl absorption to a change in the nearest surroundings [4]. Analysis of the spectra suggested that one cannot rule out the insignificant presence of double bonds in the carbon chain at the surface (because of structural defects and destructive processes) and also the existence of various functional groups–aldehyde, ketone, and carboxyl (acidic) groups [5]. The poly(vinyl chloride) spectrum has a broad band at 3410 cm^{-1}, which is likely to be caused by hydrogen bonds involving the acidic hydrogen of the polymer and the hydroxyl oxygen of water vapor captured during casting of the polymer film from a tetrahydrofuran solution.

Additionally, we analyzed poly(vinyl chloride) and poly(methyl methacrylate) by the quantum-chemical density functional theory method. Figure 1 presents the results obtained.

Figure 1a shows that the poly(vinyl chloride) mac-romolecule indeed has strong electron-acceptor sites (δ_H = 0.178), which confirms the mainly acidic nature of this polymer. However, along with them, poly(vinyl chloride) also contains chlorine atoms with excess electron density (δ_{Cl} = –0.065). Although a chlorine atom is a weak electron-donor site, it can exhibit basic properties in the presence of a strong acid. In its turn, poly(methyl methacrylate) contains mobile hydrogen atoms. Given a certain arrangement of the polymer chain, some of these atoms can be involved in the formation of an intramolecular hydrogen bond (and also intermolecular hydrogen bonds in the bulk of the polymer) (Figure 1b).

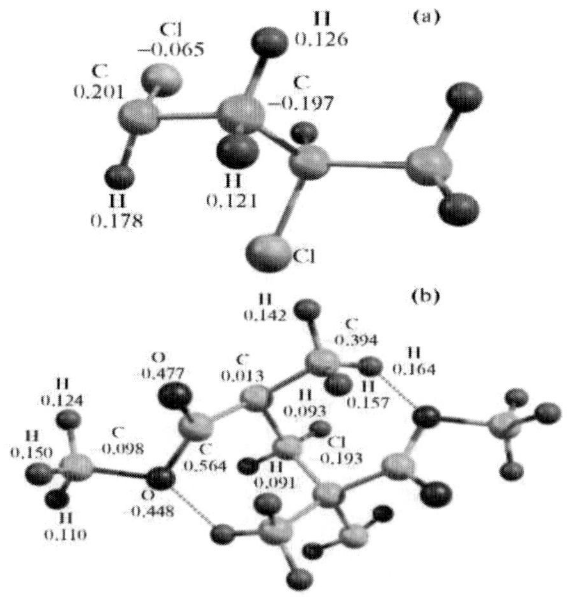

Figure 1. Electron density distribution in monomeric units of (a) poly(vinyl chloride) and (b) poly(methyl methacry-late).

All the above suggests that the surface of neither poly(methyl methacrylate) nor poly(vinyl chloride) can be considered monopole. Most likely, such a conclusion is valid for any polymer surface. Nonetheless, various polymer surfaces have long and successfully been used as probe surfaces, e.g., for determining the surface tension of liquids and oligomers [6]. A legitimate question arises as to whether it is possible and, if so, how to obtain reliable information on the γ^+ and γ^- can be considered monopole. Most likely, such a conclusion is valid for any polymer surface. Nonetheless, various polymer surfaces have long and successfully been used as probe surfaces, e.g., for determining the surface tension of liquids and oligomers [6]. A legitimate question arises as to whether it is possible and, if so, how to obtain reliable information on the γ^+ and γ values of liquids and solid surfaces. Della Volpe et al. proposed to solve this issue by solving a system of nonlinear equations [7]. We modified and simplified the method for solving such systems. The determined parameters of test liquids [8] can be used to find the sought-for values for polymers by two ways:

1. by solving systems of nonlinear equations containing the cosines of the contact angles of an unknown polymer surface with test liquids [8] or
2. by applying a graphical method for determining the γ^+ and γ^- of a polymer surface.

In the first case, the system is highly overdeter-mined, there are many equisignificant solutions, and the problem of choosing the only correct solution remains open.

In the second case, the van Oss–Chaudhury– Good equation

$$W_{LS} = \gamma_L(\cos\theta + 1)$$
$$= 2\sqrt{\gamma_S^{LW}\gamma_L^{LW}} + 2\sqrt{\gamma_S^{+}\gamma_L^{-}} + 2\sqrt{\gamma_S^{-}\gamma_L^{+}},$$

where W_{LS} is the adhesion work done by liquid *(L)* on solid *(S)*, is transformed to the form $y = ax + b$, and a straight line in the coordinates

$(A, \sqrt{\gamma_L^+/\gamma_L^-})$ is plotted, where $A = \left(\frac{W_{LS}}{2} - \sqrt{\gamma_S^{LW}\gamma_L^{LW}}\right)\Big/\sqrt{\gamma_L^-}.$

The y-intercept of the straight line is $\sqrt{\gamma_S^+}$, and the slope is $\sqrt{\gamma_S^-}$. This method gives the only value of each of the quantities γ^+ and γ^-.

Figure 2 exemplifies the graphical method for determining the surface free energy parameters of poly(vinyl chloride) and poly(methyl methacrylate). For constructing the graphs, we used the previously determined [8] surface free energy parameters of aniline, dimethyl sulfoxide, formamide, glycerol, and double distilled water. The correlation coefficients for both straight lines are 0.99. For poly(methyl methacrylate), the graphically found acid and base parameters are 0.28 and 0.06 mJ/m^2, respectively; and for poly(vinyl chloride), they are 1.54 and 0.48 mJ/m^2, respectively.

Thus, we have shown the invalidity of using poly(methyl methacrylate) and poly(vinyl chloride) polymer surfaces as probe monopole surfaces and demonstrated the possibility of correctly determining the acid and base parameters of the surface free energy of polymers by a simple and illustrative graphical method.

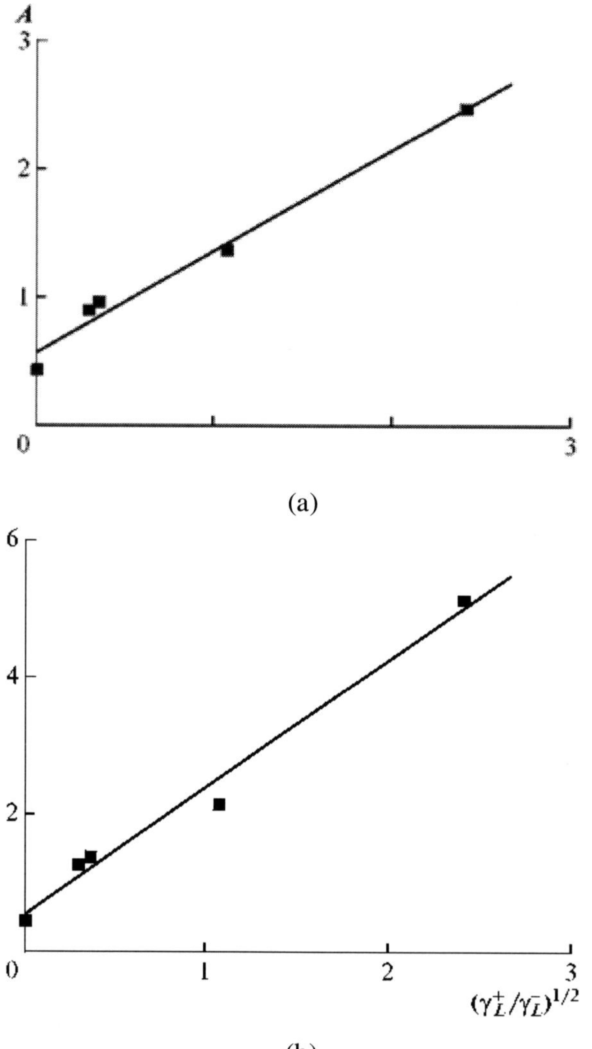

(b)

Figure 2. Graphical determination of the surface free energy parameters of (a) poly(vinyl chloride) and (b) poly(methyl methacrylate).

REFERENCES

[1] van Oss, C.J., Chaudhury, M.K., and Good, R.J., *Adv. Colloid Interface Sci.,* 1987, no.28, pp. 35–64.

[2] Berg, J.C., *Wettability,* New York: M. Dekker, 1993.

[3] Jan 'czuk, B., Bialopiotrowicz, T., and Zdziennicka, A., *J. Colloid Interface Sci.,* 1999, no. 211, pp. 96–103.

[4] Kazitsina, L.A. and Kupletskaya, N.B., *Primenenie UF-, IK-, YaMR- i mass-spektroskopii v organicheskoi khimii* (Applications of UV, IR, NMR, and Mass Spectroscopy to Organic Chemistry), Moscow: Mosk. Gos. Univ., 1979.

[5] Troitskii, B.B., Domrachev, G.A., Anikina, L.I., et al., *Vestn. Nizhni Novgorod. Gos. Univ.*, 2001, no. 1, pp. 60– 71.

[6] Chalykh, A.E., Busygin, V.B., and Stepanenko, V.Yu., *Vysokomol. Soedin.*, 1999, vol. 41, no. 11, pp. 1843– 1846.

[7] Della Volpe, C., Siboni, S., Maniglio, D., and Morra, M., *J. Adhesion Sci. Technol.*, 2003, vol. 17, no. 11, pp. 1425–1456.

[8] Starostina, I.A., Stoyanov, O.V., Makhrova, N.V., and Deberdeev, R.Ya., *Dokl. Phys. Chem.*, 2011, vol. 463, part 1, pp. 8–9 [*Dokl. Akad. Nauk*, 2011, vol. 436, no. 3, pp. 343-345]. RR

In: Quantitative Chemistry, Biochemistry and Biology
Editors: G.E. Zaikov, O.V. Stoyanov et al.
ISBN: 978-1-62948-332-0

Chapter 17

INVESTIGATION OF ETHYLENE–VINYLACETATE COPOLYMERS BINARY MIXTURES STRUCTURE

O. V. Stoyanov[1], R. M. Khuzakhanov[1], L. F. Stoyanova[1], V. K. Gerasimov[2]*, A. E. Chalykh[2], A. D. Aliev[2], M. V. Vokal'[2] and G.E. Zaikov

[1] Kazan State Technological University, Kazan, Republic of Tatarstan, Russia
[2] Frumkin Institute of Physical Chemistry and Electrochemistry, Russian Academy of Sciences, Moscow, Russia
N.M. Emanuel Institute of Biochemical Physics, Moscow, Russia

ABSTRACT

The method of differential scanning calorimetry was used to analyze the structure of binary mixtures of ethylene and vinylacetate copolymers with different contents of reagent monomers. The anomalous deviation of the degree of crystallinity from additivity in the system is found and a possible interpretation of this experimental fact is suggested.

Keywords: Copolymers of ethylene and vinylacetate (CEVA), binary mixtures of copolymers, melt adhesives, degree of crystallinity of polymers

Mixing ethylene–vinylacetate copolymers (CEVA) with different contents of vinylacetate (VA) results in unusual dependences of the properties of materials on their composition [1–3]. In particular, in the case of a small difference in the VA chain link content in CEVA in the binary mixture, the positive deviation of the composition properties from additivity is observed (in the case of a large difference, it is negative). Binary CEVA–CEVA mixtures are of interest as melt adhesives with improved adhesive properties [2, 3] and their structural characteristics were studied in detail.

* chembio@sky.chph.ras.ru.

Studies have shown that, in the case of a CEVA-14 + CEVA-22 mixture (the number corresponds to the VA content in %), the concentration dependence of degree of crystallinity (*k*) evaluated using the differential scanning calorimetry (DSC) method features a positive deviation from additive values (Fig. 1). This is a characteristic of this system, since the dependence of *k* is generally either additive, e.g., in the case of a mixture of polyethylenes of different types; as a rule, a negative deviation from additivity is observed for mixtures of different crystallizing polymers or mixtures of crystallizing and amorphous polymers. In our case, the negative deviation of *k* from additivity is observed for the CEVA-7 + CEVA-29 mixture (Fig. 1). The obtained result is typical for mixtures with limited mutual solubility. The mutual effect of the components in these systems causes a decrease in the degree of structural ordering.

To confirm the obtained results, an IR spectral study of the CEVA-14 + CEVA-22 and CEVA-7 and CEVA-29 mixtures was carried out. The 1894 cm^{-1} band was analyzed. This band is a combination of the modes of 1168, 720, and 730 cm^{-1} (λ (CH_2) + 720, 730 *R* (CH_2)) and is identified as a crystalline band in polyethylene (PE) [4–6], i.e., it characterizes the absorbance of local conformers in the crystal phase. It is shown a priori (Fig. 2) that the optical density of the 1894 cm^{-1} band for low- and high-density polyethylene (LDPE and HDPE), as well as CEVA with a different content of VA, correlates with the degree of crystallinity in the studied polyolefins. Therefore, an analysis of this band may be used for the independent estimation of the degree of structural ordering in CEVA–CEVA binary mixtures. The experimental data is presented in Fig. 3.

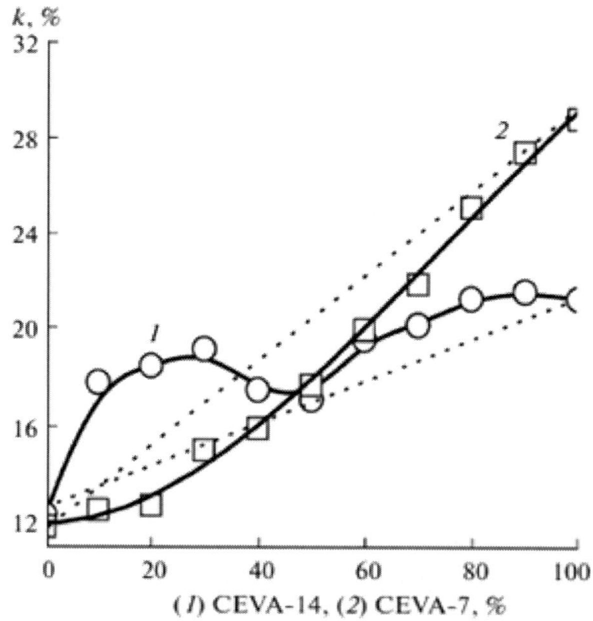

Figure 1. Dependence of degree of crystallinity (*k*) on composition of mixture: (*1*) CEVA-14 + CEVA-22 and (2) CEVA-29 + CEVA-7.

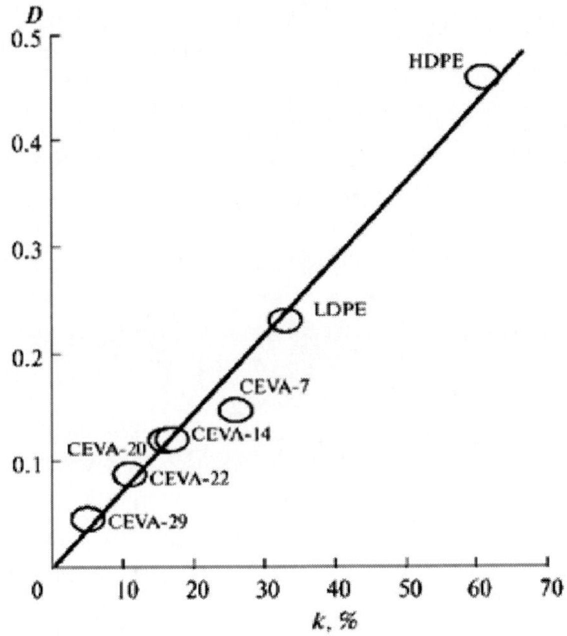

Figure 2. Correlation of (*1*) optical density (*D*) of 1894 cm^{-1} band and degree of crystallinity (*k*) of PE and CEVA.

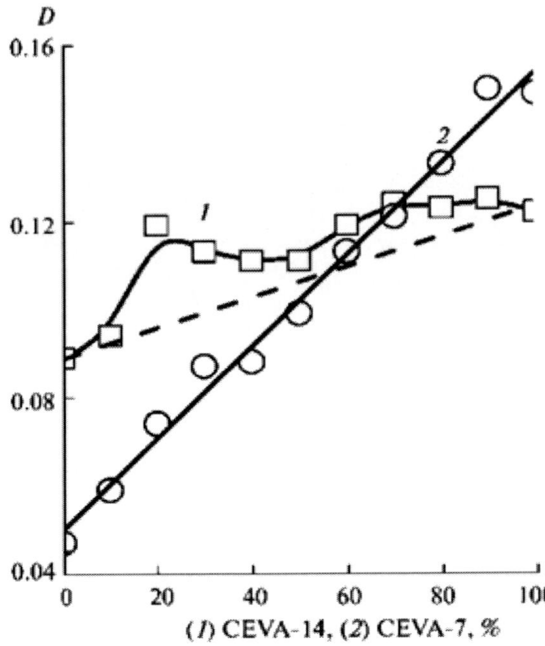

Figure 3. Dependence of optical density (*D*) of 1894 cm^{-1} band on mixture composition: (*1*) CEVA-14 + CEVA-22 and (*2*) CEVA-7 + CEVA-29.

As follows from the experimental data, the concentration dependences of *k* and *D* are consistent for both systems (to a greater degree for the CEVA-14 + CEVA-22 system). A certain difference in the character of curves for the CEVA-7+ CEVA-29 system is apparently due to a difference in the sensitivity of methods that characterize the crystallinity of the material through different perspectives. Nevertheless, one should note that both mixtures have close *k* values in the region of medium compositions. The *D* values are close in both systems in the same region.

The obtained results indicate that both methods register independently. On one hand, an increase in the degree of structural ordering with regard to the additive values in the CEVA–CEVA composition with a small difference in the content of VA chain links in the mixture components and, on the other hand, the conventional behavior of the mixture with a larger difference in the content of VA chain links in copolymers.

It is obvious that these regularities are due to a difference in the melt phase state in the course of sample formation. A study of phase equilibriums in the CEVA-30– CEVA-7, CEVA-14, CEVA-22, CEVA-14–CEVA-7, CEVA-22 systems showed that a decrease in the difference in VA content for the components of the mixture characteristically leads to an increase in the region of compatibility and a decrease in the upper critical temperature of the solution (Figs. 4, 5).

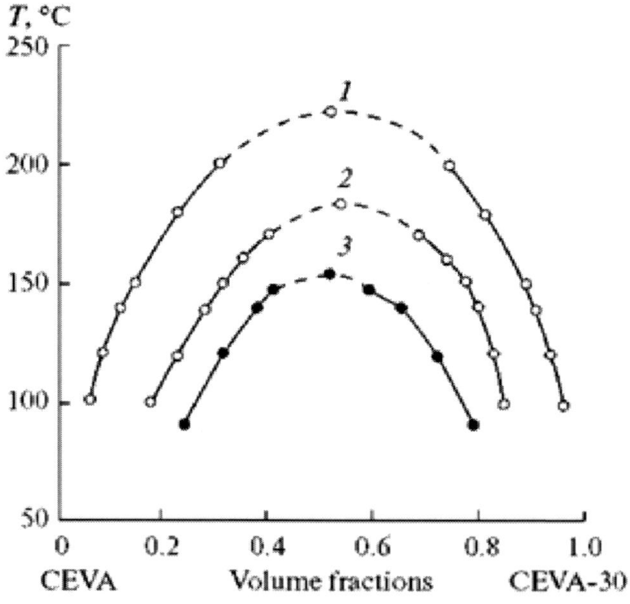

Figure 4. Phase diagrams of the systems of CEVA-30— CEVA with content of VA groups being (*1*) 7, (*2*) 14, and (*3*) 20%.

A comparison of the data on amorphous equilibrium and a change in the degree of crystallinity of copolymer mixtures leads to the following conclusions. The crystallization of copolymers occurs in the temperature range of 103–60°C. The large difference in the composition of copolymers leads to crystallization as early as in the form of a phase-separated two-phase system, where each of the phases represents a nearly pure copolymer. In this case, crystal phase separation occurs in the same way as in the case of a pure copolymer. Therefore, the degree of crystallinity of the copolymer mixture is determined by the phase ratio, which is an additive quantity.

A small difference in the copolymer composition results in the crystallization of the copolymer mixture in single-phase regions corresponding to the nonaddi-tivity of the degree of crystallinity. One may assume that heterogeneity of the composition of copolymers, which generally inhibits their crystallization, actually facilitates this process for copolymers of close compositions.

The formation of the sample phase structure in the process of mixing in the melt depending on the degree of compatibility of the binary CEVA–CEVA mixture components, which determines the regularities of the phase structure and physicomechanical properties of the obtained materials [7].

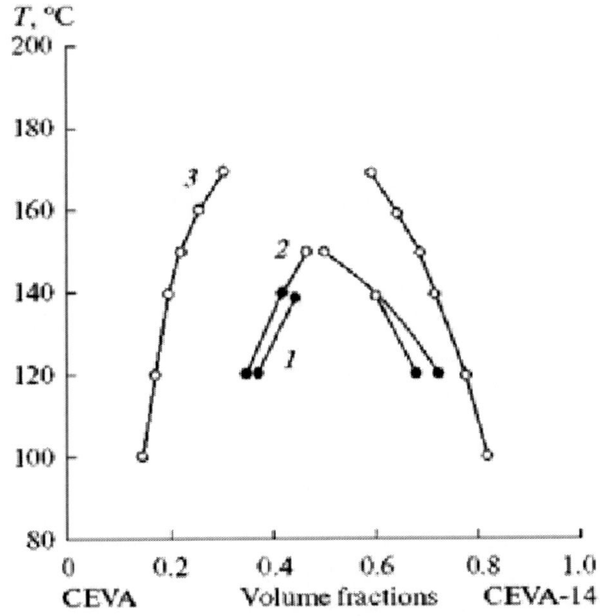

Figure 5. Phase diagrams of systems of CEVA-14–CEVA with content of VA groups being (*1*) 7, (*2*) 22, and (*3*) 30%.

References

[1] E. R. Mukhamedzyanova, R. M. Khuzakhanov, O. V. Stoyanov, and A. E. Zaikin, in Proc. IX All-Russia Conf. "Structure and Dynamics of Molecular Systems— Yal'chik-2002", (MarGTU, Yoshkar-Ola, 2002), Vol. IX, No. 2, pp. 46–49.

[2] R. M. Khuzakhanov, Ya. V. Kapitskaya, R. Ya. Deber-deev, and O. V. Stoyanov, Klei. Germetiki. *Tekhnologii,* No. 5, 15–16 (2005).

[4] R. M. Khuzakhanov, Ya. V. Kapitskaya, E. R. Mukhamed-zyanova, et al., Klei. Germetiki. *Tekhnologii,* No. 4, 24–27 (2005).

[5] P. J. Miller, J. F. Jackson, and R. S. Porter, *J. Polym. Sci. Polym. Phys.* Ed. 11, 2001–2012 (1973).

[6] R. Zbinden, Infrared Spectroscopy of High Polymers (*Mir,* Moscow, 1966; Academic Press, New York, 1964).

[7] A. V. Gribov, Theory of Spectra of Polymers (Moscow, 1977) [in Russian].

[8] R. M. Khuzakhanov, O. V. Stoyanov, E. R. Mukhamed-zyanova, et al., *Izv. Vyssh. Uchebn. Zaved., Khim. Khim. Tekhnol.* 45 (5), 103–105 (2002).

In: Quantitative Chemistry, Biochemistry and Biology
Editors: G.E. Zaikov, O.V. Stoyanov et al.
ISBN: 978-1-62948-332-0

Chapter 18

EFFECT OF MELAPHEN ON A MITOCHONDRIA OF SPROUTS OF PEAS UNDER STRESSFUL INFLUENCES BY THE AFM METHOD

***V. I. Binyukov*[1], *E. M. Mil'*[1*], *I. V. Zhigacheva*[1†], *A. A. Albantova*[1‡], *I. P. Generozova*[2‖] *and A. G. Shugaev*[2#]**

[1] Emanuel Institute of Biochemical Physics,
Russian Academy of Sciences, Moscow, Russia
[2] Timiryazev Institute of Plant Physiology,
Russian Academy of Sciences, Moscow, Russia

ABSTRACT

In our work, we investigated the effects of combined action of insufficient moisture and moderate cooling to 10-14 ° C, and the processing of peas plant growth by regulator melaphen (the melamine salt of bis (ox methyl) phosphoric acid) on the AFM images of isolated mitochondria 5 day-old seedlings of pea (Pisum sativum). Atomic force microscopy (AFM) revealed a statistically significant change in the shape of mitochondria when exposed to cold-drought and detected reducing the effect by the action of the drug melaphen at concentrations of 2 x 10^{-12}M and revealed the ability to divide.

Keywords: Atomic force microscopy (AFM), melaphen, mitochondria

---

* elenamil2004@mail.ru.
† zhigacheva@mail.ru.
‡ albantovaaa@mail.ru.
‖ igenerozova@mail.ru.
ag_shugaev@ippras.ru.

Aim and Background

In nature, plants are exposed to not one but several environmental factors. In this connection, the problem of studying the rearrangements of plant cell metabolism under exposure to several abiotic factors is of special importance. Stressors modify cellular membranes and especially mitochondrial and chloroplast membranes, initiating free radical processes that eventually lead to the activation of lipid peroxidation (LPO) [1]. It is known that plant growth and development regulators increase the resistance of plants to both biotic and abiotic stressors, including water deficiency [2, 3]. One of such growth regulators is melaphen, a melamine salt of bis(hydroxymethyl)phosphinic acid. In this study, we investigated the combined effect of moisture deficit, moderate cooling to 10–14°C, and of pea seeds with the organophosphorus plant growth regulator melaphen on the intensity of lipid peroxidation (LPO) and the atomic force microscopy (AFM) images of mitochondria isolated from five-day-old pea (Pisum sativum) seedlings.

Introduction

In the nature the plant is exposed to influence at once several factors of environment. The study of metabolic conversion in the plant cell at the combined action of several abiotic factors is actual in particular the combined effects of insufficient moisture and moderate cooling.

Energy metabolism plays an important role in the adaptive response organism. In this case, the mitochondria play a key role in the energetic, redox and metabolic processes in the cell [4]. In particular, it was found that the ambient temperature changes lead to changes in the lipid composition of mitochondrial membranes. At the same time there is a change in the number and degree of saturation of free fatty acids, which is probably a sign of the stress factor [5]. Increase the amount of free fatty acids (FFA) changes the redox state of the inner mitochondrial membrane. Depending on the redox state of the mitochondrial respiratory chain expression changes of stress genes. Under the effect of extreme temperatures, excessive salt or lack of moisture in some degree increased ROS. Thus mitochondria are extremely functional-dependent organelles.

Mitochondria in animal cells and yeast combined in a branched network, called the "mitochondrial reticulum" [6].

At the same time, mitochondria at higher plant are single and have either spherical or cylindrical form [7]. Under stress conditions (heat shock, anoxia, UV irradiation, or exposure to strong oxidants), mitochondria aggregate to form tight clusters located near the chloroplast or in other areas of the cytosol. The formation of “giant mitochondria” is accompanied by an increased generation of reactive oxygen species (ROS) by these organelles, which is prevented by antioxidants [8, 9]. The standard procedure of isolation of mitochondria in a sucrose solution leads to complete destruction of intermitochondrial contacts. For this reason, mitochondria are represented in separate vesicles 0.86–1.18 μm in diameter and 0.3–0.4 μm in height. The morphology of isolated mitochondria possibly reflects their functional state [10]. It was previously shown that the treatment of plants with strong ROS_ producing oxidants (methylviologen and *s*-_triazine) caused changes in the shape and size of mitochondria [7]. In this study, we investigated the combined effect of moisture deficit, moderate cooling to 10–14°C, and of pea seeds with the growth regulator melafen (melamine salt of bis(hydroxymethyl)phosphinic acid) on the intensity of lipid peroxidation (LPO) and the atomic force microscopy (AFM) images of mitochondria isolated from five_day_old pea (*Pisum sativum*) seedlings.

Experimental Part

Pea seeds were allowed to imbibe for 60 min either in water or in 2 × 10^{-12} M melaphen solution. After 2 days, half of the melafen_treated seedlings were transferred onto dry filter paper in an open cell, where they were kept at 14°C for two days (drought + cold + melaphen).

The second half of the melafen-treated seeds and half of the control seeds were kept for the same time in sealed cells with periodically moistened filter paper at 14°C (cold and cold + melafen). Two days later, all the seedlings were transferred to a room with temperature 22°C and placed in a sealed cell with a periodically moistened paper. The control group of seedlings was kept at 22°C throughout the experiment. On the sixth day, the number of seeds germinated was counted and mitochondria were isolated.

Isolation of Mitochondria

Mitochondria were isolated from epicotyls of six_day_old seedlings by differential centrifugation [11]. The isolation medium contained 0.4 M sucrose, 20 mM KH2PO4 (pH 7.4), 5 mM EDTA, 10 mM KCl, and 0.2% fatty acid_free BSA.

Atomic Force Microscopy

Samples for AFM studies of mitochondria were prepared as follows. The mitochondria in the isolation buffer were applied onto the surface of a silicon substrate and incubated for 1 min, after which the substrate was washed with the same buffer without BSA. Then, 2% glutaraldehyde in the BSA_free buffer was applied, and the sample was incubated for 3 min. In all procedures, the substrate was wet. Then, the surface of the substrate was washed with distilled water and air_dried. The control silicon samples were prepared in a similar way without mitochondria. Mitochondria samples for atomic force microscopy (AFM) were investigated on the device SOLVER P47 SMENA at frequency 150 kHz in tapping mode.

The volume image of mitochondria is the product of the average cross-sectional area at the height of the image in the section. Cross sections were made at the height of 30 nm. The volume image of mitochondria is the product of the average cross-sectional area at the height of the image in the section. In the analysis and processing of data were used the program Image Analyzis и Statistica 6. The figures shows the atomic force microscopy(AFM) images of mitochondria from pea seedlings.

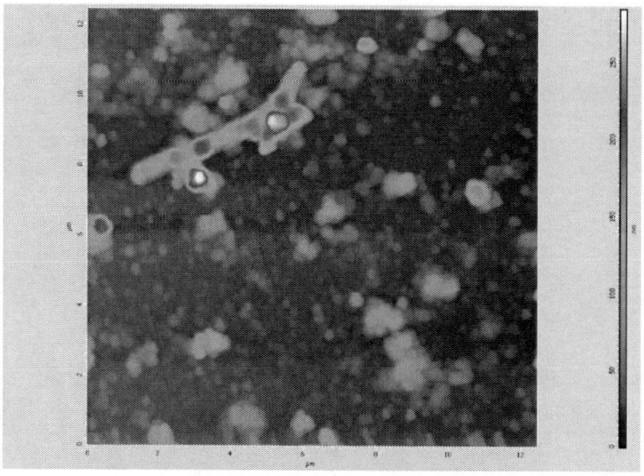

a)

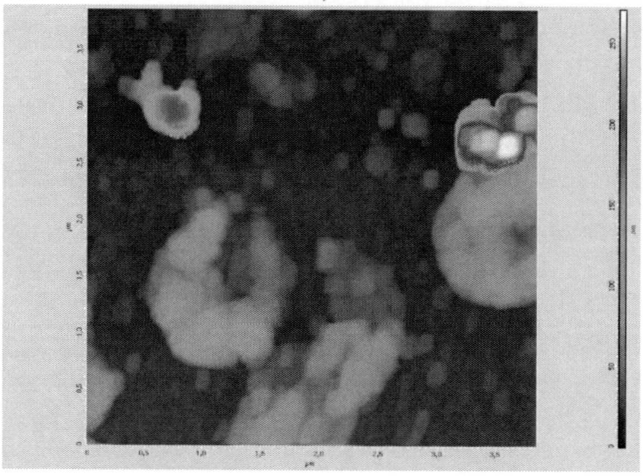

b)

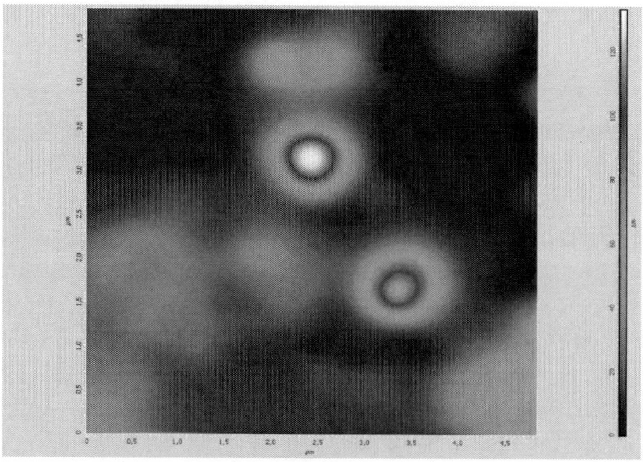

c)

Figure 1. AFM image of mitochondria from pea seedlings in control (a), in the presence of melaphen-cold-drought (b) and under the influence of cold-drought (c).

The Level of Lipid Peroxidation

The LPO level was evaluated fluorescently. After homogenization of the experimental material in a glass homogenizer with a Teflon pestle, mitochondrial membrane lipids were extracted from 3--5 mg of the mitochondrial protein using a chloroform--methanol mixture (2:1, by volume). The mitochondria:chloroform--methanol ratio was 1 : 10. After homogenization of mitochondria (1 min, 10°C) and addition of an equal volume of distilled water, the suspensions were mixed thoroughly and transferred to 12-ml centrifuge test tubes. Washing with H_2O is necessary for removal of flavin components with fluorescence maxima around 520 nm. The resulting preparations were centrifuged for 5 min at 600 g. Then, 1 ml of chloroform was withdrawn from the bottom layer and mixed with 0.1 ml of methanol. Fluorescence was measured in 10-mm quartz microcuvettes (1.4 ml) using a. FluoroMax spectrofluorometer (HoribaYvon GmbH, Germany). Chloroform (1 ml) to which 0.1 ml of methanol was added served as control. Excitation wavelength was 360 nm fluorescence, emission - 420-470 nm. The results are expressed in arbitrary fluorescence units converted to mg of mitochondrial protein.

RESULTS AND DISCUSSION

AFM images of the mitochondria of pea seedlings have undergone a two-day effects of insufficient moisture and moderate cooling, significantly changed and differed from controls.

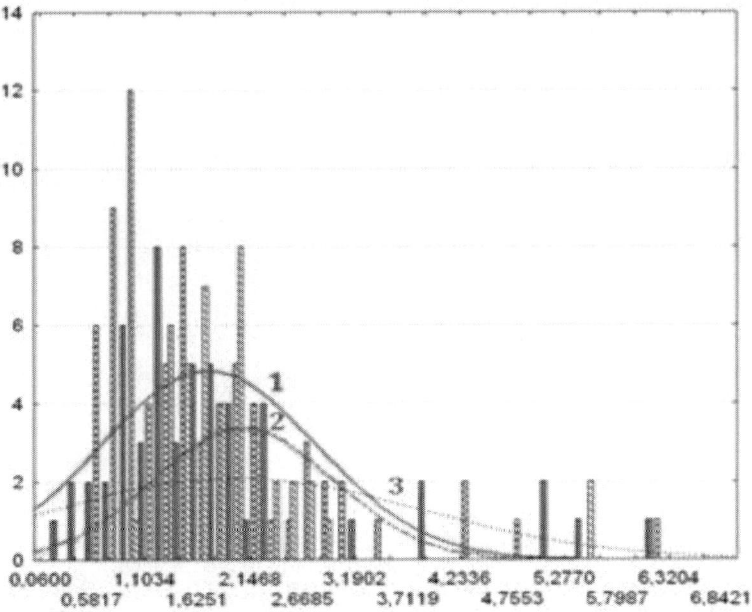

Figure 2. Frequency distribution of the mitochondria length (mkm) and the approximation of the normal distribution on the x-axis - the length of the mitochondria (mkm). The ordinate-number in sample 1-control, 2-melafen+cold + drought, 3 - cold+drought.

Was observed an increase the height of the area of the image and a number of mitochondria, and the number of dividing mitochondria significantly decreased.

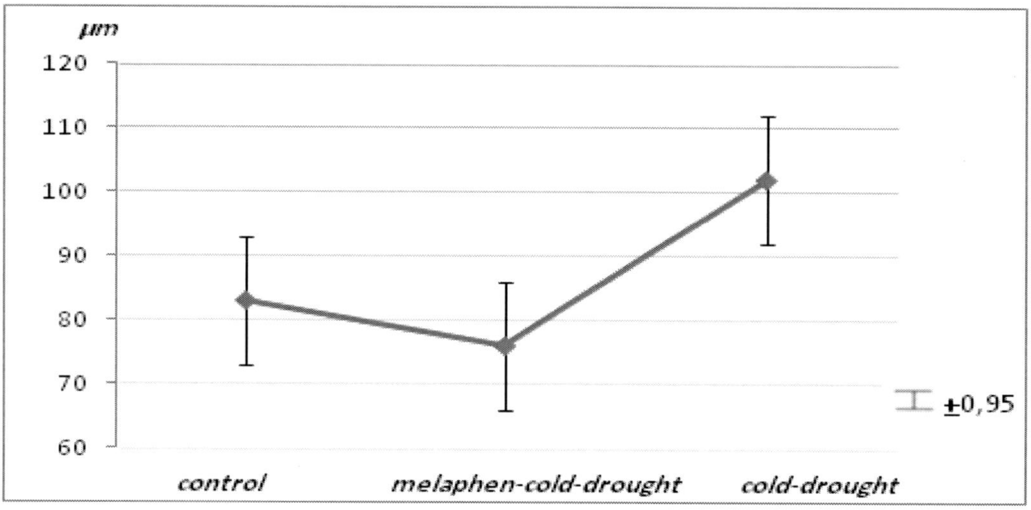

Figure 3. The figure shows the average height (µm) AFM image of mitochondria from pea seedlings in control when exposed cold-drought- melaphen and under the influence of cold-drought. On the horizontal axis - a group of plants. The vertical axis - AFM height image of mitochondria in microns.

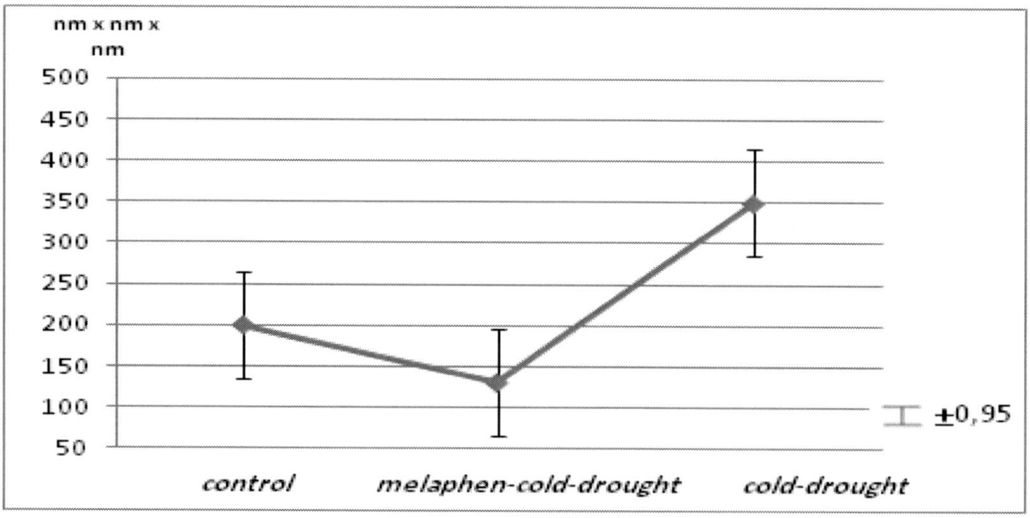

Figure 4. The figure shows the average volume (nm x μm^2) AFM image of mitochondria from pea seedlings in control when exposed cold-drought- melaphen and under the influence of cold-drought. On the horizontal axis - a group of plants. The vertical axis - AFM average volume of mitochondria in nmx$\mu m^{2.}$

Statistical analysis of the length and the amount of pre-fixed, with glutaraldehyde mitochondria indicates the appearance of single, not divisible mitochondria with greater volume and length of seedlings in the group exposed to the stresses in comparison with the control group. Similar results were reported by Logan and Scott, who treated *Arabidopsis thaliana* leaves with methylviologen [7]. The comparison of published data and the results of this experiment allowed us to assume that the combination of moderate cooling and moisture deficit probably leads to an increase in ROS generation and subsequent swelling of mitochondria in the cells of pea seedlings [13]. Indeed, simultaneous cooling and moisture deficit led to LPO activation in the mitochondrial membranes in pea seedlings. The fluorescence intensity of LPO products increased 3 and 2.5 times, respectively (Fig. 5).Soaking the seeds in a 2 x 10^{-12}M solution of melaphen prevented the changes in morphology of mitochondria. Size and shape of mitochondria approaching to the control. При этом reduced the content of LPO products in mitochondrial membranes: the fluorescence intensity of LPO products decreased almost to the control level (Fig. 4).

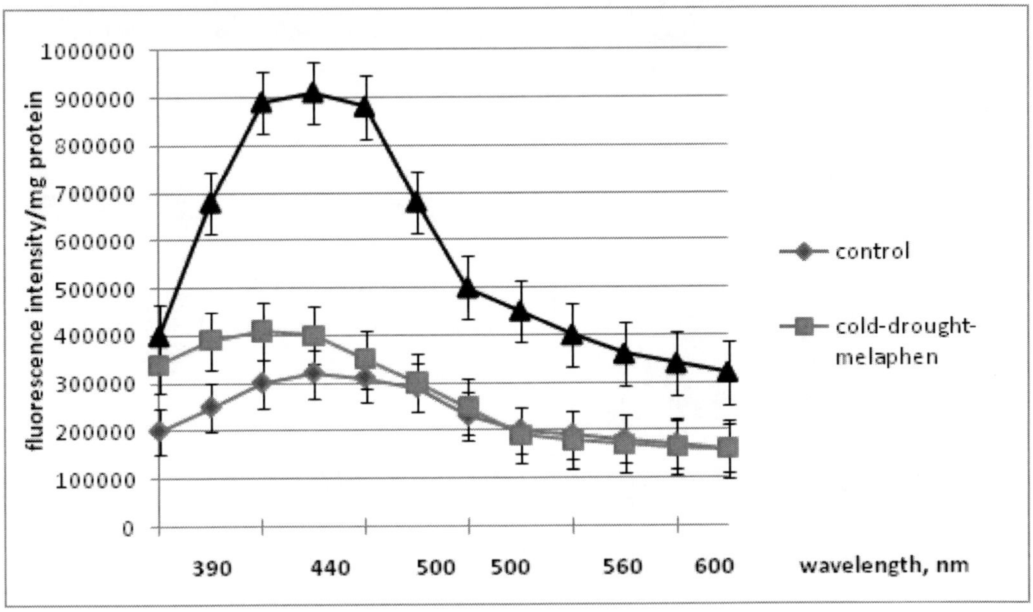

Figure 5. The fluorescence spectra of lipid peroxidation products in the membranes of the mitochondria of pea seedlings in times of insufficient moisture with moderate cooling and processing peas with melaphen.

It can be assumed that the protective effect of the drug is due to its antioxidant properties [7]. Reducing the intensity of free radical is reflected in the low intensity of lipid peroxidation and maintaining mitochondria morphology.

CONCLUSION

Science its structure melaphen similar to ATP, it can be assumed that it performs in a cells similar signaling functions.

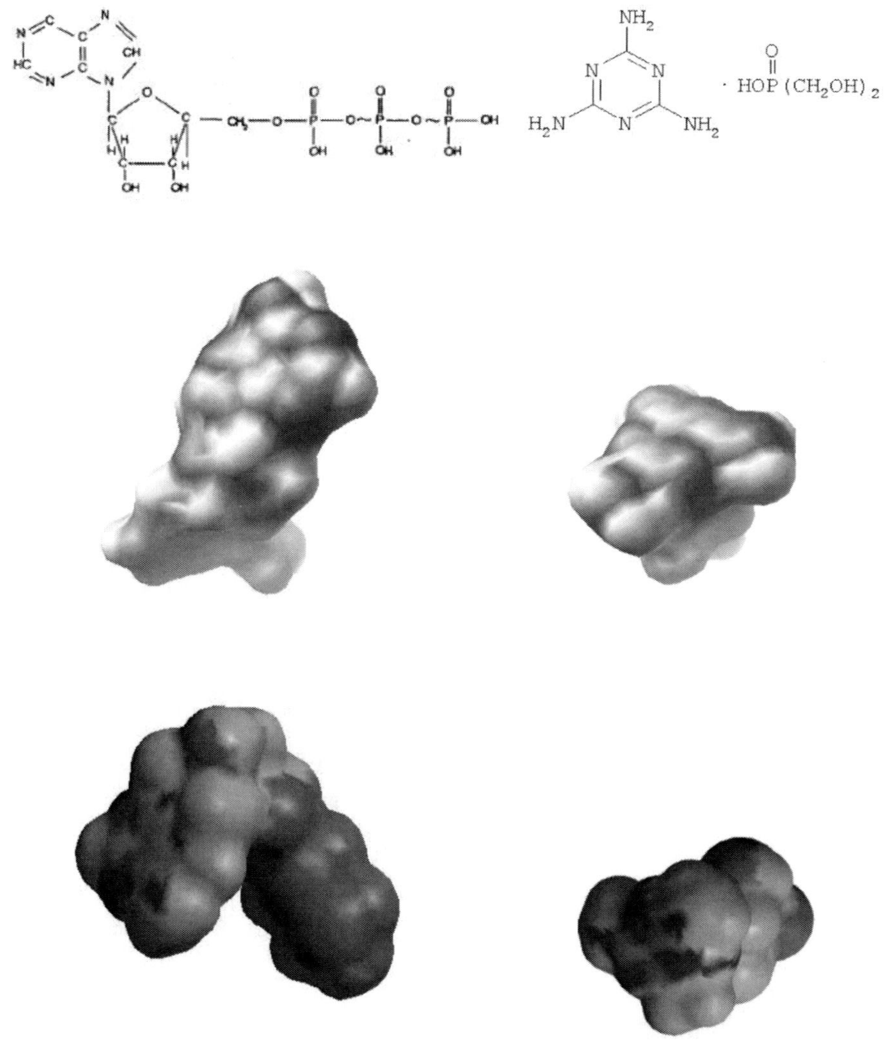

Figure 6. The structural formula of a molecule of ATP(1) and melaphen (2) and their molecular model.

It was therefore of interest to compare the molecular models melaphen and ATP. In figure presents the structural formulas and models of the surfaces of these molecules with the color of their atoms (with used MGL Tools1.5.4 and Python.5.2). Indeed, the purine moiety of ATP and melamine moiety melaphen in size, shape and position of heteroatoms pretty close, what may to some extent indicate a possible role as an artificial melaphen signaling molecule mimics of ATP.

Now has shown that ATP plays a role not only as substrate, (supplying energy) but in very small concentrations performs a signaling function, increasing signaling through ATP receptors on the outer membrane of cells, in particular, the signal for growth and division.

We can asume that melaphen like ATP, refers to those specific chemical factors that are able to regulate the growth of very low concentrations in plant cells and it can interact with the same receptors as ATP, and its phosphine groups can also amplify signals for the implementation of the metabolic processes as suggested Kashina [14].

To date, the mitochondrial K-ATP sensitive channel was found in the inner membrane of mitochondria in several laboratories. To date, the mitochondrial K-ATP sensitive channel was found in the inner membrane of mitochondria in several laboratories. It is shown opening of K-ATP sensitive channel regulate redox state of the inner mitochondrial membrane and this channel is inhibited by nucleotides [15]. Possible functions of this channel may be a reduction generation of reactive oxygen species by mitochondria and the regulation of mitochondrial volume[15,16]. If melaphen serve as the ATP, the drug treatment would not only did not reduce generation ROS by respiration chain and, consequently, the intensity of lipid peroxidation, but also would contribute to the swelling of mitochondria. Since melaphen reduces lipid peroxidation and prevents swelling of mitochondria, suggests that protective effect of the drug is related to its antioxidant properties. Melaphen preventing oxidation of unsaturated fatty acids in the membranes of mitochondria preserves the functional activity of these organelles [17].

REFERENCES

[1] Ahmad, P., M. Sarwat, Sharma S. Reactive oxygen species, antioxidants and signaling in plants// *Plant J. Biol., 2008*, vol. 51, P. 167–173.

[2] Chalova, L.I. and Ozeretskovskaya, O.L., Biokhimiya immuniteta, pokoya, stareniya rastenii (The Biochemistry of Plant Immunity, Dormancy, and Aging), Moscow: *Nauka,* 1984.

[3] Thomas, J.C., Bohnert, H.J., Salt Stress Perception and Plant Growth Regulators in the Halophyte Mesembryanthemum crystallinum// *Plant Physiol.,* 1993, vol. 103, P. 1299–1304.

[4] Skulachev V.P // Phenomenon of programmed cell death. Mitochondria, cells and organs: the role of reactive oxygen species. *Soros Educational Journal,* 2001, No 6, p. 4-10.

[5] Rodríguez M., E. Canales, O. Borrás-Hidalgo. Molecular aspects of abiotic stress in plants// *Biotecnología Aplicada* 2005, V.22 (1), P. 1-10

[6] Bakeeva L.E., Chentsov Yu.S., Skulachev V.P.// *Biochim.et biophys. acta.* 1978. V. 501. № 3. P. 349–369.

[7] Logan D.C., Leaver C.J . Mitochondria-targeted GFP highlights the heterogeneity of mitochondrial shape, size and movement within living plant cells // *J. Exp. Bot*. 2000. V. 51. P. 865–871.

[8] Scott, I., Logan, D.C., Mitochondrial morphology transition is an early indicator of subsequent cell death in Arabidopsis //*New Phytol.,* 2008, vol. 177, P. 90–101

[9] Zhang, L., Yinshu Li, Da-Xing, Caiji Gao Characterization of mitochondrial dynamics and subcellular localization of ROS reveal that HsfA2 alleviates oxidative damage caused by heat stress in Arabidopsis.// *J. Exp. Bot.,* 2009, vol. 60, P. 2073–2091/

[10] Claypoo, S.M., McCaffery, J.M., and Koehler, C.M. Mitochondrial mislocalization and altered assembly of a cluster of Barth syndrome mutant tafazzins// *J. Cell Biol.,* 2006, vol. 174, no. 3, P. 379–90

[11] Popov, V.N., Ruge, E.K., and Starkov, A.A., The influence of Inhibitors of Electron Transport on Generation of Active Oxygen Species at Succinate Oxidation by Pea Mitochondria, *Biochemistry* (Rus.), 2003, vol. 68,no. 7, P. 910-916.

[12] Fletcher B.I., Dillard C.D., Tappel A.L. Measurement of fluorescent lipid peroxidation products in biological systems and tissues // *Anal. Biochem.* 1973. V. 52. P. 1–9.
[13] Skulachev, V.P. Phenomenon programmble death. Mitochondria, cells, organs: the role of reactive oxygen species// *Soros. Obraz. Zh.,* 2001, vol. 7, no. 6, P. 1–10.
[14] Kashina O.A, Studies of the effect of organophosphorus compounds melaphen on growth and energy processes Chlorella cells// *summary of Ph.D. Kazan*, 2004. P.21.
[15] Petrussa E., Casolo V.,Braidot E., Chiandussi E., Macri F.,Vianello A. (2001). Cyclosporine A induces the opening of a potassium selective channel in high plant mitochondria// *J.Bioenerg. Biomembrane,* 2001, V. 33, P. 107-11
[16] Pastore, D., Stoppelli, M.C., Di. Fonzo N, Passarella S. (1999). The existence of the K^+ channel in plant mitochondria//. *J. Biol. Chem.,* 1999, V. 274, P. 26683-26690.
[17] I. V. Zhigacheva, , T. A. Misharina, , M. B. Terenina,, N. N. Krikunova,, E. B. Burlakova, I. P. Generozova,, A. G. Shugaev,, S. G. Fattakhov, A. I. Konovalov. Fatty Acid Composition of Mitochondrial Membranes of Pea Seedlings under Conditions of Insufficient Moistening and Moderate Cooling // *Doklady Akademii Nauk,* 2011, Vol. 437, No. 4, P. 558–560

In: Quantitative Chemistry, Biochemistry and Biology
Editors: G.E. Zaikov, O.V. Stoyanov et al.
ISBN: 978-1-62948-332-0

Chapter 19

EFFECT OF EXTERNAL INFLUENCES ON THE STRUCTURAL AND DYNAMIC PARAMETERS OF POLYHYDROXYBUTYRATE– HYDROXYVALERATE-BASED BIOCOMPOSITES

S. G. Karpova*[1*]*, A. L. Iordanskii*[2]*, A. A. Popov*[1]*,
N. G. Shilkina*[1]*, S. M. Lomakin*[1]*,
***M. A. Shcherbina*[4]*, S. N. Chvalun*[3] *and A. A. Berlin*[2]**

[1] Emmanuel Institute of Biochemical Physics,
Russian Academy of Sciences, Moscow, Russia
[2] Semenov Institute of Chemical Physics,
Russian Academy of Sciences, Moscow, Russia
[3] Enikolopov Institute of Synthetic Polymer Materials,
Russian Academy of Sciences, Moscow, Russia
[4] Russian Scientific Center "Kurchatov Institute," Moscow, Russia

ABSTRACT

The EPR probe, DSC, and WARDX methods are applied to study the molecular dynamics and structure of 3_hydroxybutyrate–3_hydroxyvalerate copolymer (PHBHV, 5 mol %) and its mixtures with segmented polyester urethane (SPEU) upon ozone oxidation and hydrothermal treatment. It is shown that time limited (3 and 5 h) treatment of the mixed compositions with water at 40°C has no noticeable effect on the dynamics of rotation of the probe (EPR) and the crystallinity of PHBHV (DSC). However, hydrothermal treatment at 70°C causes an increase in the molecular mobility of the probe in the system, accompanied, according to XRD and DSC analyses, by increases in the degree of crystallinity and density of PHBHV crystallites. The temperature of fusion of mixed compositions also increases. During ozonation, the mobility of the probe slows gradually in PHBHV all the time, whereas in mixed compositions, a slowdown is

* e-mail: karpova@sky.chpc.ras.ru.

observed only up to 3 h of oxidation time, after which, up to 5 h of ozonation, the mobility increases, along with decreases in the fraction of the amorphous phase and in the density of the crystallites.

Keywords: Biodegradable composition, crystallinity, probe EPR, DSC, X-ray analysis, ozone, water

INTRODUCTION

In recent years, the emphasis in the field of biodegradable materials has been placed on their use in gene and cell engineering for developing bioartificial or hybrid organs and tissues. One of the key problems in the formation of cell tissues in vitro for subsequent implantation is to obtain a biodegradable polymer scaffolds with the required physicochemical and biomedical properties. Properties suitable for the production of biodegradable scaffolds are exhibited by poly-hydroxybutyrate–hydroxyvalerate (PHBHV), a natural copolymer. However, moderate mechanical properties and a relatively high cost of PHBHV seriously limits its use in this area [1]. In [2, 3], the physicochemical and transport properties of biodegradable thermoplastic matrices prepared from poly(3-hydrox-ybutyrate) intended for targeted delivery of drugs were studied. The hydrolytic degradation of PHBHV for a long period (~3 months) was examined in [4, 5]. Segmented polyester urethane (SPEU) is also widely used as a biomedical material for artificial heart components, artificial blood vessels, catheters, and matrices for drug delivery. It was demonstrated [6] that, when exposed to an aqueous medium, the supramolecular structure of a SPEU surface changed so as to minimize the surface energy via the reorientation of polar rigid segments in domains. The effect of ozone on SPEU was studied in [7]. In particular, using ozone oxidation, the authors grafted acrylic acid and sulfobetaine to polyurethane films, a treatment that reduced thrombus formation on the surface. The literature reports studies of the structural dynamic and diffusion characteristics of SPEU under deformation, orientational drawing, and ozone oxidation [8, 9]. An optimal way to solve the problem of reducing the cost and, more importantly, to improve the physicochemical characteristics of original PHBHV is to blend it with other biomedical polymers, such as amorphous cellulose acetate butyrate [10]. These blends exhibit improved mechanical properties, in particular, ultimate strain and rigidity. Similarly, one can expect an improvement in the service characteristics of PHBHV upon blending with SPEU, the most widespread polymer in cardiology. The properties of PHBHV–SPEU compositions and their behavior in aggressive environments are currently unknown. Therefore, in the present study, we investigated the influence of temperature in the aquatic environment as well as the impact of ozone oxidation on the structural and dynamic parameters of PHBHV–SPEU compositions with different ratios of polymer components.

OBJECTS AND METHODS

In this work, we studied mixed compositions of natural biodegradable copolymer polyhydroxybutyrate–hydroxyvalerate (PHBHV) and segmented polyester - urethane (SPEU).

SPEU is characterized by $Mw = 2.29 \cdot 105$, $Mn = 5.3 \cdot 104$, and ρ = 0.97 g/cm3; PHBHV, by $Mw = 2.4 \cdot 105$, $Mn = 1.5 \cdot 105$, and ρ = 1.25 g/cm3. The ratios of the components in the PHBHV–SPEU mixture were 100: 0, 40 : 60, 50 : 50, 60 : 40, and 0 : 100 wt %. The procedure for preparation of composite films was described in detail earlier [11].

The molecular mobility was studied a paramagnetic probe by determining the correlation time τ that characterizes the rotational mobility of the probe as described in [12]. The probe was the 2,2,6,6-tetramethylpiperidine-1-oxyl radical. The radical was introduced into the film from the vapor phase or by adding dropwise its solution in tetrahydrofuran to the stirred polymer solution.

The ozone oxidation of the samples was carried out using an ozone–oxygen mixture with ozone concentration of 4 · 10–5 mol/L. Thermal analysis of the samples was performed on the DSC 204_F1 instrument (Netzsch) in an argon atmosphere at a heating rate of 10 K/min. The mean measurement error was 2%. Transmission XRD patterns were recorded of a Bruker Advance D8 diffractometer (Cu*K*αradiation). The lattice parameters were determined from the corresponding interplanar spacings by using the Bragg–Wulf equation:

$$n\lambda = 2d\sin\theta,$$

where n is the order of reflection, λ = 1.5408 E is the wavelength of the incident X-rays, and 2θ is the diffraction angle.

The average effective crystallite size *Lhkl* in the crystallographic direction *hkl* was determined from the integral halfwidth of the line profile of the corresponding X-ray reflection using the Selyakov–Scherrer formula:

$$\Delta_{hkl}(2\theta) = \lambda / L_{hkl}\cos\theta$$

where $I(2\theta)$ and I_{max} are, respectively, the intensity of the line at some angle 2θ and its maximum intensity,

$$\Delta = 1/\, I_{max}\int I(2\theta)d2\theta.$$

In the diffractometric formulation, the numerator is the area under the reflection profile, whereas the denominator is its maximum ordinate. Mathematical processing of the profiles of the investigated reflections was performed with the Microcal™ Origin™ 8.0 software package.

Results and Discussion

Molecular Mobility in PHBHV and Its Mixed Compositions during Treatment with Water and Ozone

Using various methods of introducing the TEMPO radical into films prepared from PHBHV–SPEU compositions leads to a significant difference in its rotational mobility. Figure 1 shows the dependence of the correlation time of the probe introduced into film samples by sorption from the vapor phase (1) and by adding it to the polymer solution (2) on the SPEU concentration in the mixture. Comparison of these results shows that the effective molecular mobility of the probe introduced from the solution is significantly lower than the mobility of the probe adsorbed from the vapor phase. When polymer films are formed from a solution, the probe penetrates into the denser areas of PHBHV [13], whereas when sorbed from the vapor, it is distributed mainly in the amorphous regions of the studied polymers, where the segmental mobility is higher. It should also be noted that, in the first case, the polymer composition of the mixture has a greater impact on the correlation time than in the case where the radical is sorbed (Fig. 1).

When the radical is introduced into PHBHV–SPEU mixtures from the vapor phase, it concentrates primarily in the elastic regions of SPEU and the amorphous areas of PHBHV, remain almost completely outside the denser areas of PHBHV. Note that, in the initial polymers, PHBHV and SPEU, the mobilities of the sorbed radical differ threefold, whereas the mobilities of the probe introduced through the solution into the same polymers differ by almost an order of magnitude (7.5 times). As expected, the rotational speed of the probe in highly crystalline PHBHV is significantly lower than in SPEU. Thus, the method of introduction of the radical into PHBHV–SPEU films via a cosolvent, tetrahydrofuran, is more informative, since in this case, the dynamics of rotation of the probe is more sensitive to structural organization of the polymer system.

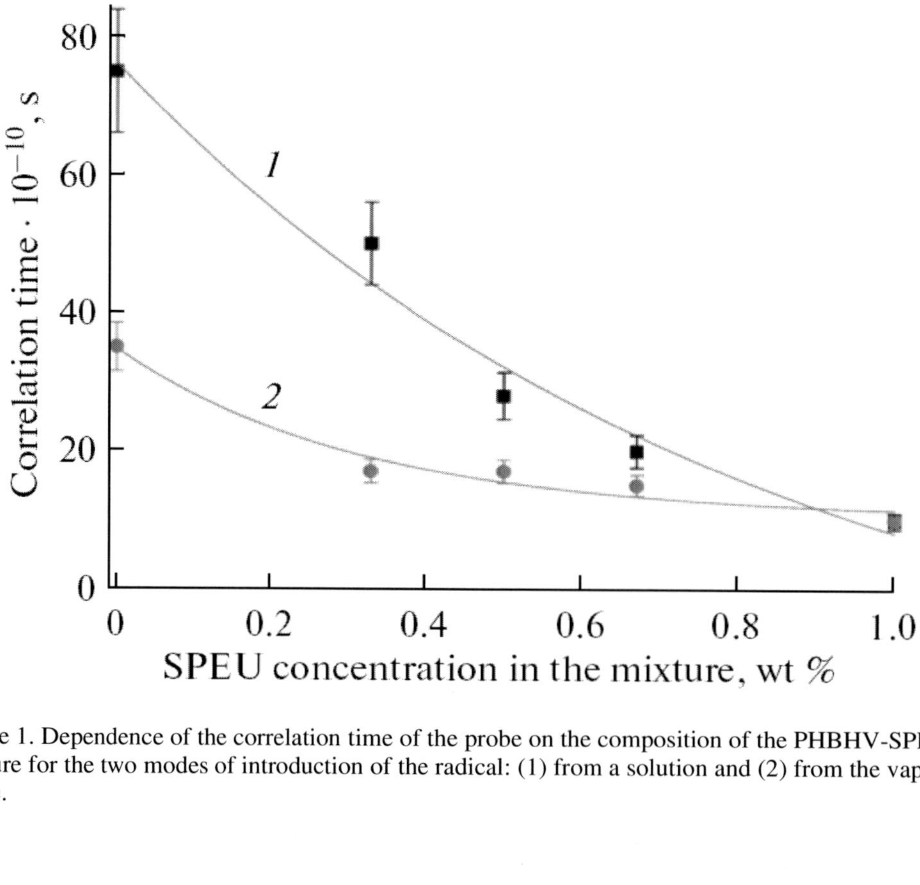

Figure 1. Dependence of the correlation time of the probe on the composition of the PHBHV-SPEU mixture for the two modes of introduction of the radical: (1) from a solution and (2) from the vapor phase.

Proceeding from our problem, let us examine how ozone oxidation influences the molecular mobility of the radical. Figure 2 (curve *1*) shows that, in individual PHBHV, the parameter τ increases approximately linearly with the time of expo - sure to ozone. In the mixed compositions, the correlation time τ grows only at the initial stage of oxidation (up to ~2–3 h), then passes through a maximum and decreases.

With the introduction of the radical from a solution, we ob -served a change in the rotational mobility at a fixed time of exposure to ozone (for 2 h; Fig. 3, curve 4). That the mobility of the radical, introduced from the vapor phase (Fig. 2) and a solution (Fig. 3) slowing down upon ozone exposure can be explained as follows. At the initial stage of the oxidation process, the degradation of the chain has not yet developed, but the oxygen containing groups produced form intermolecular hydrogen bonds, which leads to an increase in chain rigidity and, consequently, to an increase in the parameter t. We observed similar phenomena in crystalline PP and PE [4, 5]. At high degrees of exposure to ozone, when polymer chain cleavage becomes intense, the mobility of the probe increases. The observation that, in individual PHBHV, the molecular mobility decreases over the entire period of oxidation can be explained firstly by the fact that, in the intercrystallite regions of PHBHV, the fraction of straight ened chains is high enough, and, after oxidation, these chains tend to be more oriented, so that the parameter describing the degree of chain linearity [16] $\beta = h/L = \sqrt{2} /3k$ increases (here, h is the distance between the ends of the chain, L is the contour length of the macromolecule, and k is the number of segments). It is known that each polymer has its own value of β^* above which the chains tend to straighten and below which, to fold. Second, because of a high packing density and low mobility, it is sterically difficult to fold the chain, a factor that leads to a decrease in the molecular mobility at all times of ozone oxidation.

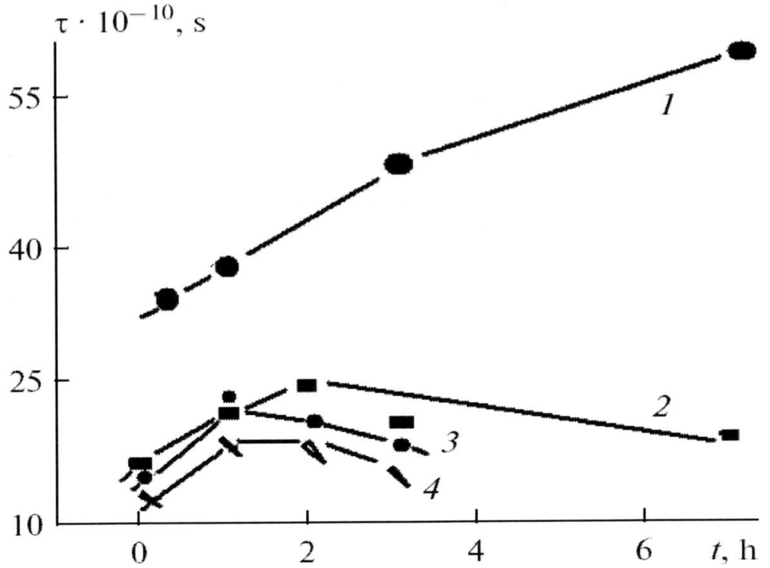

Figure 2. Dependence of the correlation time on the ozonation time for (1) PHBHV, (2) 60:40 PHBHV-CPEU, (3) 40:60 PHBHV – SPEU, and (4) SPEU.

The results obtained make it possible to draw the fundamental conclusion that the ozone oxidation largely influences the structure of PHBHV rather than that of SPEU, apparently due to a higher fraction of strained (straightened) chains of the biopolymer at the crystal–amorphous region boundary.

Additional information can be obtained from measurements of the activation energy of compositions exposed to ozone for 2 h (Fig. 3b). The figure shows that, like τ, the activation energy increases after ozone oxidation, which is particularly noticeable for the PHBHV samples (the radical introduced from the vapor phase). That the energy barrier to the rotation of the probe in individual PHBHV and in the mixed compositions is indicative of a slowdown in the motion of bio -polymer segments in the intercrystallite regions due to the accumulation of oxy -gen-containing groups, a process that leads to an increase in the density of the structure.

Hydrothermal effect on molecular mobility was studied at two different tem -peratures: 40°C, close to physiological temperature, and 70°C, adopted in a number of studies as the standard temperature for accelerated hydrolytic stability tests of biopolymers. Figure 3 shows that the heating of the polymer composites at 40°C in water for 3.5 h does not significantly reduce the molecular mobility of the probe in the mixture and does not alter the crystallinity of PHBHV, as demonstrated using DSC. Indeed, at this temperature, changes in the degree of crystalllinity of PHB and PHBHV begin to manifest themselves only after a sufficiently long hydrothermal treatment, more specifically, after several weeks of testing in phosphate buffer at 37°C [4, 5].

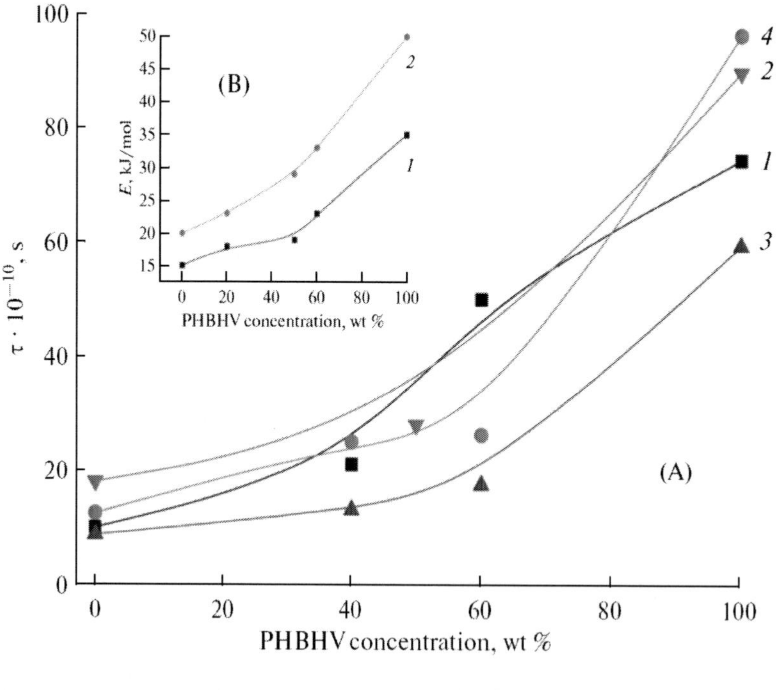

Figure 3. (A) Correlation time as a function of the composition of the mixture for (1) hydrothermal treatment at (2) 40^{0}C and (3) 70^{0}C and for (4) ozonation for 2h. (B) Dependence of the activation energy on the composition for (1) the initial mixtures and (2) after ozonation.

For PHBHV and mixtures thereof, treatment in the aquatic environment at 70°C leads to an increase in the mobility of the probe compared to the original PHBHV–SPEU compositions (Fig. 3). These changes in the EPR spectra can be explained by the plasticizing effect of water on PHBHV macromolecules, so that their segmental mobility in the intercrystallite space increases, and the relaxation processes proceed faster, which manifests itself as an increase in the frequency of rotation of the probe.

Note also that the difference between the rotational mobility of the probe in initial, oxidized, and hydrothermally treated films increases with the PHBHV concentration in the mixture (Fig. 4). This result is used to a high degree of straitened chains in the morphous phase of the PHBHV component (a high density ρ = 1.25 g/cm3) and degree of crystallinity).

Differential Scanning Calorimetry of the Compositions after Ozone Oxidation and Hydrothermal Treatment

DSC curves for PHBHV in the region of fusion (150–190°C) showed two endothermic peaks: (1) a high-temperature peak, which is responsible for the fusion of more perfect crystals, which appears as a result of primary crystallization [17] and (2) a low temperature peak, associated with the fusion of less structured crystals formed by secondary crystallization [18] (Fig. 5). Analysis of the DSC curves for individual PHBHV and the PHBHV–SPEU compositions in the native state (i.e., the original films prepared from a solution) showed that, as the content of SPEU in the mixture increases from 0 to 60%, of the enthalpy of fusion of PHBHV remains constant. Consequently, the degree of crystallinity in terms of PHBHV concentration in the mixture also remains approximately constant (Table 1). This result indicates that the interaction between the polymer components during the evaporation of the solvent and formation of the composite film from a solution is minimal, i.e., the presence of polyurethane macromolecules does not prevent the crystallization of the biopolymer. A 3°C shift of the peak of fusion of PHBHV to lower temperatures is indicative of the dilution of biopolymer molecules by SPEU molecules rather than the interaction of the polymer components.

During the second fusion, PHBHV and mixtures thereof behave in a different way. In this case, the degree of crystallinity (enthalpy of fusion) of the 40: 60 PHBHV–SPEU mixture decreases by 30%. Therefore, in contrast to the previous case (first heating), PHBHV crystallization is retarded by SPEU segments interacting with biopolymer molecules through the formation of a network of hydrogen bonds between the oxygen of PHBHV ester groups and the hydrogen of SPEU urethane groups. The constancy of the position of the maximum of the high - temperature fusion peak on the temperature scale, i.e., the temperature of fusion (TM) indirectly indicates the preservation of the perfection of the structure of PHBHV crystal both in the mixture and in the individual state.

Consider the effect of ozone oxidation on the thermophysical behavior of the system. Analysis of DSC curves for PHBHV showed that, after exposure of native specimens in an ozone atmosphere for 3 h, their crystallinity increases, decreasing after 5 h (Table 1). PHBHV–SPEU compositions obtained from a solution were oxidized for 2 h. Table. 1 shows that ozone oxidation does not affect the crystallinity of the pure biopolymer, but slightly increases it for PHBHV–SPEU mixtures, from 59 to 64%. By analogy with earlier studied polymers, PE, PP, and others [14, 15], and in accordance with the dynamics of the mobility of the probe (see the previous section), the observation of the growth and subsequent decline of the degree of crystallinity can be explained by changes in the ozone oxidation mechanism over time: at first stage, the crystalline phase is built up, being partially destroyed at later stages.

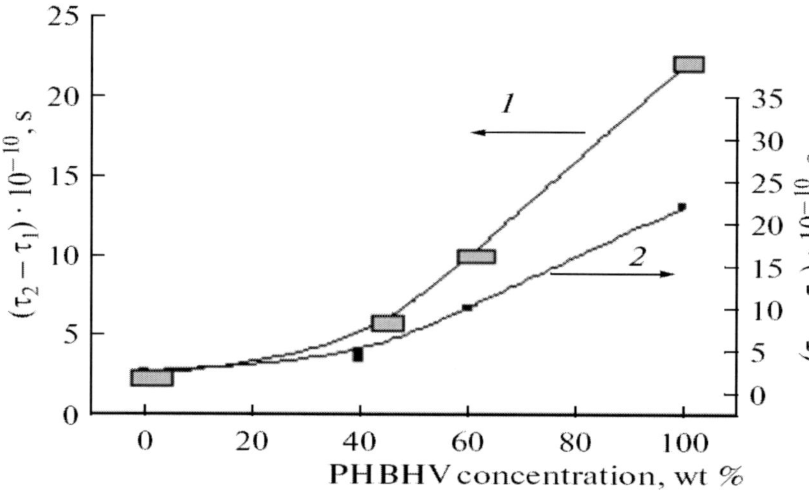

Figure 4. Dependence of the difference in the correlation times for compositions (1) ozonized for 2 h and (2) subjected to hydrothermal treatment on the PHBHV concentration in the mixture.

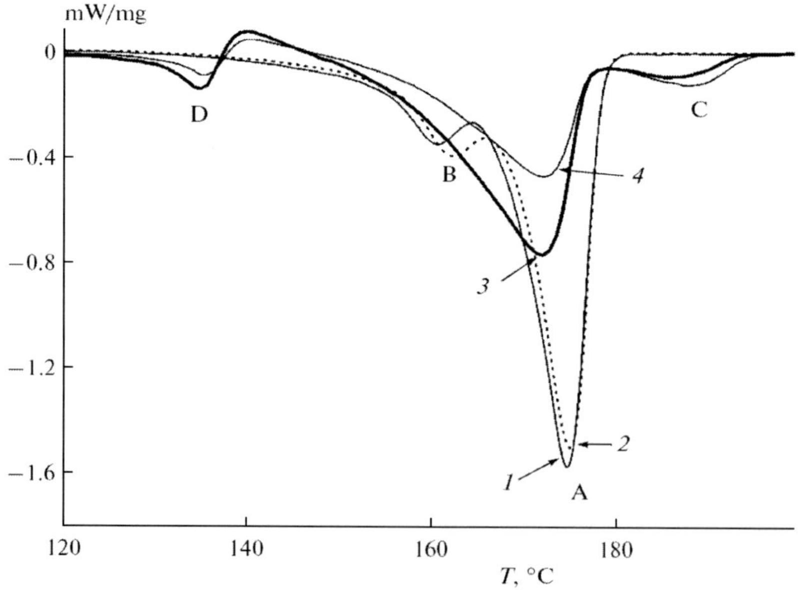

Figure 5. DSC curves of fusion of (1,2) films of PHBHV and its mixtures with SPEU in proportion of (3) 60:40 and (4) for (1) a dry PHBHV and (2-4) samples kept in water at 70^{0}C for 3.5h and then dried in air at 25^{0}C for 48h. The main thermal transitions are denoted as A, B, C, D (see the text). The arrows indicate the main peaks of fusion in the thermograms for each of the four samples.

It was shown [19, 20] that, the most strained areas of the macromolecules are most reactive with respect to ozone. In the case of PHBHV, for first several hours, ozone also affects the most strained areas of the macromolecules, which are located mainly in the intercrystallite regions of PHBHV, near the end faces of the crystallites. Given the high crystallinity of PHBHV and a fairly narrow space between the lamellae, the degree of orientation of interdomain chains is high, with an extended conformation being most probable, but after exposure to ozone (rupture of intermolecular bonds), they may take part in building crystalline regions, as evidenced by a decrease in the fraction of the amorphous phase and a small increase in the crystallite size (XRD data, Tables. 2). During further expo - sure of the samples to ozone (5 h), the oxidation process reaches the surface of the PHBHV crystals, thereby reducing the overall crystallinity and, consequently, decreasing the temperature of fusion of the crystallites (Table 1), from 175 to 171°C, which also is also indicative of an increase in the defectiveness of the crystallites after 5 h of ozone oxidation.

Let us now consider the effect on ozone oxidation on the mixed compositions. Note first that all the thermograms of the PHBHV–SPEU samples feature small endothermic and exothermic peaks at 135 and 140°C, with specific heats of 4.5 and 9.3 J/g, respectively. The origin of these peaks is not fully understood; one can only assume, by analogy with the data for PHB [21], that cold crystallization and the melting of these crystals occur. However, the fact that this effect is observed only in the presence of the second component, not for individual PHBHV, allows us to put forward an alternative explanation. The first peak (135°C) represents the melting of imperfect crystallites in the interphase layers at the PHBHV–SPEU boundary, whereas the subsequent exothermic peak, at 140°C, is associated with the displacement of SPEU chains from the interphase zone, which occurs at the initial stage of PHBHV crystallization.

Comparison of DSC curves for PHBHV samples with different thermal histories, namely native films (first heating) and the same films obtained from the melt (second heating) showed that, all other things being equal, the most noticeable change in the crystallinity occurs in the native films obtained directly from a solution. As a result of intermolecular interactions, the samples subjected to melting exhibit no extreme time dependence, with their crystallinity reducing markedly, from 59 to 41%, only at the longest exposure (5 h, Table 1).

The use of polymer composites as medical implants and therapeutic systems for a controlled release of drugs requires additional information on the effects of water on the physicochemical properties of polymer systems. Therefore, the next stage of the work was to study the effect of water molecules on the dynamic and thermal behavior of PHBHV–SPEU compositions at different ratios of the components. Note that the effect of water on the polymer composition for 5 h under conditions close to physiological, i.e., at pH 7.4 and40°C, causes no significant structural and dynamic changes. However, the aqueous medium at elevated temperature (70°C) is a specific aggressive environment for the PHBHV–SPEU system. Conducting experiments with biodegradable polymers for studying the rapid degradation at 70°C is quite common and well-known procedure proposed, in particular, as the standard test for the rate of hydrolysis of PHB, PHBHV, and their homologues [22].

Treatment of the samples in water at 70°C for 3.5 h acts differently on individual PHBHV and compositions thereof. In the first case, the biopolymer demonstrates the stability of the crystalline phase: according to DSC analysis, it remains unchanged, retaining the same degree of crystallinity as the original, not treated in water polymer (59%, see Table 1). More over, there is solid evidence that the quality of the crystallites after hydrothermal treatment becomes higher.

Table 1. Degree of crystallinity and temperature of fusion for compositions subjected to ozone oxidation and hydrothermal treatment as obtained by the DSC method

Mixtures	Fusion					
	first heating			second heating		
	1	2	3	1	2	3
PHBHV 100%	58	175	161	59	172	165
PHBHV–SPEU 60 : 40%	62	172	–	57	171	163
PHBHV–SPEU 40 : 60%	58	172	–	41	172	166
Ozonation for 2 h						
PHBHV 100%	59	175	161	58	172	169
PHBHV–SPEU 60 : 40%	62	172	–	57	173	167
PPHBHV–SPEU 40 : 60%	64	172	–	52	174	170
Ozonation for 3 h						
PHBHV 100%	65	175	162	59	173	170
Ozonation for 5 h						
PHBHV 100%	50	171	–	41	170	–
Hydrothermal treatment at 70°C for 3.5 h						
PHBHV 100%	59	177	162	59	174	163
PPHBHV–SPEU 60 : 40%	58	172	–	54	170	160
PHBHV–SPEU 40 : 60%	50	168	–	15	169	–

Notes: (1) The degree of crystallinity corresponding to the mean peak of fuson; (2) high-temperature peak of fusion of crystallites, ^{0}C; (3) low-temperature peak of fusion, ^{0}C.

Table 2. Parameters of the crystal lattice of PHBHV after ozone oxidation and hydrothermal treatment

PHBHV	Interplanar spacing d_{020}	Crystallite size l_{020}	Interplanar spacing d_{110}	Crystallite size l_{110}
Initial	6.639	215	5.269	160
Ozone, 2 h	6.630	230	5.263	160
Ozone, 5 h	6.646	240	5.278	185
Water	6.606	255	5.245	190

Indeed, after hydrothermal treatment, *TM* shifts to higher temperatures, from 175 to 177°C (for the native sample) and from 172 to 174°C (after second heating).

According to XRD data, the hydrothermal treatment of PHBHV films leads to an increase of the crystallinity of the samples, improvement of their crystalline structure, and a significant increase in the crystallite size (Table 2). All these results can be explained by the plasticizing effect of water molecules on PHBHV, so that their segmental mobility in the intercrystallite space increases, the relaxation processes occur faster, crystallites are built up at the expense of the surrounding polymer segments (which gain additional mobility), and quality of the cry -stallline regions improves.

The DSC curves of PHBHV–SPEU compositions hydrothermally treated at 70°C exhibit a moderate decrease in crystallinity, from 59 to 50% for native samples, and a sharp decrease in crystallinity in the same conditions for samples subjected to fusion, from the same 59 to 15%. The decrease in the crystallinity of PHBHV is accompanied by a decrease in the melting temperature: from 174°C for the initial sample to 170°C for the 60 : 40 PHBHV–SPEU mixture and to 168°C for the 40 : 60 mixture. Thus, these results show what a strong influence the second component of the mixture (segmented polyurethane) has on the molecular dynamics and structure of PHBHV. In most cases, SPEU molecules interact with molecules of the biopolymer to form intermolecular bonds, thereby creating energy and steric hindrances to the crystallization of the PHBHV component. The results presented here are of practical interest, showing that compositions of the two proposed medical polymers exhibit higher resistance to hydrothermal influence when the PHBHV component is predominant in the mixture.

XRD Analysis of Compositions after Ozone Oxidation and Hydrothermal Treatment

X-ray diffraction analysis showed that the PHBHV samples had a relatively high degree of crystallinity. The diffractograms of the PHBHV films at 25°C exhibited at least five narrow reflections corresponding to an orthorhombic lattice with lattice constants of a = 5.74 E, b = 13.24 E, and c = 5.98 E (Fig. 6). These values are quite close to the published data for PHB (a = 5.76 E, b = 13.20 E, c = 5.96 E. [23, 24]. A small discrepancy in the values of a, b and c at an average standard deviation of measurements of ±0.005 can be attributed to the presence of a small number of hydroxyvalerate comonomer units in the PHB molecule (~5 mol %), which leads to a slight change in the unit cell size of the lattice [24].

Treatment of samples with ozone or water vapor leads to certain changes in the supramolecular structure of the samples, more specifically, an increase in the crystallinity of the samples, a significant increase in the crystallite size, and, in the case of hydrolysis, to a perfection of their crystalline structure, which appears to reduce the principal interplanar spacings and d020 and d110. More over, the ratio of the integrated intensity of crystalline reflections to the total integrated intensity is 0.66 for the initial samples, 0.72 and 0.93 for the samples treated with ozone for two and five hours, respectively, and 0.84 for the samples treated with water. Note that the density of the crystallites decreases slightly upon ozone oxidation, but increases by 1% upon treatment with water. This led us to conclude that ozone oxidation is accompanied by the etching of the amorphous phase, which results in the observed increase in the degree of crystallinity. Hydrothermal treatment is accompanied by a significant increase in the mobility of molecular chains that contribute to a perfection of the crystal structure of the material. The crystallite size and the corresponding lattice constants are presented in Table. 2. The diffractograms of X-ray scattering are shown in Fig. 6.

CONCLUSION

Thus, in the present work, the effects of thermal treatment in the aquatic environment and of ozone oxidation on the structural and dynamic parameters of PHBHV–SPEU mixed compositions with different ratios of the polymer components was for the first time investigated. The findings allowed us to draw the fundamental conclusion that the ozone oxidation largely affects the structure of PHBHV rather than that of SPEU, apparently due to a higher fraction of strained chains of the biopolymer in the crystal–amorphous phase boundary region.

Depending on the treatment duration, the material of these areas can partially convert to augment the crystalline phase, thereby increasing the stiffness of the polymer matrix. In contrast, at high ozone exposures, when the intense cleavage of polymer chains and intermolecular bonds, including hydrogen bonds, begins to dominate, which act as effective cross-links, the chain mobility and, consequently, the rotational speed of the probe increases.

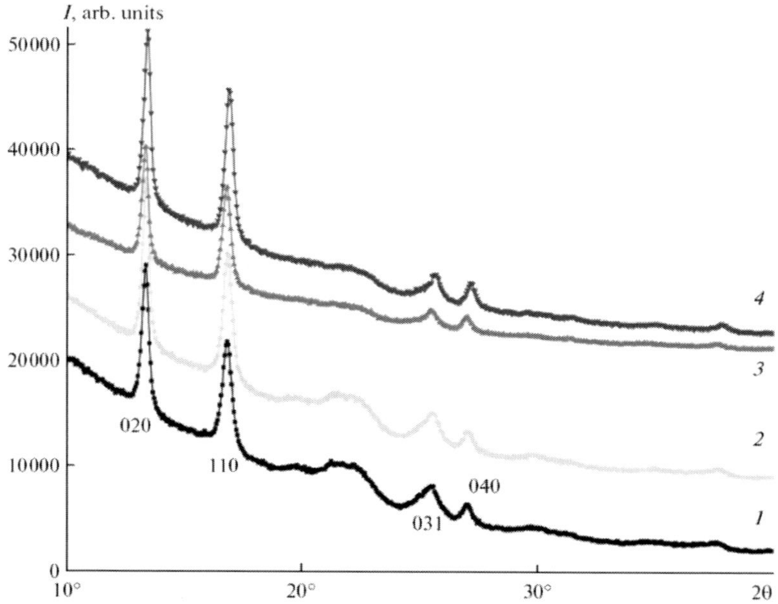

Figure 6. X-ray diffractograms for (1) the initial samples and samples treated with ozone for (2) 2h and (3) 3h and with (4) water vapor.

Exposure of the samples to water for 3.5 h at 70°C acts differently on individual PHBHV and compositions thereof. The individual biopolymer shows stability of the crystalline phase: according to DSC measurements, it does not change, while according to XRD analysis, hydrothermal treatment leads to an increase in the crystallinity of the samples, perfection of their crystalline structure, and significant increase in the crystallite size. All these results can be explained by the plasticizing effect of water molecules on PHBHV, so that their segmental mobility in the intercrystallite space increases, the relaxation processes become faster, crystallites are built up at the expense of the surrounding polymer segments (which gain additional mobility), and the crystalline regions become more perfect.

The results obtained are of practical interest, showing that compositions of the two proposed medical polymers exhibit higher resistance to hydrothermal impacts when the PHBHV component is predominant in the mixture.

Acknowledgments

This work was supported by the Department of Chemistry and Materials Science of the Russian Academy of Sciences within the framework of the program "Creating a New Generation of Macromolecular Structures" and the Presidium of the Russian Academy of Sciences within the framework of the program "Fundamental Science for Medicine" (grant no. 5P).

References

[1] L. Mandelkern and A. Allon, *J. Polym. Sci.* 4, 447 (1966).

[2] M. I. Shtil'man, *Polymers for Medical Biological Purposes* (Akademkniga, Moscow, 2006) [in Russian].

[3] R. Shi, D. Chen, Q. Liu, Y. Wu, X. Xu, L. Zhang, and W. Tian, *Int. J. Mol. Sci.* 10, 4223 (2009).

[4] A. P. Boskhomdzhiev, A. P. Bonartsev, E. A. Ivanov, T. K. Makhina, V. L. Myshkina, et al., *Plast. Massy,* No. 8, 13 (2009).

[5] A. P. Bonartsev, A. L. Iordanskii, G. A. Bonartseva, et al., *Plast. Massy,* No. 3, 6 (2010).

[6] M. Manno, A. Emanuele, V. Martorana, P. San Biagio, and D. Bulone, *Biopolymers* 59, 51 (2005).

[7] Y. Youling, A. Fei, Z. Xiaopeng, Z. Wei, S. Jian, and L. Sicong, *Colloids Surf. B: Biointerfaces* 35, 1 (2004).

[8] S. G. Karpova, A. A. Popov, E. E. Zhukova, M. I. Shtil'man, and A. L. Iordanskii, *Plast. Massy,* №. 12, 23 (2012).

[9] S. G. Karpova, A. A. Popov, A. V. Khvatov, N. S. Klenina, and A. L. Iordanskii, *Khim. Fiz.* № 2, 30 (2011).

[10] S. G. Karpova, A. A. Popov, S. N. Chvalun, Yu. A. Zubov, and G. E. Zaikov, *Vysokomol. Soedin.* A, No. 7, 1404 (1986).

[11] S. G. Karpova, N. N. Blinov, A. A. Popov, and G. E. Zaikov, *Vysokomol. Soedin. A,* No. 10, 2201 (1989).

[12] N. Suttiwijitpuhdee, H. Sato, J. Zhang, and T. Hashimoto, *Polymer* 52, 461 (2011).

[13] A. L. Zhul'kina, E. L. Ivantsova, A. G. Filatova, and A. L. Iordanskii, *Crystallogr. Rep.* 54, 464 (2009).

[14] A. A. Popov, S. G. Karpova, E. V. Podkopaeva, A. L. Kovarskiim, et al., *Vysokomol. Soedin.* B 22, 868 (1980).

[15] I. I. Kamaev, A. L. Aliev, A. L. Iordanskii, and A. M. Wasserman, *Polymer* (UK) 42, 515 (2001).

[16] S. G. Karpova, A. A. Popov, S. N. Chvalun, Yu. A. Zubov, and G. E. Zaikov, *Vysokomol. Soedin.* A, No. 7, 1404 (1986).
[17] S. G. Karpova, N. N. Blinov, A. A. Popov, and G. E. Zaikov, *Vysokomol. Soedin.* A, No. 10, 2201 (1989).
[18] S. N. Chvalun, Candidate's Dissertation in Physics and Mathematics (*Phys. Tech. Inst.*, Dolgoprudnyi, Moscow, 1981).
[19] R. Pearce and R. H. Marchessault, *Polymer* 35, 3990 (1994).
[20] G. Hauser, J. Schmidtke, and G. Strobl, *Macromolecules* 31, 6250(1998).
[21] A. A. Popov, N. Ya. Rapoport, and G. E. Zaikov, *Oxidation of Oriented and Strained Polymers* (*Khimiya,* Moscow, 1987) [in Russian].
[22] A. N. Ozerin, Candidate's Dissertation in Chemistry (*NIFKhI im. Karpova*, Moscow, 1977).
[23] M. L. Di Lorenzo, P. Sajkiewicz, P. La Pietra, and A. Gradys, *Polym. Bull.* 57, 713 (2006).
[24] A. P. Bonartsev, A. L. Iordanskii, G. A. Bonartseva, and G. E. Zaikov, in *Environmentally Degradable Materials Based on Multicomponent Polymeric Systems*, Ed. By C. Vasile and G. E. Zaikov (Brill, Leiden, Boston, 2009), Ch. 2, pp. 35–75.
[25] P. J. Barham, A. Keller, E. L. Otun, and P. A. Holmes*, J. Mater. Sci.* 19, 2781 (1984).
[26] M. L. Di Lorenzo, M. Raimo, E. Cascone, and E. Martuscelli, *J. Macromol. Sci.: Phys.* B 40, 639 (2001). SPELL: 1.

In: Quantitative Chemistry, Biochemistry and Biology
Editors: G.E. Zaikov, O.V. Stoyanov et al.
ISBN: 978-1-62948-332-0

Chapter 20

COMPOSITE MATERIALS BASED ON LDPE WITH THE ADDITION OF WOOD FLOUR AND RUBBERS OF DIFFERENT NATURE

A. V. Koroleva[1*], Yu. K. Lukanina[1], A. V. Khvatov[1], N. N. Kolesnikova[1], P. V. Pantiukhov[1], A. A. Olkhov[2] and A. A. Popov[1]

[1] Federal State Institution of Science
N.M. Emanuel Institute of Biochemical Physics, Moscow, Russia
[2] M.V. Lomonosov Moscow State University of Fine Chemical Technology, Moscow, Russia

ABSTRACT

Currently the world's consumption of plastics is growing in this regard there are difficulties in disposing of large amounts of household plastic waste. The solution is to develop biodegradable polymers with a specific expiration date. This work is dedicated to the creation of biodegradable composite material based on LDPE, wood flour and rubbers of different nature.

Keywords: Biodegradation, LDPE, wood flour, natural rubber, EPDM

AIM AND BACKGROUND

The research carried out in this paper, aimed at establishing the possibility of creating composite materials based on polyethylene and natural rubber with biodegradable additive assessment of their compatibility, mechanical and rheological properties as well as their propensity to biodegradation in soil [1-2]. The basis for creating biodegradable composite

* korolevaanna83@gmail.com.

served as LDPE, natural additive - wood flour and rubbers of different nature. To study the workability (adaptability) of the resulting composition, an experiment was carried out on receipt of the melt flow index. In order to determine the physical and mechanical properties of the composite material, indicators were obtained such as tensile modulus and elongation at break. To study the biodegradability of the composite in the environment, field tests were conducted in soil and the water absorption was determined.

INTRODUCTION

Currently, there is a lot of work aimed at creating biodegradable composites based on polyolefines and natural supplements [3-7]. It is used as a component of a composite material of natural rubber due to the instability of microbiology is new and interesting direction. From a microbiological damage natural rubber faced in the initial stages of its production, as natural rubber latex contains 30-35% of the hydrocarbons in the form of rubber particle size of 0.15-0.5 mm. There are cases of microbial damage vulcanized natural rubber in the operation of underground electrical cables, rubber insulated from NR, sealing materials in contact with the ground, sea and tap water, sewage. Synthetic rubbers are a large group of elastomers that differ in chemical composition and biological stability. The general rule for synthetic rubber, as well as for other polymers, is the growth of biological stability by increasing the length of the macromolecular chain [8]. Among the least biostable synthetic rubber is EPDM.

EXPERIMENTAL PART

The objects of study were selected mixture of low density polyethylene (LDPE) with wood flour (WF), natural rubber (NR) and ethylene-propylene rubber (EPDM). WF content is 40 wt.%, Rubber injected at 10 and 20 masses.%. As composite materials based on LDPE blending performed on a laboratory mixer at a temperature of 140 °C for 5 minutes and then get the film samples in a laboratory press. The sample thickness was 100 ± 10 microns.

In order to identify the strength characteristics of the obtained mixtures were conducted physical and mechanical tests in a laboratory tensile machine PM-10 (Table 1, Fig. 1). Testing, training and payment module samples were assayed according to GOST 9550-81.

To study the rheological properties of the compositions, a method for determining the melt flow rate was used (MFR) (GOST 11645-73). The experiment was conducted in a laboratory type IIRT capillary viscometer at 190 ° C and a load of 2.16 kg, the dwell time in the chamber 5 minutes. The data are presented in Table. 1. Crushed sample was placed in a heated, covered on one side, the camera device, stand 5 minutes to complete melting of the material. After a time delay, the cap removed, the melt flow was under the load of 2.16 kg. After some time the knife cut off part of the sample (at least 3) and weighed, defining and calculating the average weight melt flow for 10 minutes.

Determination of water absorption of composite materials was carried out in accordance with GOST 4650-80 for 24 hours at 23 ± 2 ° C. The data presented in Fig. 2.

To assess the biodegradation in the soil samples were placed in a combined land, prepared in accordance with GOST 9.060-75. Moisture and soil acidity were kept constant.

Data on changes in the mass of film samples for 8 months of exposure are presented in Table 2.

RESULTS AND DISCUSSION

From these data it is clear that the introduction of the filler significantly reduces the elastic modulus and elongation at break of the material. This is due to the fact that wood flour is fine addition [9]. Flour particles in the matrix of polyethylene are defects, stress concentrators, which results in uneven load distribution in the matrix. Upon loading the sample is detachment of the filler particles from the matrix pores are formed [10], which are then expanded, leading to premature failure of the composite.

Table 1. Characteristics of the samples

Samples	Melt flow rate, g/10min	Modulus, MPa	Elongation at break%
LDPE	2	212	400
LDPE 60-40 WF	1	113	9
LDPE 50-10 NR-40WF	1	22	8
LDPE 40-20 NR-40WF	8	33	8
LDPE 50-10 EPDM-40 WF	-	39	10
LDPE 40-20 EPDM-40 WF	-	48	8

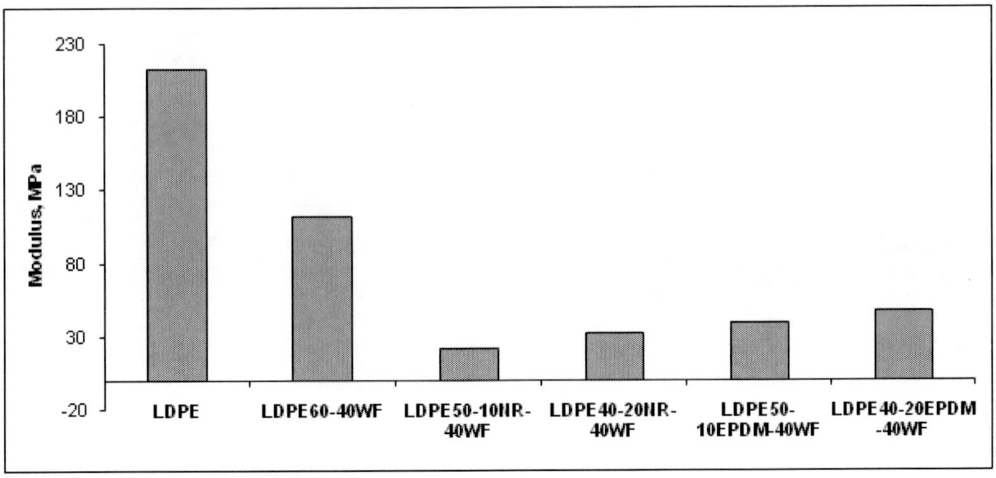

Figure 1. Tensile Modulus, MPa.

Introduction of WF in LDPE results in a slight increase in the viscosity of the melt. The addition of 10% NR is not conducive to change the value of the indicator, while 20% NR plasticizes the polymer composite during the heating process, which improves its workability. When adding EPDM observed the opposite result, as in the processing of cross-linked EPDM [11], which leads to a sharp increase in the viscosity of the melt, as a consequence of the material does not flow.

The addition of 40% wood flour increases the rate of 9. With the introduction of rubber (natural and synthetic), water absorption increases due to amorphization of the polymer matrix.

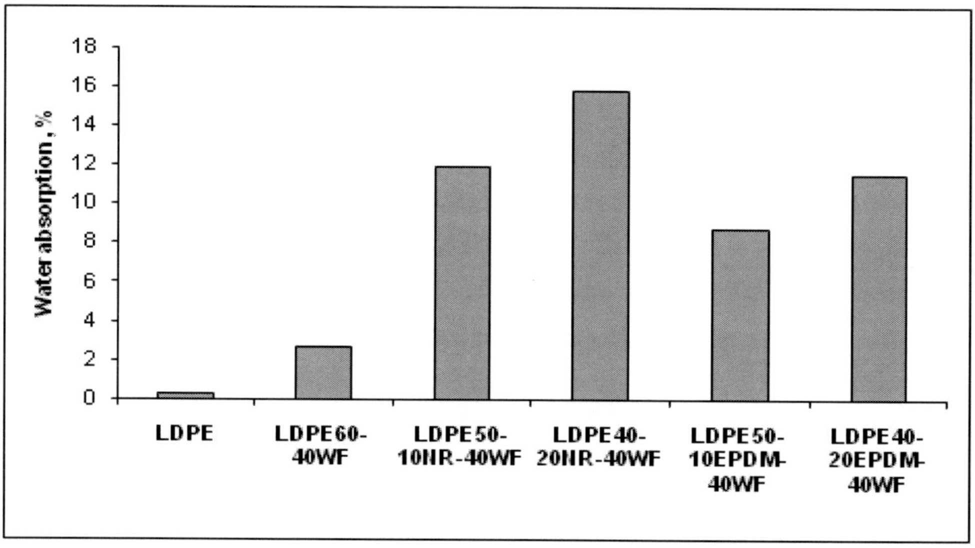

Figure 2. Water absorption in 24 hours.

Table 2. Assessment of the degree of decomposition of film samples in the soil

Samples	The mass change (%)* for 2 months	The mass change (%) * 3.5 months	The mass change (%) * 8 months
LDPE	0	0	0
LDPE 60-40 WF	+4	-6	-10
LDPE 50-10 NR-40WF	- 6	-17	-20
LDPE 40-20 NR-40WF	- 5	-24	-33
LDPE 50-10 EPDM-40 WF	+2	-15	-17
LDPE 40-20 EPDM-40 WF	+ 10	-13	-20

"+" means an increase in the mass of the sample under study, the symbol
"-" means a decrease in the mass of the studied sample.

It should be noted that during the action of the soil at the initial stage, the number of competing processes: water absorption, washing additives and degradation. The first leads to an increase in mass of the sample, the second and third - to decrease. This data only indicates which process dominates in the case of each sample, but cannot serve as a basis for predicting future impacts of soil. Weight loss of a sample of pure PE is zero, but with the introduction of wood dust exposure in 3.5 months is a loss of mass of the sample. The introduction of rubber leads to an acceleration of the process of biodegradation in the soil. It should be noted that the addition of natural rubber has a 10% weight loss increases the filling factor of 2, and at 20% - 3 times in 8 months of exposure.

CONCLUSION

As a result of these investigations, the possibility of biodegradable material creation based on LDPE and rubbers of different nature was shown. In the course of analyzing the obtained rheological parameters, it can be concluded about how the data processing of composites. For example, for samples with NR should choose extrusion and injection molding, as the main methods of processing. Whereas, for the sample with EVA suitable method pressing.

It should be noted that the introduction of LDPE and rubber wood flour increases the porosity of the polymer matrix. It causes high values of water absorption data of composite materials and facilitates the penetration of microorganisms, accelerating their growth and development in the process of biodegradation in the soil, and thus accelerates the fragmentation pattern.

REFERENCES

[1] A.A Popov, A.V. Koroleva «Biodegradable polymer composites based on polyolefin and natural polymers», *Modern Problems in Biochemical Physics:* New Horizons, 2012, pp.123-132.

[2] M.V. Bazunov, Y. Prochukhan, "Disposal of waste polymers", *Journal of Bashkir University,* 2008. T. 13. Number 4. page 880.

[3] J.K. Lukanina, A.V. Khvatov, N.N.Kolesnikova, A.A.Popov. Structure and properties of biodegradable polymer composite materials. *Progress in chemical and biochemical physics, kinetics and thermodynamics,* 2008, p. 209-218.

[4] Yu.K. Lukanina , N.N. Kolesnikova, A.V. Khvatov, A.N. Likhachev, A.A. Popov Influence of the structure of a composite material's polymeric template on the development of micromycetes. *Handbook of Chemistry, Biochemistry and Biology: Nova Science Publishers* 2010, pp. 357-363

[5] A.V. Khvatov, O.V. Shatalova, A.V. Krivandin, Yu.K. Lukanina, A.A. Popov. Structure and mechanical properties of biodegradable materials based on LDPE and natural supplements. *Deformacia I Razrushenie materialov.* 2012, № 8. 45-48.

[6] Yu.K. Lukanina, N.N. Kolesnikova, A.V. Khvatov, A.N. Likchachev, A.A. Popov. Influence of polypropylene structure during micromycete growth on their composition. *Journal of the Balkan tribological association.* 2012, Vol.18, № 1, p. 142-148.

[7] Yu.K. Lukanina, N.N. Kolesnikova, A.N. Likhachev, A.V. Khvatov, A.A. Popov. Effect of the polymer matrix structure the on the micromycetes development on mixed compositions of polyolefins and cellulose// *Plasticheskie massi,* 2010, №11, c. 56-59.

[8] "The burning, destruction and stabilization of polymers", Ed. Zaikova GE, Publisher *"Fundamentals and Technologies,"* Moscow 2008., P.73-79.

[9] Yu.S. Lipatov "Physical chemistry of filled polymers", publishing house *"Chemistry",* 1977. P159

[10] S.L. Bazhenov, A.A. Berlin, A.A.Kulkov, V.G. Oshmyan. "Polymer composites. Strength and technology", *Publishing House" Intelligence ",* Dolgoprudny 2010., Page 147.

[11] "Encyclopedia of Polymer", publishing house *"Soviet encyclopedia",* Moscow 1977., V. 3. str.1011.

In: Quantitative Chemistry, Biochemistry and Biology
Editors: G.E. Zaikov, O.V. Stoyanov et al.
ISBN: 978-1-62948-332-0

Chapter 21

FUNGICIDAL INFLUENCE OF IONS OF CU^{2+} ON MICROORGANISMS AS A WAY OF BIOCORROSION PROTECTION OF METAL CONSTRUCTIONS

Yu. K. Lukanina [1,*], N. N. Kolesnikova [1], A. V. Khvatov [1], A. N. Likhachev [2] and A. A. Popov [1]

[1]N.M.Emanuel Institute of Biochemical Physics,
Russian Academy of Science, Moscow, Russia
[2]M.V. Lomonosov Moscow State University, Moscow, Russia

ABSTRACT

In the course of metal constructions exploitation by one of determinatives of their durability are protection processes against soil corrosion. The results of microbiological investigations of the effect of increasing concentrations of Cu^{2+} ions on the inhibition of growth and Aspergillus niger, Fusarium moniliforme, Penicillium chrysogenum, Penicillium cyclopium, Trihoderma viride, Actinomycetales sp. were studied.

Keywords: Corrosion protection of metals, soil corrosion, micromicets corrosion

AIM AND BACKGROUND

Corrosion of metals – destruction owing to chemical or electrochemical environment interaction – the main scourge of all metal designs and constructions. Distinguish various kinds of corrosion of metal: by the character of destruction – continuous and local, by the form courses – chemical and electrochemical, by the course conditions – atmospheric, soil, sea, biocorrosion, contact, slot-hole, corrosion by wandering currents, an external current,

---

[*] julialkn@gmail.com.

etc., by the features of conditions of operation – corrosion of the chemical and oil equipment, pipelines, the power equipment, courts, building constructions, materials, etc [1-4].

The most destructive is soil corrosion. Annual losses of metal owing to soil corrosion reach 4 %. Different purpose pipelines, tanks, piles, poles, cables, well casing, all kinds of steel structures operating in the soil are exposed to the soil corrosion [2-6]. Soil corrosion always proceeds on the electrochemical mechanism (the only exception is very dry soils) [7]. Electrochemical corrosion arises during metal contact with surrounding electrolytically conducting medium. An original cause of electrochemical corrosion is thermodynamic instability of metals in environments surrounding them. The oxygen depolarization proceeds at the cathode. On the cathode also can pass and hydrogen depolarization (only in the conditions of acidic soils) [8].

Tremendous influence on soil corrosion of metals is rendered by microorganisms existing in soil. The soil corrosion of metal constructions caused by microorganisms, carries the name of biological (biocorrosion). Waste products of various microorganisms which are present in water, a soil, intensify corrosion process. Microorganisms action on metals can variously occur. First of all, corrosion of metals can cause aggressive microorganisms metabolites — mineral and organic acids and the bases, enzymes and others. They create the corrosion active environment in which in the water presence take place electrochemical corrosion. Colonies of microorganisms can create on the metal surface outgrowths and mycelium films under which take place corrosion as a result of electric potentials difference between metal and microorganisms [9].

The different methods are applied for protection of metal against soil corrosion. The basic methods of protection are: protective coating and insulation products, the creation of the built environment, electrochemical protection, and the use of special methods of installation [10-11].

In the practice of protecting metals from soil corrosion is often applied cathodes protection [12-13]. This significantly reduces (minimizes) the rate of soil corrosion. To protect drilling platforms, welded metal bases, underground pipes to be connected as a cathode to an external power source. The anode used auxiliary inert electrodes. The negative pole of the direct current is connected to the pipe, and the positive - with the anode. Thus, for cathode protection is a vicious loop current, which flows from the positive pole of the power supply cable to the isolated ground to the anode. The anode grounding current spreads along the ground and into the protected piping, then flows through the pipeline, and the pipeline for insulated cables back to the negative terminal of the power supply. Electrical current coming out of the anode in the form of positive ions of the metal, so the anode is gradually destroyed, and the pipeline is protected from destruction.

In this paper we conducted comprehensive research the possibility of using copper salt solution to protect against biological corrosion under the action of fungi and actinomycetes. These results will allow us to create an adjustable directional protection structures from biocorrosion.

EXPERIMENTAL PART

For the definition of a range of the concentration of ions of the copper suppressing growth of microorganisms four solutions of copper (II) sulfate with following normal concentration of ions of copper have been prepared:

Four copper (II) sulfate solution with the following concentrations of copper ions to determine the range of concentrations of copper ions that inhibit the growth of microorganisms were prepared:

1) 10^{-3} H: 5,78 pH;
2) 10^{-2} H: 5,14 pH;
3) 10^{-1} H: 4,33 pH;
4) 1 H: 3,6 pH.

Mycological analysis of the ions concentration effect was performed on the following test cultures in the complex soil microbiota:

Micromycetes: Aspergillus niger, Fusarium moniliforme, Penicillium chrysogenum, Penicillium cyclopium, Trihoderma viride.
Actinomycetes: Actinomycetales sp.

The considered complex of microorganisms is dominating in agrobiocenosis and causes biological corrosion of metal designs. Inoculation suspensions test cultures were carried out in Petri dishes on medium wort agar at 25°C. Each Petri dish in a culture medium were cut 4 holes, and places the products with different concentrations of copper ions. Zone suppression agent of growth and development of the fungus was measured as the distance from the center of the hole with drug to fungal spawn Figure 1.

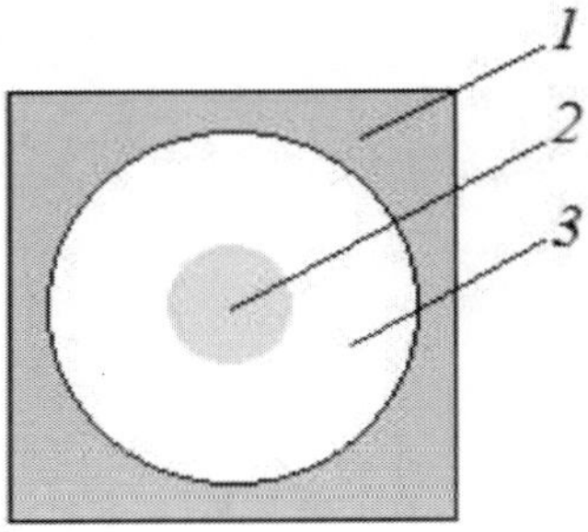

Figure 1. Suppression zone of fungi growth, where 1 - fungi spawn, 2 - hole with the drug, 3 - zone suppression.

RESULTS AND DISCUSSIONS

The table shows the values of the suppression the zones radii of microbial growth under the influence of drugs with different concentrations of copper ions.

Table 1. Radius of a suppression zone of microorganisms (from the c hole center to the fungi spawn on day 5 exposure, ± 2mm)

Concentration Cu^{2+} / Microorganisms	10^{-3} N	10^{-2} N	10^{-1} N	1 N
Aspergillus niger	0	0	5	25
Fusarium moniliforme	0	0	0	30
Penicillium chrysogenum	0	0	0	20
Penicillium cyclopium	0	0	7	20
Trihoderma viride	0	0	5	20
Actinomycetales sp.	10	15	20	35

On the first and second day under the influence of drugs with different concentrations has slowed growth of spores and poor development of mycelium. On day 5 preparations with a concentration of 10^{-3}, 10^{-2} and 10^{-1} H exhibit fungistatic properties when exposed to all strains of micromycetes (i.e. delay and stop the growth of the fungus at the initial stage of exposure, followed by the resumption of growth), and a preparation with the concentration of 1H - fungicides (i.e., exposure leads to the death of the fungus).

The most sensitive microorganisms presented to the copper ions were Actinomycetales sp., growth inhibition is observed even at a concentration of 10^{-3} N (Figure 2-3). As the concentration of copper ions change the character of the inhibition is a power function ($Y = 33 \bullet X^{0,18}$).

At 15 and 20 days of incubation changes the pigmentation (color) fungi Penicillium chrysogenum and Penicillium cyclopium, while the inhibition observed on Day 5, do not cancel for any fungus.

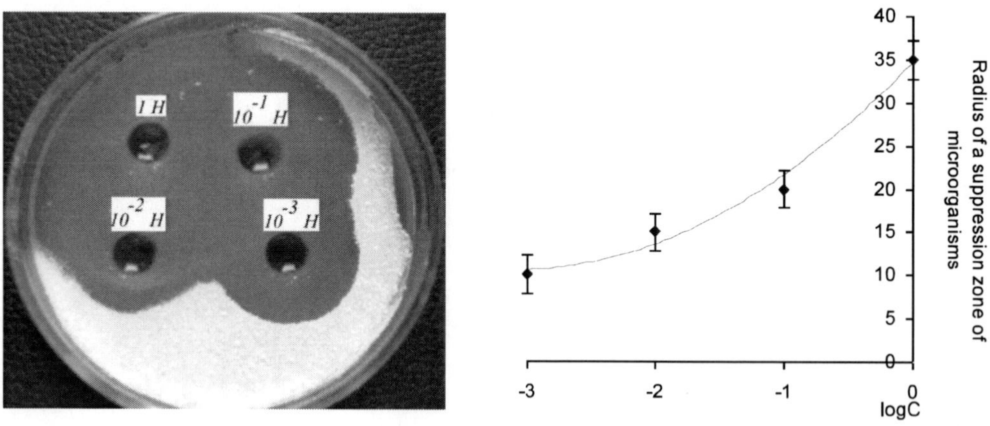

(a) (b)

Figure 2. Suppression zones of growth Actinomycetales sp. Depending on concentration of ions of copper, a) photography, b) graph.

Its widely known that the development process of mycological corrosion is as follows: the first stage - education mikrocolony, their growth, the emergence of metabolic products and local accumulation of electrolytes with excess hydrogen ions, the second stage - the emergence of electrochemical cells on the surface of the metal, and the third stage - the emergence of the cathode (anode) depolarization the capture of protons (electrons) fungi and their metabolic products.

Inertia fungi to low concentrations of metal ions in the first day due to the occurrence of the metal ions sorption by the cell wall of the fungus, which is protective [14]. Getting inside the cells, they inhibit the respiratory chain enzymes, and splits oxidation and oxidative phosphorylation, resulting in the cell dies. With increasing concentration of copper ions in the preparation process is intensification of the death of the fungus. It should be noted that the relationship is nonlinear and is specific to different organisms. During the experiment, such fungi as Penicillium chrysogenum, Penicillium cyclopium, Trihoderma viride demonstrate high adapt capacity to the adverse conditions (drug concentration was 10^{-1}N) as the formation of stable colonies. At low concentrations of the drug it is also possible for micromicetes to use metal ions as a source of minerals, energy or electron acceptors.

CONCLUSION

As a result of these studies, the uses of solutions of copper salts as an effective method of protection against biological corrosion of metal structures of different shapes and metals have been found. The adapt capacity of some micromycetes in presence of metal ions in small concentrations was observed. Optimal parameters for the electrochemical method of protection against biological corrosion of metal objects, caused by fungi and actinomycetes are the concentration of copper ions in the solution is not less than 0.1 N.

REFERENCES

[1] Semenova I.V., Florianovich G.M., Khoroshilov A.V. *Corrosion and Corrosion Protection* // M. Fismatlit, 2002, 335 p..

[2] Vigdorovich V.I., Shel N.V., Krylova A.G. Features of the atmospheric metals corrosion // *Bulletin of the Tambov University.* Series: Natural and Technical Sciences. 2001,T. 6, № 3, Pp. 279-289.

[3] Semenov S.A., Gumargalieva K.Z., Zaikov G.E. Characteristics of the processes and characteristics of damage of materials engineering by microorganisms in operation // *Bulletin of MITHT.* 2008, T. 3, № 2, Pp. 1-21.

[4] Alehova T.A., Alexandrova A.V., Zagustina N.A., Novozhilova T.Y., Romanov S.Y. Mikrosokopicheskie fungi on the Russian segment of the International Space Station / / *Mycology and phytopathology.* 2009, T. 43, № 5, P. 377.

[5] Mikhailov A.A., Strekalov P.V. Modeling of atmospheric corrosion of metals and types of dose-response functions // *Corrosion: Materials, protection.* 2006, № 3, Pp. 2-13.

[6] Mikhailov A.A., Strekalov P.V. Modeling of atmospheric corrosion of metals and types of dose-response functions (continued) / / *Corrosion: Materials, protection.* 2006, № 4, Pp. 2-10.

[7] Karpov V.A., Kovalchuk Y.L., Kharchenko U.V., Beleneva I.A. Microaccretions impact on marine corrosion of metals and destruction of protective coatings // *Corrosion: Materials, protection.* 2011, № 3, Pp. 11-18.

[8] Andreuk E.I., Bilai V.I., Koval E.Z., Kozlova I.A. Microbial corrosion and its agents. Kiev: *Nauk. Dumka*, 1980, 286 p.

[9] Zaikina N.A., Duganova N.V. The formation of organic acids, fungi isolated from samples of affected biological corrosion. // *Mycology and phytopathology.* V.9, № 4, 1975, p.303.

[10] Glazkov V.I., Zinevich A. M., Kotik V.G., et al *Corrosion long metal structures.* M.: Nedra, 1969.

[11] *Protection against corrosion, aging, biological damage of machinery, equipment and structures: a guide: in 2 volumes* / ed. A.A. Gerasimenko. Moscow, 1987.

[12] Glazov N.P. Particular corrosion protection of steel underground piping // *Physical chemistry of surface protection materials.* 2004, T. 40, № 5, Pp. 522-528.

[13] Sulimina E.Y. Investigation of pulsed current cathode protection on the technology of small-diameter pipeline. // *Oil and Gas business.* 2011, № 12, Pp. 52-58.

[14] Lebedeva E.V., Nazarenko A.V., Kozlova I.V., Tomilin B.A. Effect of increasing concentrations of copper on soil fungi are. // *Mycology and phytopathology.* 1999, № 33, 257-263.

In: Quantitative Chemistry, Biochemistry and Biology
Editors: G.E. Zaikov, O.V. Stoyanov et al.
ISBN: 978-1-62948-332-0

Chapter 22

INTERFACIAL RHEOLOGY OF HEN EGG WHITE LYSOZYME-5-METHYLRESORCINOL MIXTURES AND THEIR FOAMING PROPERTIES

E. I. Martirosova*[*] *and I. G. Plashchina
Emanuel Institute of Biochemical Physics RAS, Moscow, Russia

ABSTRACT

The capacity of methylresorcinol (MR) to affect the hen egg white lysozyme (LYS) surface activity and viscoelastisity of its adsorption layers at air/solution interface is compared with the effect of MR on the foam formation ability and foam stability. It was shown that the MR introduction increases multiplicity and stability of foams, and the concentration of surfactant had an important role.

Keywords: Lysozyme, methylresorcinol, complex dilatation modulus, phase angle, foam formation ability, foam stability

INTRODUCTION

Mixtures of proteins and surfactants are often used in many technological applications, including food and pharmaceutical industries, cosmetics, coating processes, and so on. In many of these applications protein–surfactant mixtures are used in the manufacture of the various dispersions. These dispersions contain two or more immiscible phases (aqueous, oil and/or gas phases) in the form of foams and emulsions. Dispersions are inherently unstable systems because of their large interfacial area (Maldonado-Valderrama, Patino, 2010).

The stability of these systems is generally achieved through a protective interfacial layer around the particles (emulsion droplets or foam bubbles). The properties of this interfacial

[*] E-mail: ms_martins@mail.ru.

layer are governed by the composition and structure of the adsorbed material and in turn would determine the properties of the dispersion (Dickinson, 1996).

A wide use of lysozyme in medicine for the treatment of various infections and in the food and cosmetic industry to prevent bacterial contamination of the stuffs leads to production of the resistance of microorganisms to lysozyme action. In recent years, a great deal of attention has been devoted to the investigations aimed at studying lysozyme modifications, affecting lysozyme properties while retaining its enzymatic activity. One of the ways to solve this problem is modification of lysozyme structure with using weak nonspecific interaction with MR. It is known numeral effects of MR on the structure and functions of lysozyme. In particularity, in concentration range of 10^{-7}–10^{-3}M MR the specific and nonspecific enzymatic activities raise, its substratum specificity extends and temperature range of lysozyme catalysis expands (Petrovskii et al., 2009).

Recently we have found out that MR due to diphylic character of its structure is capable to self-organization in a solution and can behave itself similarly to nonionic surfactants (Martirosova, Plashchina, 2012). MR changes the adsorption behavior of LYS in their mixed solutions. MR was shown to change thermodynamic affinity of lysozyme to the solvent with the resulting the protein being more surface active and reological properties of its adsorption layers at the air/solution interface are higher in the wide range of MR concentrations (Martirosova et al., 2012). For this reason, we suppose, that using of lysozyme in mixture with MR can be favorably for production of antibacterial pharmacological products of foam type with enhanced physical stability and period of using while increasing its enzymatic activity and wider spectrum of action.

The aim of this paper is to investigate the effects of MR on the air/water interfacial properties (interfacial dilatational complex modulus and interfacial pressure) of lysozyme and comparing the effect on its foam formation ability (volume, multiplicity, stability of foams).

MATERIALS AND METHODS

A sample of hen egg white lysozyme (Sigma-Aldrich, USA) with activity 20,000 U/mg and molecular mass 14,445 Da was used.

Alkyl-substituted hydroxybenzenes, 5-methylresorcinol (5-Methylbenzene-1,3-diol) (Sigma-Aldrich, USA) with molecular mass 124,14 g/mol (anhydr) was taken.

Tetradecane (Sigma-Aldrich, USA) with molecular mass 198,39 g/mol was used.

All reagents for 0.05M phosphate buffer (pH 6.0) preparation in Milli-Q water were of analytical grade.

Solution Preparation

Lysozyme protein powder was dissolved in buffer for 2 hour and were centrifuged at 20 000g (Beckman 21, Germany) during 1 hour at room temperature.

The lysozyme concentration was determined by measuring the absorbance at 280 nm and using a specific extinction value 26.4 for $A_{1cm}^{1\%}$ (Page et al., 1995). The operating solutions were prepared by mixing equal volumes of protein and methylresorcinol solutions within 40-50 min just before using in experiment.

FOAM FORMATION

The initial feed solution volume was 4 mL. The protein solution with constant concentration of 1.0 wt % and MR solutions with varying (0-1.0 wt %) concentrations were used. The comparatively high lysozyme concentration was necessary to obtain fast and sufficient amounts of foam in experiment.

Foam formation was performed by using of homogenizer Heidolph DIAX 900 (Germany). Air was introduced through a porous of homogenizer under constant air flow rate of 23,000 r/min, within 1.5 min. Experiments were conducted at the temperature 25°C. The responses selected for evaluating the experimental plan were the foam volume and multiplicity. Evaluation was carried out in the reaction vessel after the completion of foam formation. Foam volume was set by the geometric measurement. Foam multiplicity was calculated using the formula:

$$L = Vn / Vg,$$

where L is the foam multiplicity; Vn is the height of foam column and Vg is the height of the initial solution column.

Foam stability was evaluated using the parameter of half-life. Determined by visual observation is the foam half-life, i.e. the time for the foam height to reduce to 50% of its initial value. Time of the foam destruction was fixed by timer.

The triple preparations were made for each study in all experiments.

Surface Dilatational Properties Measurement

The surface rheological parameters of adsorbed LYS -MR films at the air-water interface were measured with using of an automatic drop Tracker tensiometer (ITC Concept, France). Adsorbed films were formed over 60 000 s to guarantee steady-state. Sample solutions included a lysozyme (0.075 mg/mL) and varying 5-methylresorcinol (0.16-67.2 mM) concentrations in 0.05M phosphate buffer, pH 6.0 at 25°C.

The surface dilatational modulus (E) and the phase angle (θ) were measured as a function of time with constant amplitude ($\Delta A/A$) of 3% and angular frequency (ω). The range of angular frequencies used was 0.007-0.625 rad/s. The sinusoidal oscillation for surface dilatational measurement was made with 3-5 oscillation cycles followed by a time of 3-5 cycles without any oscillation. The average standard accuracy of the surface pressure is roughly 0.1 mN/m. The reproducibility of the results (for at least two measurements) was better than 0.5%.

RESULTS AND DISCUSSION

Proteins are often used as foaming agents because of their ability to unfold at the interface, thus creating films with high surface elasticity and also steric resistance against coalescence of films. Although they decrease the interfacial tension and hence reduce the driving force for disproportionation, often quite high protein concentrations are required for the formation of stable foams. For this reason, mixing with low-molecular weight surfactants can be very efficient and can increase the quality of the foam (Alahverdjieva et al., 2008).

Destabilization of foams is a dynamic process that includes disproportionation, coalescence in addition to drainage of the thin film between bubbles. In addition to the bulk phase viscosity, all of these processes involve interfacial film properties (Bos, van Vliet, 2001; Foegeding et al., 2006; Murray, 2007; Sanchez, Nino, 2008; Wilde, 2000). The greater stabilizing effect may be attributed to a greater enhancement of the local viscosity in a foam lamellae which tends to inhibit film drainage, as well as to increased thickness of the mixed adsorbed layer which tends to enhance steric stabilization and inhibit bubble coalescence (Dickinson et al., 1996)

Early the dynamic surface tension measurements performed in phosphate buffer solutions (pH 6.0, I = 0.05 M) at constant lysozyme concentration, 5.1×10^{-6} M, and varying concentrations of MR, have shown that the LYS-MR mixture is characterized by more high rate of adsorption and more low quasi-equilibrium surface tension, than pure LYS. MR critical micelle concentration in lysozyme presence is 32.7 mM (Martirosova et al., 2012).

The effect of MR on the rheological properties of lysozyme adsorption layers at the air/solution (0.05 M phosphate buffer, pH 6.0) interface at 25°C was studied using a dynamic drop tensiometry and dilatometry methods. The quasi-equilibrium adsorption layers (the formation time of 60 000 - 70 000 sec) were subjected to compression/dilatation deformation sinusoidally in the field of linear viscoelasticity.

The dependences of complex viscoelastic modulus (**E**) of mixed adsorption layers , as well as its elastic (real part, $\mathbf{E_{rp}}$) and viscous (imaginary part, $\mathbf{E_{ip}}$) components as functions of MR concentration at the smallest 0.007 rad/s and the biggest 0.62 rad/s frequency values of the applied deformation are presented in Figure 1 and 2. In Figure 3 the dependences of phase angle on MR concentration at the same conditions are shown.

The complex modulus of elasticity (**E**) and its elastic component $\mathbf{E_{rp}}$ increases with growth of MR concentration from 0.16 to 33 mM (CMC_{MR}). The viscosity component ($\mathbf{E_{ip}}$) and phase angle increase with MR concentration up to 3.21 mM and then decreases up to 33 mM. The rheological behavior of the adsorption layers as well of intact lysozyme as modified ones is solid-like ($\mathbf{E=E_{rp}}$, $\mathbf{E_{rp}} >> \mathbf{E_{ip}}$, where $\mathbf{E_{ip}}$ is elastic component, and low frequency dependence of complex modulus **E**).

As seen, MR can both increase and decrease the viscoelastic parameters of LYS adsorption layers depending on concentration. The effect is more expressed at low frequencies of deformation. Therefore, changing the MR concentration in LYS solution in a wide range allows us to modify interfacial properties and the type of its rheological behavior.

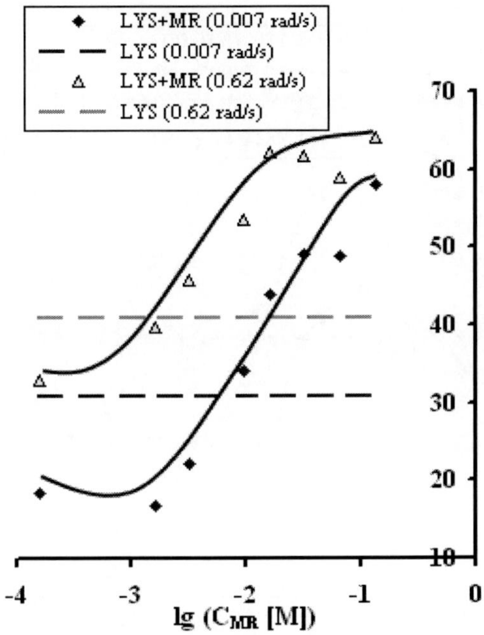

Figure 1. Dependence of Lysozyme complex elasticity modulus and it real part on MR concentration at frequency 0.007 and 0.62 rad/s. The drop lines indicate the interval of change corresponding parameters for pure LYS.

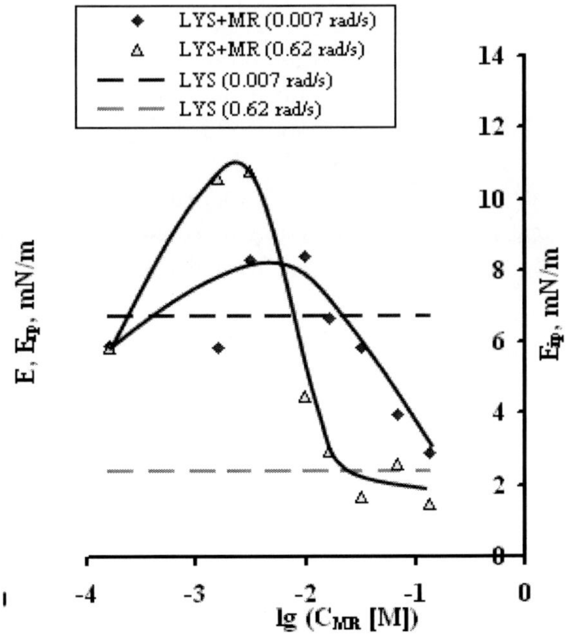

Figure 2. Dependence of Lysozyme imagine part of complex elasticity modulus on MR concentration at frequency 0.007 and 0.62 rad/s. The drop lines indicate the interval of change corresponding parameters for pure LYS.

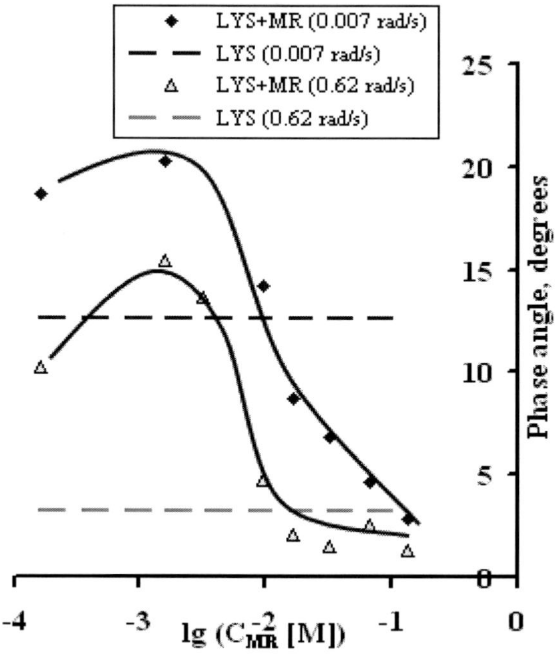

Figure 3. Dependence of Lysozyme phase angle on MR concentration at frequency 0.007 and 0.62 rad/s. The drop lines indicate the interval of change corresponding parameters for pure LYS.

Based on our results, and taking into account that there are positive associations established between the foam drainage half-life and the interfacial properties (the interfacial pressure and the interfacial elasticity) (Xin Yang, 2010), we have chosen the MR concentration for testing the foam formation properties of Lys in mixed solutions: lower and higher CMC_{MR} 32.7 mM (Tabl. 1).

At first the experimental studies the amount of foam and its multiplicity_was evaluated.

Thus, it was found that the foaming of lysozyme solution in the 30 and 100-fold molar excess of MR increases the foam total amount in 2.2 and 1.8 times, respectively, and the foam multiplicity increases in 1.8-2.3 times (tabl. 1).

In the second stage of studies the foams stability was measured, which is proportional to the rate of destruction.

Table 1. The results of volume and multiplicity foams produced with lysozyme, depending on the concentration of MR

MR concentration, mM	Foam volume, mL	**Multiplicity**
0	4.0	0.9
19.5	8.8	2.2
65.0	7.2	1.8

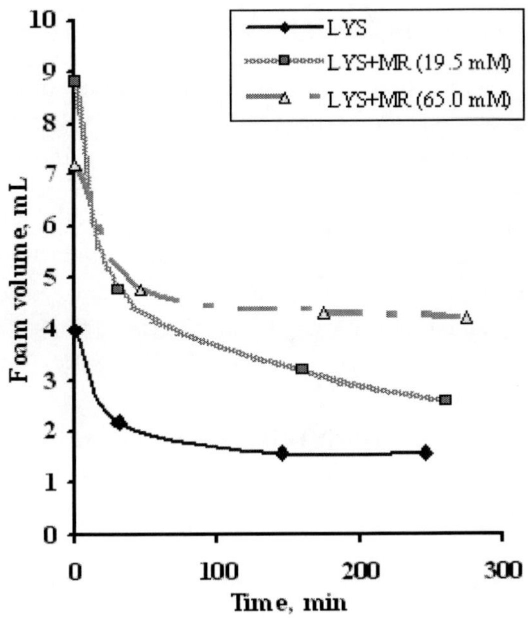

Figure 4. Dynamic curves of foams destruction.

Figure 4 shows the dynamics of the foam destruction. It implies that the half-life is the same, 45-50 minutes, for control and the sample containing 19.5 mM MR. Half-life of sample containing 65.0 mM MR is more than 4.5 hours (data not shown). The foam trend destruction of the lysozyme-MR mixture (19.5 mM) has the character, unusual for the two other samples. This is due to the transition state MR from the molecular to associative form in this concentration. It was shown that MR alone has a critical micelle concentration equal 16.6 mM and in mixture with LYS – 32.7 mM (Martirosova, Plashchina, 2012).

There, these studies have shown that the MR introduction increases the multiplicity and stability of foams. The concentration of the surfactant has a crucial importance.

It should to be noted that for all samples the main part of the liquid consumed into the foam formation follows back during the first half-hour of storage. Thus, the amount of remaining liquid in the foam was in 1.5-2.0 times higher in MR-containing samples than in the control. Also in these samples, the sizes of foam bubbles are much less, and the density of packing is higher compared with the control (data are not shown)

Foams stability can play an important role in the liquid viscosity between the films. This is due to the fact that the rate of the liquid drainage is slowed with the increasing viscosity. However, at the present time it is generally accepted that the surface viscosity can not be the determining factor for the foams stability, although the correlation between these properties exists.

CONCLUSION

The main factors which determine the foam formation ability and physical stability of the foams in mixed Lys-MR solutions are surface activity and complex dilatation modulus. With the increase of MR concentration in certain ranges, a surface activity and complex dilatation modulus of interfacial layers increased and the phase angle decreased; this means that the viscoelasticity of interfacial layers became higher. Under these conditions, the foam volume and multiplicity as well as the stability of foams were growing. This effect may be used for the creation of a mixed protein-MR system foam type for pharmaceutical applications with an improved physical stability and a wide range of antibacterial actions.

REFERENCES

Alahverdjieva V.S., Khristov Khr., Exerowab D., Miller R. Correlation between adsorption isotherms, thin liquid films and foam properties of protein/surfactant mixtures: Lysozyme/C_{10}DMPO and lysozyme/SDS Colloids and Surfaces A: Physicochem. *Eng. Aspects.* (2008) 323, 132–138.

Bos, M.A., & van Vliet, T. Interfacial rheological properties of adsorbed protein layers and surfactants: a review. *Advances in Colloid and Interface Science,* (2001) 91, 437–471.

Dickinson E., Izgi E. Foam stabilization by protein-polysaccharide complexes. *Colloid and Surfaces.* (1996) 113, 191-201.

Foegeding, E.A., Luck, P. J., Davis, J.P. Factors determining the physical properties of protein foams. *Food Hydrocolloids,* (2006) 20, 284–292.

Maldonado-Valderrama J., Patino J.M.R. Interfacial rheology of protein–surfactant mixtures. *Current Opinion in Colloid & Interface Science,* (2010) 15, 271–282.

Martirosova E.I., Plashchina I.G. Adsorption behavior of methylresorcinol and its mixtures with lysozyme at air-water interface. In Biochemistry and Biotechnology. *Nova Science Publishers.* N.-Y. (2012) 95-104.

Murray, B.S. Stabilization of bubbles and foams. *Current Opinion in Colloid and Interface Science,* (2007) 12, 232–241.

Page C.N., Vajdos F., Fee L., Grimsley G., Gray Th. How to measure and predict the molar absorption coefficient of a protein. *Prot. Sci. (*1995) 4**,** 2411-2423.

Petrovskii A.S., Deryabin D.G., Loiko N.G., Mikhailenko N.A., Kobzeva T.G., Kanaev P.A., Nikolaev Yu.A., Krupyanskii Yu.F., Kozlova A.N., El'-Registan G.I. Regulation of the functional activity of lysozyme by alkylhydroxybenzenes. *Microbiology.* (2009). 78 (2), 146-155.

Rodrıguez Patino, J.M.R., Sanchez, C.C., Rodrıguez Nino, M.R. Implications of interfacial characteristics of food foaming agents in foam formulations. *Advances in Colloid and Interface Science,* (2008) 140, 95–113.

Wilde, P.J. Interfaces: their role in foam and emulsion behavior. *Current Opinion in Colloid and Interfaces Science,* (2000) 5, 176–181.

Xin Yang, E. Allen Foegeding. Effects of sucrose on egg white protein and whey protein isolate foams: Factors determining properties of wet and dry foams (cakes) *Food Hydrocolloids* (2010) 24, 227–238.

In: Quantitative Chemistry, Biochemistry and Biology
Editors: G.E. Zaikov, O.V. Stoyanov et al.
ISBN: 978-1-62948-332-0

Chapter 23

STRUCTURE AND THERMOPHYSICAL PROPERTIES OF BLENDS BASED ON ISOTACTIC POLYPROPYLENE AND LOW-DENSITY POLYETHYLENE

E. E. Mastalygina[1,2*], S. G. Karpova[1], N. N. Kolesnikova[1], U. K. Lukanina[1] and A. A. Popov[1,2]

[1]Department of Biological and Chemical Physics of Polymers, Emanuel Institute of Biochemical Physics of Russian Academy of Sciences, Moscow, Russia

[2]Department of Chemistry and Physics, Plekhanov Russian University of Economics, Moscow, Russia

ABSTRACT

In this paper, the structure and thermophysical properties of isotactic polypropylene (PP) and low-density polyethylene (LDPE) blends were studied by differential scanning calorimetry (DSC) and electron paramagnetic resonance (EPR). The experiment shows non-additive changes of the thermal behavior and segmental mobility in the amorphous regions of immiscible PP/LDPE blends in relation to the properties of PP and LDPE initial homopolymers. Based on the results of the study, the probable process of the structure formation for PP/LDPE blends of various compositions was described.

Keywords: Blend, isotactic polypropylene, low-density polyethylene, crystallization, segmental mobility.

AIM AND BACKGROUND

Despite the fact that numerous publications are devoted to the subject of polypropylene-polyethylene blends, many questions still remain unclear. Moreover, the results of several

* E-mail: elena.mastalygina@gmail.com.

studies published in the literature contradict one another. A number of researchers consider that the blends of PP and PE are absolutely immiscible, so PP and PE form independent phases in the blend. Consequently, the structure and properties of the PP/PE blends depend on the composition additively [1-4]. Other authors suppose that the PE component influences the behavior and morphology of PP-PE blends, while the PP component has no impact on the properties of the blend [5-7]. Another point of view is that PP and PE have a mutual influence on the crystallization process and the formation of the phase structure and, consequently, the properties of the blend [8,9].

For example, Martuscelli et al. revealed three different types of crystallization behavior of PP/HDPE blends at three ranges of crystallization temperatures [1]. Below 125°C, simultaneous crystallization of PP and HDPE was observed. From 125°C to 127°C, two crystallization peaks were observed with PP crystallizing first. Above 127°C, HDPE does not crystallize, and the crystallization of PP was observed only from a melt mixture of PP/HDPE. Wenig & Meyer [2] and Bartczak et al. [3] have determined that the presence of PE did not influence the PP spherulite growth rate or the process of crystallization. In contrast, Ramsteiner et al. [5] found that this growth rate was influenced by the presence of PE. Rybnikar [4] established that the morphology of PP is only temperature-dependent; it remains unaffected by the presence of another polymer. Other authors reported that the shape, size and orientation of LDPE occlusions remain undisturbed during the crystallization of PP spherulites. On the other hand, LDPE occlusions introduce major changes into the internal structure of PP spherulites [6]. Similar results were obtained for the PP/LDPE fibers in the study [7]. According to the authors, the LDPE component has a greater effect on the total crystallinity of the PP/LDPE blends than the PP component. Another standpoint was confirmed by the studies [8, 9]. It was found that LDPE acts as a structure-modifying agent, while PP plays the role of a strengthening (reinforcing) filler.

The aim of this work was to study the effect of the composition of PP/LDPE blends on the phase structure, in particular the quality and quantity of crystallites and amorphous regions of both components in the blends. The thermophysical and dynamic parameters of the studied systems were obtained using DSC and EPR-spectroscopy.

INTRODUCTION

Creation of new materials by blending polymers implies controlled modification of the component structures. Materials with a set of new features can be created owing to the mechanical mixing of different polymers. In addition, the production of new polymer blends does not require changes in the process of polymers production [10]. Polyolefins is the biggest group of commercial polymers; hence, they are the main component of plastic waste [11]. The blends based on the most popular sorts of polyolefins – low density polyethylene (LDPE) and isotatic polypropylene (PP) – are interesting from the practical and ecological points of view.

It is important to note that certain properties of these blends are better than those of the parent homopolymers. Adding PE to PP allows improvement of the low temperature impact behavior of PP [12]. In turn, the PP content contributes to the increase of toughness, high modulus and heat resistance to PE [13]. The PP/LDPE blends can be used as engineering plastics, consumer items for household use, goods with improved performance, etc.

Moreover, PP/LDPE blends are the products of plastic recycling. Since separation of PP and PE in solid domestic waste (SDW) is very expensive, many recycling companies sell mixtures of reworked PE and PP ("polymer mix") [13]. Thus, the study of the structure of blends of these materials will increase the field of application of recycled plastics. This is the reason for why PP-LDPE blends are the main subject considered in this paper.

EXPERIMENTAL PART

Isotactic polypropylene (PP) ($M_w = 3.5\times10^5$, $M_n = 8\times10^4$, $M_w/M_n = 4.4$) and low-density polyethylene (LDPE) ($M_w = 2\times10^5$, $M_n = 4\times10^4$, $M_w/M_n = 5$) were the objects under investigation. The PP contents in the blends are 0, 5, 10, 20, 30, 40, 50, 60, 70, 80, 90, 95, 100 wt%.

The blend preparation was produced by mechanical mixing of PP and LDPE in a laboratory mixer at a temperature of 190°C for 5 min in the nitrogen atmosphere to prevent oxidation of LDPE. Isotropic films were fabricated by pressing with cellophane substrate at a temperature of 190°C and pressure of 7.8 MPa, followed by quenching in water (25°C). The thickness of the PP/LDPE films was 130±10 μm.

The degree of crystallinity χ and the melting temperature T_m of crystallites of each component in the blend were determined by DSC measurements performed on a differential scanning microcalorimeter DSM-10mA operated at a temperature scan rate of 8°C/min [14]. The temperature scale was gauged by Indium (T_m = 156.6°C, ΔH = 28.44 J/g).

The average melting temperatures T_m were measured by the maximums of the melting peaks in DSC thermograms. The melting heats ΔH of samples were calculated from the melting peak areas limited by a baseline. The degree of crystallinity χ_{PP} and χ_{LDPE} was calculated using the following formula:

$$\chi = \frac{\Delta H_i}{\Delta H_0} \times 100 \tag{1}$$

where ΔH_i – melting heat calculated relatively to the i-polymer content (PP or LDPE) in the blend, ΔH_o(PP) = 147 J/g – melting heat of a completely crystalline PP [15], ΔH_o(LDPE) = 295 J/g - melting heat of a completely crystalline LDPE [16].

The amorphous regions were studied using EPR spectroscopy, which allows the estimation of the molecular mobility in disordered regions [17]. A stable nitroxyl radical, 2,2,6,6-tetramethylpiperidine-1-oxyl (TEMPO) was used as a paramagnetic probe. The radical was introduced into the films from vapor at a temperature of 30°C. The rotational mobility of the radical probe was determined by the correlation time τ_c. The following formula is used to calculate τ_c:

$$\tau_c = 6{,}65 \times 10^{-10}\left(\sqrt{\frac{I_+}{I_-}} - 1\right) \times \Delta H \tag{2}$$

where I_+ and I_- are intensities of the first and third peaks in the EPR spectrum; ΔH – half-width of the first peak in the EPR spectrum [17].

The dependence of τ_c on the blend composition is adjusted for isotropic and oriented samples. The orientation stretching was produced by local heating at 70ºC (PP/LDPE = 0:100–50:50) and at 90°C (PP/LDPE = 50:50–100:0). The oriented films with a draw ratio λ = 4 were used.

Results and Discussion

The PP and LDPE are thermodynamically incompatible polymers forming crystalline lattices of different types. The structures of the two polymers in a blend are formed separately because of their different melting temperatures. PE crystallizes into a more stable orthorhombic crystalline lattice, while PP – into a monoclinic crystalline lattice [18]. Our task was to study the crystallization process of both polymers taking into account the influence of either component on the structure formation of the blend.

Figure 1 shows the dependence of the degree of crystallinity χ of each component of the blend on the PP/LDPE blend composition. The degree of crystallinity of LDPE changes slightly (χ_{LDPE} = 19.4±1.5%) across the full range of the compositions. An increase in the PP content causes a small decrease of χ_{LDPE}. Thus, the presence of PP does not have a significant effect on the crystallization process of LDPE.

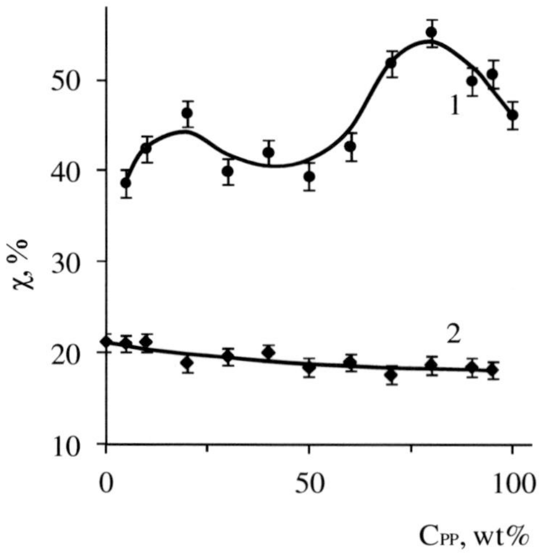

Figure 1. Effects of the PP-LDPE blend composition on the degree of crystallinity χ of PP (1) and LDPE (2).

The degree of crystallinity of PP is changed more significantly, depending on the composition of the blend. Higher values of χ_{PP} are observed for the blends with the LDPE content from 10 to 30 %, with the maximum of χ_{PP} equal to 55.3% (PP/LDPE = 80:20). Apparently, since LDPE has a lower solidification temperature as compared to PP, the process of crystallization of the PP phase takes place in the presence of molten LDPE. A small amount of molten LDPE in the blend acts as a plasticizer, so the process of crystallization of PP proceeds more perfectly. The increase of the LDPE content by more than 30 wt% prevents the process of crystallization of PP (the volume fraction of the crystalline phase of PP decreases). Blends with the LDPE contents from 30 to 60 wt% and below 5 wt% have a lower value of χ_{PP} (χ_{PP} = 41±1.5%). It seems that the phase inversion occurs in the range of compositions containing PP from 30 to 60 wt%.

As the result, both components of the blend have a mutual influence on the crystallization process and the formation of the phase structure and, consequently, the properties of the blend. However, the LDPE component has a greater effect on the process of the structure formation of PP/LDPE blends than the PP component.

Table 1. The values of melting temperatures of PP and LDPE for PP/LDPE blends as a function of weight percent of PP

C_{PP}, wt%	0	5	10	20	30	40	50	60	70	80	90	95	100
T_m (PP), °C		159	163	162	162	162	162	162	162	163	162	162	164
T_m (LDPE), °C	106	106	106	106	106	106	106	106	106	106	106	104	

The measusrement error is ± 5%

The melting temperatures T_m(PP) and T_m(LDPE) for the blends of different compositions vary within the narrow limits of 160.8±3.1°C and 104.1±0.8°C, respectively (Table 1). It means that the lattice perfection of PP and LDPE is practically constant for the entire range of compositions. The minimum values of T_m(PP) and T_m(LDPE) are observed at the lower content of the corresponding polymer in the blend. Thus, the addition of one polymer to the other causes an insignificant reduction in the value of T_m as compared with the corresponding value for the homopolymer.

Figure 2 presents the dependence of the correlation time τ_c for isotropic and oriented (λ=4) samples on the blend compositions. The value of τ_c characterizes the molecular mobility of macrochains in an amorphous phase.

The value of τ_c increases with the increase in the PP content, i.e. the mobility of macrochains in the system decreases. For isotropic samples, there is a small increase of the τ_c value (from 2.9×10^{-10} s to 5.6×10^{-10} s) up to 60–70 wt% of PP in the blend and then the value of τ_c increases sharply with the maximum value (14.7×10^{-10} s) for PP. For oriented samples, there is a far less dramatic increase in the value of τ_c in passing from LDPE to PP (τ_c increases from 12.3×10^{-10} s to 18×10^{-10} s).

Comparison of the values of τ_c for oriented and isotropic samples of the blends of the same composition indicates that a higher content of LDPE in the blend is associated with a greater difference between the values of τ_c for the isotropic and oriented samples. Thus, the τ_c value for an isotropic sample in LDPE is 4 times greater than that for an oriented sample, while in PP, the difference between the values of τ_c is approximately 3×10^{-10} sec.

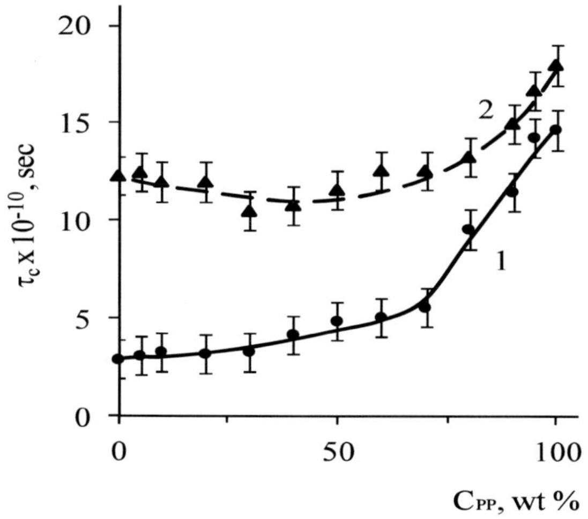

Figure 2. Effect of the PP-LDPE blend composition on the correlation time τ_c for isotropic (1) and oriented (2) samples.

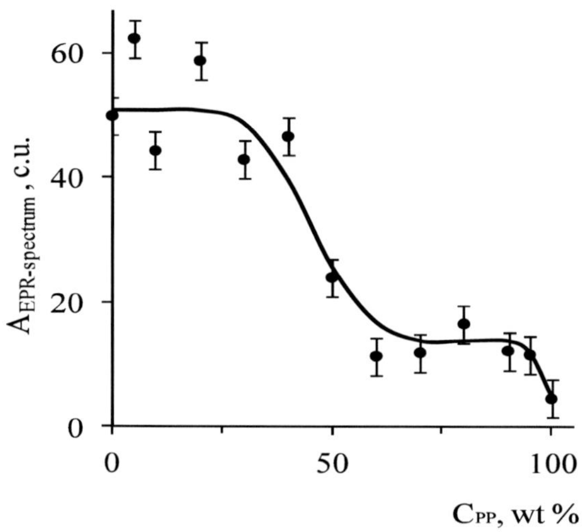

Figure 3. Effect of the PP-LDPE blend composition on the area *A* of the EPR-spectrum calculated as an amorphous region.

Using the area of the EPR spectrum allows the estimation of the amount of stable nitroxyl radicals sorbed by samples of the blends of different composition under identical conditions. The dependence of the areas of the EPR spectra for the blends of various compositions on the content of PP in the blend is presented in Figure 3. The area of the EPR spectrum decreases with the increase of the PP content in the blend. Therefore, the TEMPO radical has a better solubility in the LDPE amorphous regions than in the PP ones. A sharp decrease in this value is over a range of compositions containing from 40 to 60 wt% of PP. This indirectly suggests the occurrence of the phase inversion and the formation of the PP continuous phase in this range of compositions.

These data confirm the fact that LDPE has looser amorphous regions as compared to PP. LDPE tends to sorb up to 60–70 wt% of the PP content from the probe of the blend. Given that the correlation time of a radical rotation is an integral characteristic of the PP/LDPE blend, the value of τ_c grows rapidly only when a certain amount of the radical penetrates into PP. It explains the type of the dependence between the area of the EPR spectrum and the blend composition presented in Figure 2.

Figure 4 demonstrates the relationship between the activation energy E_A of the radical probe rotation and blend compositions. A slight increase in E_A is observed at the initial part of this graph (PP content up to 60 wt%). Further, the value of E_A grows more rapidly, increasing from 41 kJ/mol for the composition with PP/LDPE = 60:40 to the maximum value of 56 kJ/mol for PP. The increase in the E_A value indicates that the amorphous regions gain stiffness. The dependence of the E_A value on the radical probe rotation presented here suggests that the LDPE continuous phase transforms to the discrete phase in the range of 40–60 wt% of PP.

Conclusion

Based on the data obtained in the study, it appears possible to describe the probable process of the structure formation for different PP/LDPE blends. After a gradual addition of PP to LDPE, the former begins to form a dispersed phase in the continuous phase of LDPE up to the PP content in the blend of about 40 wt %. Phase inversion with the formation of the PP continuous phase is observed in the range of compositions with the PP content from 40 to 60 wt%. This results in an increase in the volume fraction of the crystalline phase of PP and the overall stiffness of the system. Molten LDPE in the blend acts as a structure-modifying agent, so the crystallization process of PP proceeds more perfectly.

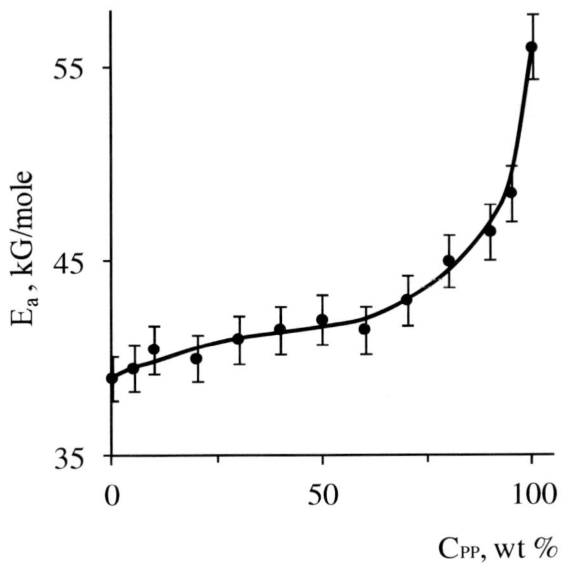

Figure 4. Effect of the PP-LDPE blend composition on the activation energy E_A of the radical rotation.

REFERENCES

[1] Martuscelli E., Pracella M., Avella M., Greco R., Ragosta G. Properties of polyethylene-polypropylene blends - crystallization behavior. // *Makromolekulare Chemie-macromolecular chemistry and physics. 1980.* Vol. 181. P. 957-967.

[2] Wenig W., Meyer K. Investigation of the crystallization behaviour of polypropylene - polyethylene blends by optical microscopy. // *Kolloid-Zeitschrift & Zeitschrift für Polymere. 1980*. Vol. 258, No. 9. P. 1009-1014.

[3] Bartczak Z., Galeski A., Pracella M. Spherulite nucleation in blends of isotactic polypropylene with high density polyethylene. // *Polymer.* 1986. Vol. 27. P. 537-543.

[4] Rybnikar F. Crystallization and morphology in blends of isotactic polypropylene and linear polyethylene. // Journal of Macromolecular Science. Part B. Physics. 1988. Vol. 27. P. 125-129.

[5] Ramsteiner F., Kanig G., Heckmann W., Gruber W. Improved low temperature impact strength of polypropylene by modification with polyethylene. // Polymer. 1983. Vol. 24. P. 365-370.

[6] Galeski A., Pracella M., Martuscelli E. Polypropylene spherulite morphology and growth rate changes in blends with low density polyethylene. // Journal of Polymer Science. *Part B. Polymer Physics.* 1984. Vol. 22. P. 739-747.

[7] Ujhelyiova A., Marcincin A., Legen J. DSC analysis of polypropylene-low density polyethylene blend fibres. // *Fibres & Textiles in Eaten Europe.* 2005. Vol.13. No.5 (53). P. 129-132.

[8] Livanova N.M., Popova E.S, Ledneva O.A., Popov A.A. Properties and phase structure of the poly(propylene)-low-density poly(ethylene) blends. // *Polymer Science. Ser. A.* 1998. Vol. 40. No. 1. P. 41-46.

[9] Prydilova O., Petronyuk U., Tertyshnaya U., Ledneva O., Popov A. The correlation of structure and properties of PE-PP blends. // *Chemistry and computer modeling.* 2002. No. 11. P. 103-106.

[10] Paul D. R. and Newman S. *Polymer Blends.* Academic Press, New York. 1978. P.189-204.

[11] James L. White, David Choi Polyolefins: Processing, Structure Development and Properties. Hanser-Gardner Publications. *Cincinnati.* 2005. P. 1-2.

[12] J. Karger-Kocsis. *Polypropylene Structure, blends and Composites: Vol. 2 Copolymers and Blends.* 1994. P. 57-59.

[13] Nwabunma D., Kyu T. *Polyolefin blends.* John Wiley & Sons, Inc. Canad*a*. 2007. P. 3-4, 59-66.

[14] Bershtein V.A., Egorov V.M. *Differencialnaya scaniruyushya kalorimetriya v fizikohimii polimerov* (Differencial scanning calorimetry in physycochemistry of polymers). L.: Himiya. 1990. P. 10-25.

[15] Vunderlih B. Fizika makromolekul (Physics of macromolecules). M.: Mir. 1984. P.124.

[16] Godovskiy U.K. *Teplofizicheskie metody issledovaniya polimerov* (Thermophysical methods of polymers investigation). M.: Himiya. 1976. P. 67.

[17] Vasserman A.M., Kovarskiy A.L. Spinovye metki i zondy v fizikohimii polimerov (Spin labels and probes in physicochemistry of polymers). M.:*Nauka,* 1986. P. 5-40, 124-152.

[18] Kuleznev V.N. Smesi polimerov. Structura i svoystva (Polymer blends. Structure and properties). M.:*Himiya.* 1980. P. 78-90.

In: Quantitative Chemistry, Biochemistry and Biology
Editors: G.E. Zaikov, O.V. Stoyanov et al.
ISBN: 978-1-62948-332-0

Chapter 24

INVESTIGATING THE EFFECT OF FUNGI ON THIN FILMS OF POLYETHYLENE, POLYETHYLENE–POLY-3-HYDROXYBUTYRATE AND POLYETHYLENE–D_2W

Yu. V. Tertyshnaya*, S. Dempsey[1] and A. A. Popov

Emanuel Institute of Biochemical Physics,
Russian Academy of Sciences, Moscow, Russia,
[1]Durham University, Durham, UK

ABSTRACT

In the work influence of fungi on films of PE – PHB, PE – d_2w and pure PE without the additional power supply was investigated.

It was established that films of PE – PHB were attacked by fungi more intensively, than other samples.

Tests in the climatic chamber were conducted and it was revealed that with growth of PHB content the sizes of cracks increase. This fact is coordinated with the data on degradation of films of PE – PHB in soil.

Keywords: Polyethylene, poly-3-hydroxybutyrate, degradation, fungi, climatic chamber

AIM AND BACKGROUND

The main idea of current paper is focused on investigation the effect of fungies on films of pure polyethylene and polyethylene composites and study of biodegradation of the samples in soil.

* E-mail: yersinia@bk.ru

INTRODUCTION

PHB (Polyhydroxybutyrate) is an example of PHA polyhydroxyalkanoate), first isolated in 1925 by the French Microbiologist Maurice Lemoigne. The most common form of PHB is poly-3-hydroxybutyrate, denoted P3HB. Unlike its petrochemical alternatives, PHB is based on renewable resources and can be completely metabolized by a wide variety of bacteria found in a range of environments, making it truly biodegradable. A problem posed by non-biodegradable polymers is efficient waste disposal – in 1992 the total volume fraction of plastics in landfill sites had reached 20%, and the proportion continues to rise. PHB holds several advantages over other biodegradable polymers, for example its water insolubility, good oxygen permeability, UV resistance and biocompatibility.

PHB is produced by various microorganisms as an energy storage molecule under conditions of stress. Much current research is dedicated to finding an economically efficient method to produce a high yield in a short amount of time. Studies show successful PHB accumulation from a variety of substrates, including *Bacillus sp, Pseudomonas cepacia, Halomonas boliviensis* [1, 2, 3]. The type of microorganisms and feedstock used the method of extraction and the down-processing techniques determine the molecular weight and physical properties of PHB.

The biodegradation of PHB has been extensively studied recently. Previous research investigates the degradation of PHB and its copolymers in several environments, such as soils, compost, fresh water and sewage sludge. The microorganisms which act to digest the polymer have been isolated, and include various strains of *Penicillium* and *Aspergillus*. General consensus is that PHB shows increased degradation in soil environments, although good results in sewage sludge and compost are also reported [4, 5].

Of increased interest are PHB blends. PHB can be blended with many commercially available plastics such as polyethylene (PE). This produces a material with all the beneficial properties of the PE (e.g. cheap, readily available) with the biodegradable properties of the PHB. Investigation is required to determine the necessary blend parameters, as well as to determine microorganisms capable of biodegradation. This is the motivation for this paper. Similar products on the market include d_2w, produced by Symphony Environmental. d_2w is an additive that changes the molecular structure of a plastic at the end of its useful life, producing a compound which can be bio-assimilated into the environment.

This paper presents a study on the suitability of various polymer films (PE, PE/PHB blend and PE/ d_2w blend) to the growth of fungi strains, including *Penicillium*, *Aspergillus* and *Trichoderma*. Subsequent comparison of fungal growth allows conclusions to drawn about the biodegradability of the samples.

EXPERIMENTAL PART

Source of Polymers

Three polymers were tested: polyethylene (PE), PE with added d_2w, and the PE/PHB blends. The PHB (of specific structure $H[-O-CH(CH_3)CH_2COO-]_nOH$ and average molecular weight 2.5×10^5 g mol^{-1}) was sourced from Biomer biopolyesters, who use a selected soil bacterium to grow the PHB in an aqueous medium. The PE and d_2w blend was sourced from a readily available plastic carrier bag, where the d_2w was produced by Symphony Environmental.

Preparation of Polymer Film

PHB (2, 4, 8, 16, 32% mass) and PE were blended and formed into a thin film. All polymer films were cut into strips of approximate dimensions 10 mm by 100 mm. The mass of each sample was recorded.

Microbial Strains and Incubation

Water suspensions containing spores of *Aspergillus Niger*, *Penicillin Chrysagenum* and *Trichoderma Viride* were prepared. 5-6 drops of each suspension were placed onto the polymer strips. The strips were placed in an incubator held at room temperature (20°C +/- 2°C).

Observation of Samples

Samples were removed from the incubator after 14 days and 28 days. An Olympus CX41 optical microscope was used to observe the samples, under 400 x magnifications. Photographs of the sample were taken using a Canon digital camera.

The climatic measurement was occurred using 'Xentest 150-S" chamber at T=35°C with the day-night program.

Thermal Analysis

Thermal analysis was performed using a DSC differential scanning calorimeter at a heating rate of 8 °C min^{-1}, for the PE + d_2w sample. The DSC calibration was done with Indium, which has a known melting point of 156.61°C.

Both heating and cooling curves were obtained, and the endotherms produced allowed calculation of the melting point, the degree of crystallinity and the temperature of crystallization.

The degree of crystallinity α is calculated using the equation

$$\alpha = \frac{\Delta H}{\Delta H^o m} \text{ x } 100\ \%,$$

where ΔH is the melting enthalpy of the sample and ΔH°_m is the hypothetical melting enthalpy of a purely crystalline sample.

RESULTS AND DISCUSSION

First of all we hade to investigate the supermarket's sample. The IR-spectroscopy and DSC methods were used. The IR-spectra with some characteristic lines was obtained: 720, 1470, 2850 и 2920 sm^{-1}, so this polymer was polyethylene (Figure 1) [6]. The melting temperature (T_m) 134 ^0C and the crystallinity degree (α) 75% obtained by DSC method confirmed that.

For our investigation the sample 4 wt.% of PHB was taken because the d_2w content approximately was about 5 % [7] and we could compare the samples.

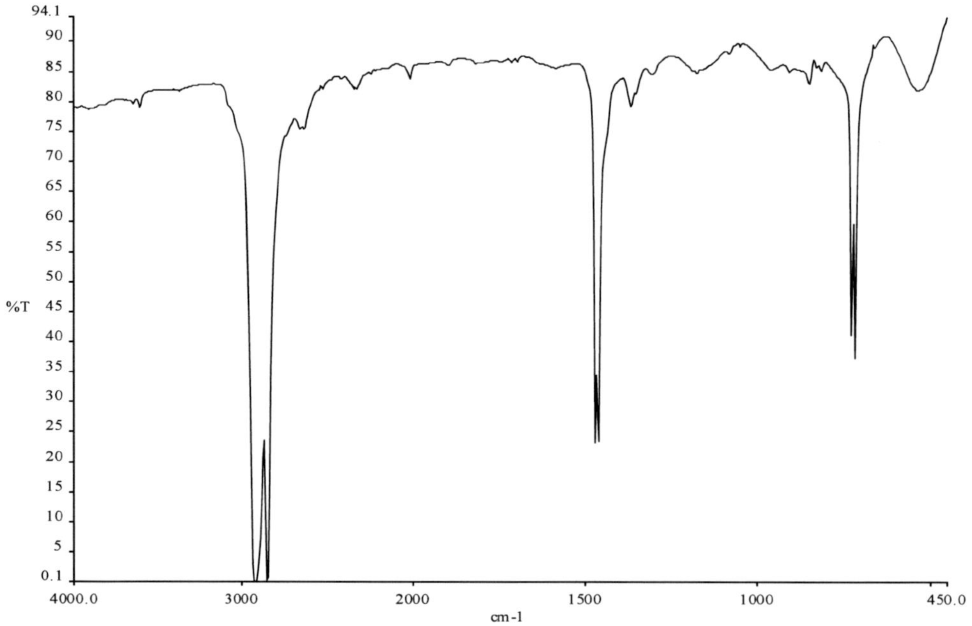

Figure 1. IR-spectra of the film sample from a supermarket.

Next step was a fungies contamination:

After 14 days:

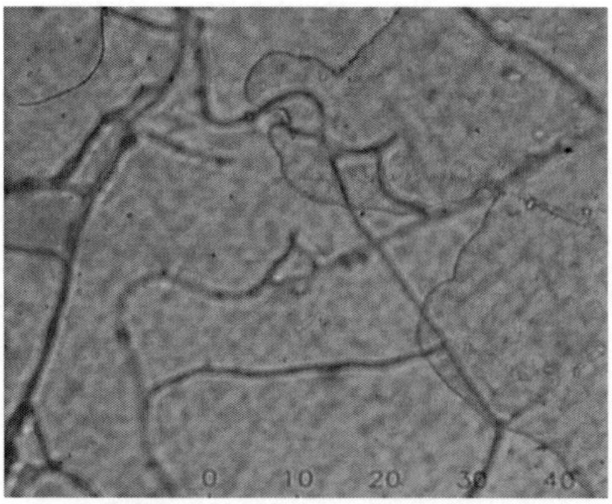

PE

Light microscope image of PE strip with *Penicillium Chrysogenum*. We observe the presence of a developed network of mycelia, indicating that the fungi has grown. The PE has started to biodegrade under the influence of the *Penicillium Chrysogenum*.

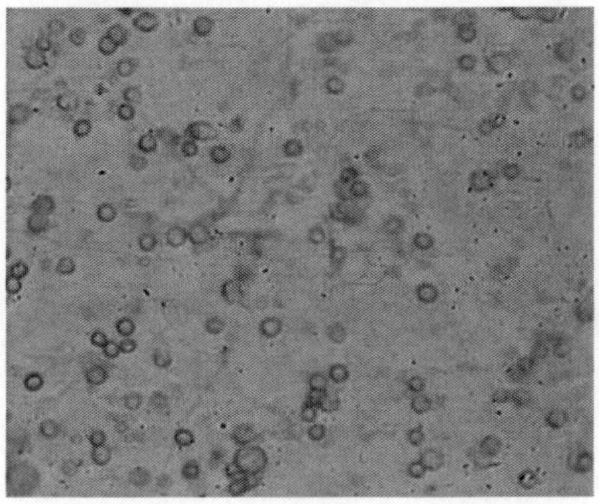

PE + PHB

Image of PHB (4%) and PE (96%) polymer strip, with *Trichoderma Viride*. We observe the early stages of mycelia growth from the spores, signalling the beginning of biodegradation.

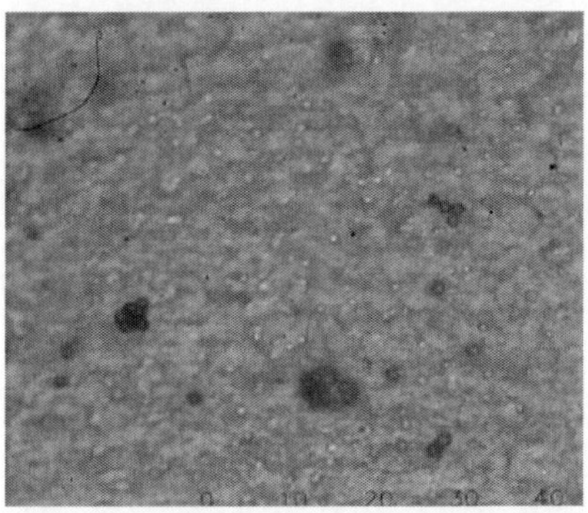

PE + d_2w

Light microscope image of PE + d_2w with the presence of *Aspergillus Niger*. No mycelia are observed, indicating that the *Aspergillus Niger* fungi has not grown on the PE + d_2w sample.

After 28 days:

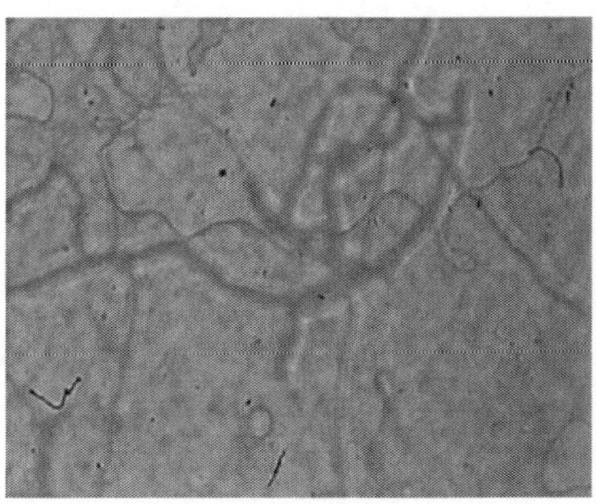

PE

For the PE sample we observed the best growth of *Penicillium Chrysagenum*. It was strange because there was no power source, but as it was industry polymer and it contained stearats which could be a power source for a while.

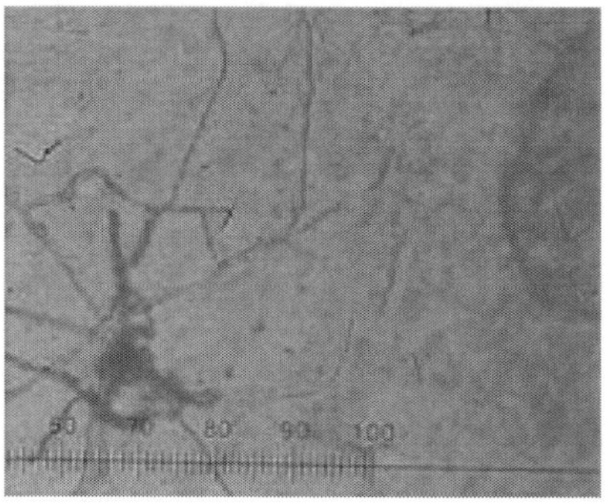

PE + PHB

We observe similar growth to that shown after 14 days. The most advanced mycelium growth is found on the sample of PE + 4% PHB, with the *Penicillium Chrysogenum*, and also the *Trichoderma Viride*. We observe a more complex mycelium structure, with many branches.

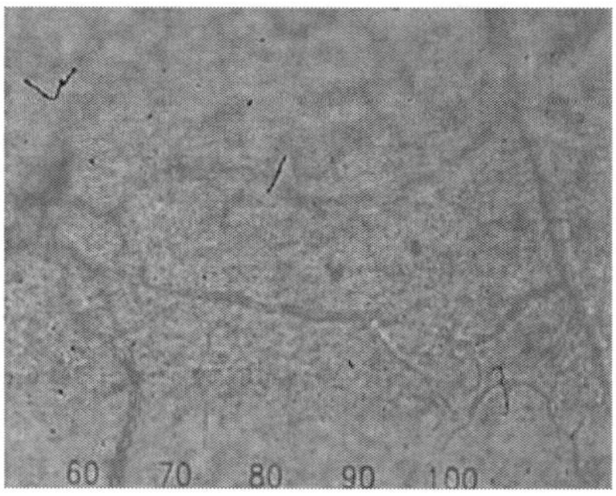

PE + d_2w

The best growth on the PE + d_2w occurred for the *Trichoderma Viride*. We observe less mycelium growth on this sample for *Penicillium Chrysogenum*, and hardly any growth for the *Aspergillus Niger*.

The growth of three fungi strains *Aspergillus Niger*, *Penicillium Chrysagenum* and *Trichoderma Viride* on thin film samples of PE, PE+d_2w and PE+4% PHB was investigated. The samples were subjected to water suspensions of the three kinds of spores, and incubated for 14 days and 28 days. Comparison of light microscope images allowed us to determine the extent of mycelia growth. It was found the best mycelium growth occurs on PE + PHB for both *Penicillium Chrysagenum* and *Trichoderma Viride*.

After fungi contamination the degradation in soil of pH= 6.0-6.2 at room temperature (20°C +/- 2°C) was studied. Biodegradation was monitored every 3 months for approximately 12 months by measuring residual mass. For this, the buried samples were recovered, washed with distilled water and dried at room temperature before being weighed and then buried again in their respective trays. The residual mass was calculated as the ratio between the final and initial weights. The samples of pure PE and PE-PHB (2 wt.%) had no mass changes even after 12 months. The significant weight loss was observed for samples with 16 and 32 wt.% of PHB: 6 and 10 % respectively. These data are confirmed the climatic chamber measurement.

As a result of climate processes of ageing in polymer composites can precede processes of structure formation, which result in various flaws. It may be gaps of macromolecules, the emergence of cracks in the volume and on the surface of the material. From the table 1 one can see that the craze size increased with growing PHB content.

Table 1. The craze sizes of PE – PHB films

PE – PHB wt.%	Length of crazes, μm	Width of crazes, μm
100:0	No	No
98:2	No	No
96:4	60-1000	0.5-10
92:8	180-1000	3-10
84:16	200-1000	15-30
68:32	200-1000	30-45

The presence of defects in the samples, as a rule, leads to the growth of colonies of microorganisms, when the material loses its physic-mechanical characteristics and the integrity of the matrix. Based on the above, it can be assumed that in natural conditions to the extent of the impact of environmental factors on the surface of the film will be formed cracks, the presence of which will contribute to the penetration of the dispute mold fungi and other microorganisms. Microorganisms, eating PHB, will be developed, to allocate metabolites and destroy the film, fragmenting them up to a size commensurate with the size of soil particles.

REFERENCES

[1] Vishnuvardhan Reddy S, Thirumala M. and Mahmood S.K. A novel Bacillus sp. accumulating poly (3-hydroxybutyrate-co-3-hydroxyvalerate) from a single carbon substrate // *Journal of Industrial Microbiology & Biotechnology*. 2009.V. 36. P. 837.

[2] Ramsay BA, Ramsay JA, Copper DG. Production of poly-β-hydroxyalkanoic acid by *Pseudomonas* cepacia // *Appl. Environ. Microbiol.* 1989. V. 55. P. 584.

[3] Quillaguaman J, Delgado O, Mattiasson B, Kaul RH. Poly (β-hydroxybutyrate) production by a moderate halophile, *Halomonas boliviensis LC1* // *Enzyme Microb. Technol.* 2006. V 38. P.148.

[4] Manna A., Paul A.K. Degradation of microbial polyester poly(3-hydroxybutyrate) in environmental samples and in culture // *Biodegradation*. 2000. V. 11. P. 323.

[5] Tertyshnaya Yu. V., Feofanova E.S., Popov A.A. Biodegradation of the films poly-3-hydroxybutyrate and its blends in soil and water. Modern problems in biochemical physics. Nova Science Publ., Inc. 2012. P. 117.

[6] Dechant J., Danz R., Kimmer W., Schmolke R. Ultrarotspektroskopische untersuchungen an polymeren // *Akademie-Verlag*. Berlin. 1972. P.472.

[7] Esthappan S. K., Kuttappah S. K., Joseph R. Effect of titanium dioxide on the thermal ageing of polypropylene // *Polymer. Degrad. and Stab.* 2012. V. 97. Issue 3. P. 615.

In: Quantitative Chemistry, Biochemistry and Biology
Editors: G.E. Zaikov, O.V. Stoyanov et al.
ISBN: 978-1-62948-332-0

Chapter 25

DETERMINATION OF THE IDENTITY OF THE SIGNS OF NATURAL POMEGRANATE JUICE

E. A. Barishovets* and L. G. Eliseeva#
Plekhanov Russian University of Economics,
Moscow, Russian Federation

ABSTRACT

In this research we have carried out comparative studies on the establishment of chromatographic profiles of anthocyanins in fresh pomegranates (freshly squeezed juices) and comparative tests with chromatographic profiles anthocyanins of pomegranate juice from different manufacturers with the purpose of establishing the effectiveness of the method of identification for the detection of forged products.

Keywords: Identification, anthocyanins, pomegranate juice, high performance liquid chromatography, antioxidants

AIM

Identification of criteria of valuing for the evaluation of pomegranate juice's authenticity on the basis of an assessment of qualitative composition of anthocyanins pomegranate.

BACKGROUND

Pomegranate juice is widely used for the prevention and treatment of many diseases. This fruit has useful properties because of high content of vitamins, including vitamin K, folic

* E-mail: e.barishovets@yandex.ru.
# E-mail: eliseeva-reu@mail.ru.

acid, vitamin C, vitamins of B group, vitamin E and unique complex of mineral substances. In this connection it is important to identify the authenticity of drinks from the pomegranate to prevent the distribution of counterfeit pomegranate juice.

INTRODUCTION

Juices from fruits and vegetables are an essential part of a healthy person's diet. The juice has advantages, which, first of all, are necessary to the consumer: natural and fresh taste, convenient packaging. It is important that in addition to good taste and aroma it can effectively quench the thirst and has a high nutritive value due to active substances of fruits and berries, which are necessary for people. A particular importance in the line of juices is given to products with a range of functional properties that determine the important characteristics aimed at the prevention of human diseases. In this field special attention is paid to the pomegranate juice. Useful properties of pomegranate due to a high content of vitamins, including vitamin K that is necessary for blood clotting (coagulation), folic acid (one pomegranate has approximately 25% of the recommended daily norm), vitamin C (about 30% of a daily rate of consumption), vitamins of B group, vitamin E and unique complex of mineral substances. However, according to scientists the main advantage of the pomegranate is the presence of antioxidants, the number of which is several times higher than in any other kind of fruit. The number of these phytonutrients contained in pomegranates are the polyphenols (tanning substances, phytoncides, catechins, and phenol carbonic acid: chlorogenic, neo chlorogenic, p-coumaric, protocatechuic, ellagic acid, anthocyanins, leucoanthocyanins); they determine the color and the taste of juice, possess strong antioxidant, P-vitamin activity. Paramount importance of anthocyanins can be seen in the contents of which vary from 34,0 to 76.5 mg per 100 ml of juice. Pomegranate juice is widely used for the prevention and treatment of many diseases. In this connection it is important to identify the authenticity of drinks from the pomegranate to prevent the distribution of counterfeit pomegranate juice.

We have carried out monitoring of the quality of more than 15 types of pomegranate juice, presented in the retail network of Moscow and represented in different price ranges. It has been established that more than 70% of pomegranate juice in the trading network is forged. Many manufacturers are trying to make profit reducing the cost of raw material, respectively deteriorating the quality of the product, inventing new methods of falsification.

The aim of this study is the identification of criteria of valuing for the evaluation of pomegranate juice's authenticity on the basis of an assessment of qualitative composition of anthocyanins pomegranate. Several dozen of different anthocyanins were extracted from plants, but for fruit and berries their characteristics are about ten. It is typical for fruits the presence of six most common anthocyanins: delphinidin, cyanidin, malvidin, peonidin, pelargonidin, petunidin, which differ from each other due to radicals R1 and R2. Carbohydrate residues in the molecules of anthocyanins are connected with aglycone (anthocyanidin) through oxygen hydroxyl group in the 3d position of anthocyanidin. Qualitative composition and quantitative ratio of anthocyanins are very specific for each type of plant. This index is stable for every kind of fruit, therefore, is considered to be the hallmark of a particular botanical species. In a pomegranate delphinidin and cyanidin prevail.

We have carried out comparative studies on the establishment of chromatographic profiles of anthocyanins in fresh pomegranates (freshly squeezed juices) and comparative tests with chromatographic profiles anthocyanins of pomegranate juice from different manufacturers with the purpose of establishing the effectiveness of the method of identification for the detection of forged products. The comparison of chromatographic profiles according to the size of the peaks, their mutual arrangement and time for holding signals is carried out.

In the study there were used the following commercial samples: 5 samples that were marked - 100% pomegranate juice of direct extraction produced without the use of dyes and preservatives. There was a test of the following brands of juice such as: «Packmal», 100% pomegranate juice; «Grunel», natural 100 % pomegranate juice of higher grades; «Nar», pomegranate juice, unclarified; «Grante», pomegranate juice of direct extraction; «Noyan», pomegranate juice of direct extraction; 3 samples of recovered juice with the addition of the regulator sugar taste: «Red price», pomegranate juice restored with sugar; «Miri Pak», pomegranate restored with the addition of sugar; «Swell», restored pomegranate. Freshly squeezed juice has been squeezed out exactly on the day of the test in order to preserve the composition of anthocyanins pomegranate. Indicators of the quality of fresh juice were taken as a reference.

It is supposed, that packaged juice, as well as in the restored one and in juices of direct extraction, anthocyanin profile should be the same as the profile of fresh juice.

EXPERIMENTAL PART

Determination of the identity of the signs were carried out on the wave length of 520 nm by high performance liquid chromatography (HPLC). There was used a column of 150 mm x 4.6 mm, sorbent Nucleosil 100 angstroms, C18, precolumn with the same sorbent, at room ambient temperature. The mobile phase A was 0,05% aqueous solution of trifluoroacetic acid, and the mobile phase B acetonitrile for liquid chromatography.

The program elutes the consumption of mobile phase 1 ml/min

Time, min	A	B
0	90%	10%
3	90%	10%
15	30%	70%

Recorded time of chromatograms - 20 minutes

A spectrophotometric detector, wavelength detection of 520 nm. Regarding standards of dyes there have been used commercially available samples of dyes. Samples were filtered through the paper filter and then handled in a centrifuge (5000 vol/min 15 min). The volume of injection is 20 mcl.

To establish the impact of the botanical varieties of pomegranate and region of its vegetation we have investigated chromatographic profiles of different pomegranate fruit, grown in Turkey, Israel, Uzbekistan, and Azerbaijan (Figure 1, 2 and 3).

Consider the pomegranate juice (originated from different countries) chromatograms.

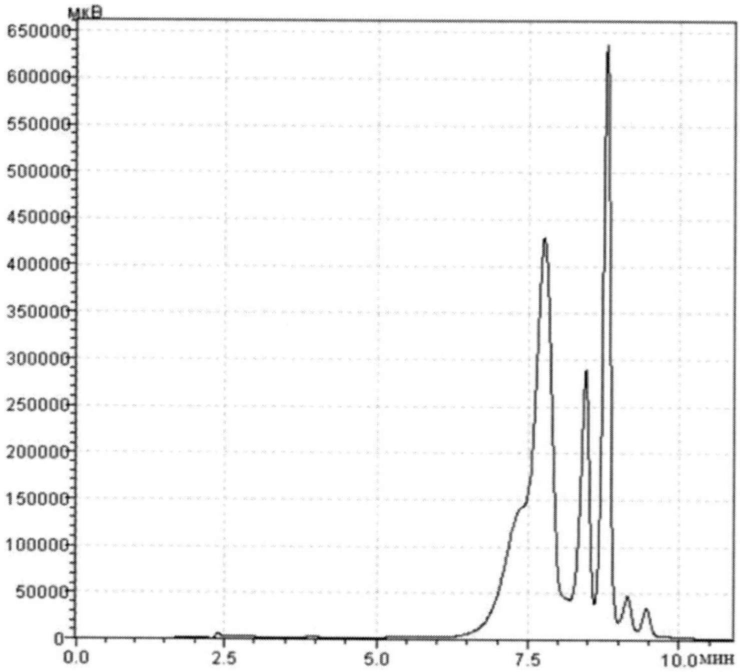

Figure 1. Chromatographic anthocyanins profile of freshly squeezed pomegranate juice, production of Turkey.

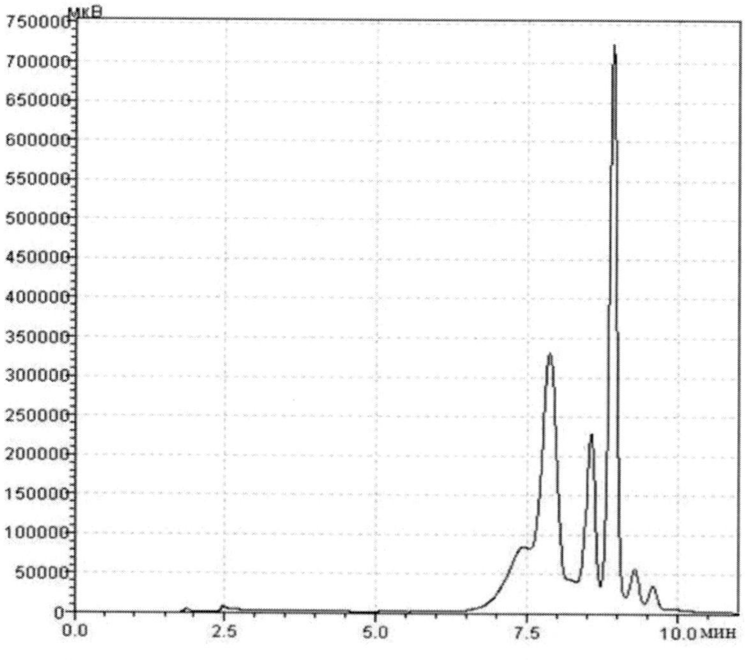

Figure 2. Chromatographic anthocyanins profile of freshly squeezed pomegranate juice, production of Israel.

According to the above chromatograms it can be concluded that the profiles of the peaks of colored (painted) substances of various pomegranate sorts almost do not differ.

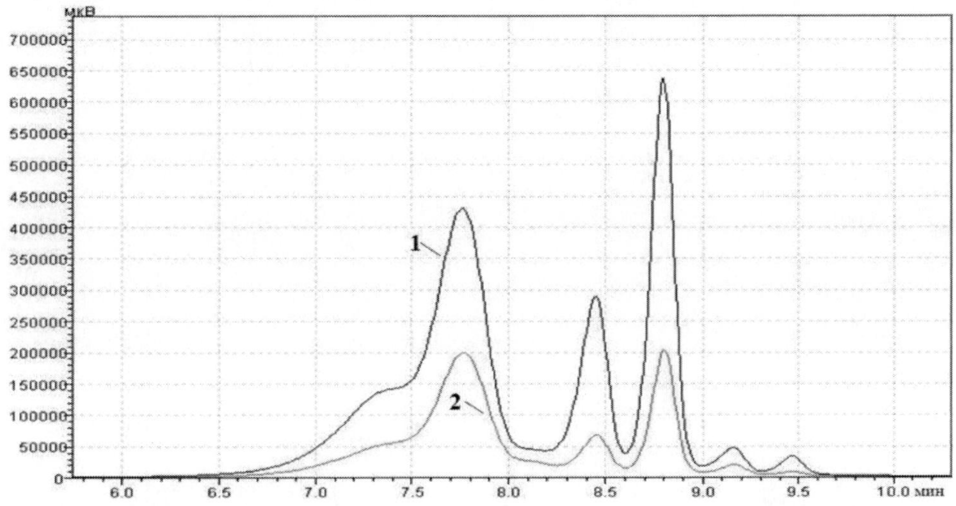

Figure 3. Chromatographic anthocyanins profile of freshly squeezed pomegranate, production of Uzbekistan (1) and Azerbaijan (2).

As it can be seen from the profiles of the qualitative composition of anthocyanins in all of the investigated samples, pomegranate was identical to the installed fluctuations in the number-of individual types of anthocyanins depending on the region of origin, but not in the anthocyanins. Consequently, these chromatographics profiles can be used as a reference in establishing the authenticity of pomegranate juice sold in the trading network. Further, we will draw a comparison of fresh juice and purchased juice chromatograms.

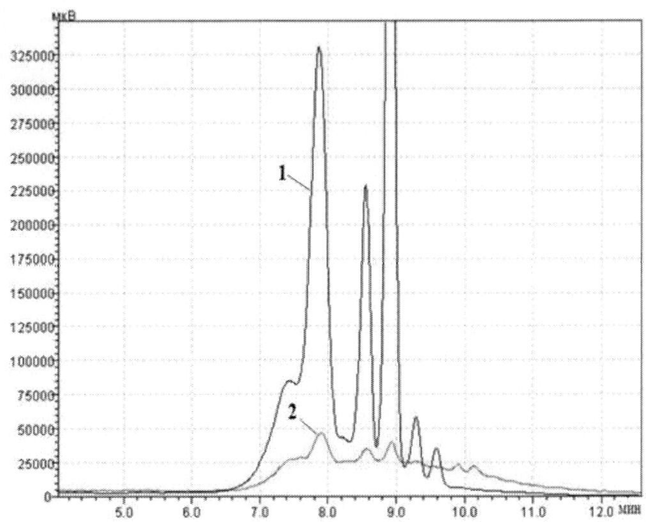

Figure 4. Chromatographic anthocyanins profile of freshly squeezed pomegranate juice (1) and pomegranate juice direct extraction «Grante» (2).

It is obvious that anthocyanins are present in these juices; the amount of them is 7 or more times lower, than in the benchmark; this fact allows us to assume that juice «Grunel» was diluted in the process of production.

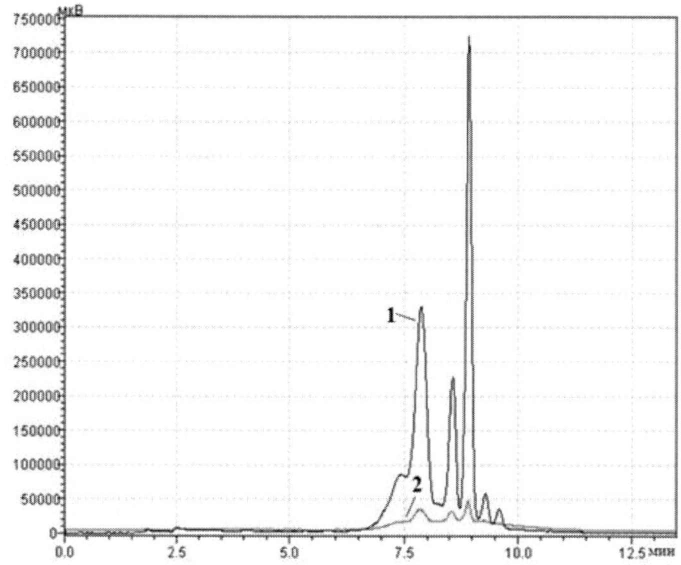

Figure 5. Chromatographic anthocyanins profile of freshly squeezed pomegranate (1) and juice «Dumbbell»(2).

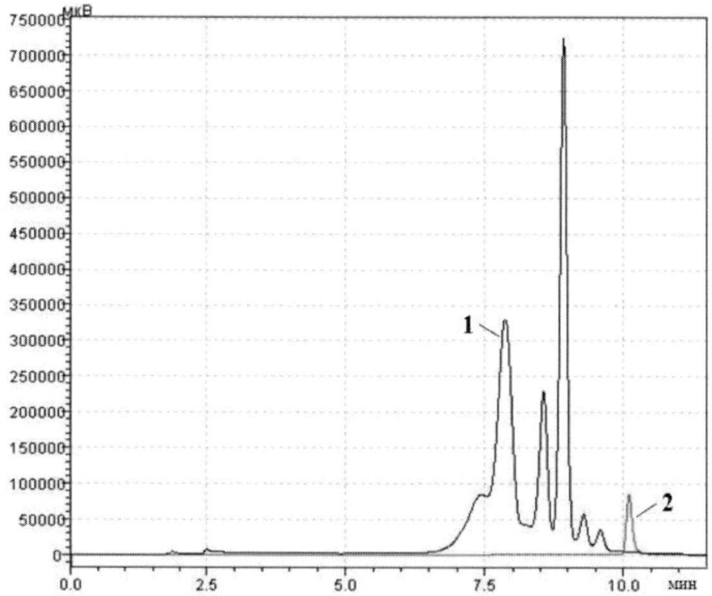

Figure 6. Chromatographic anthocyanins profile of freshly squeezed pomegranate (1) and juice «Packman» (2).

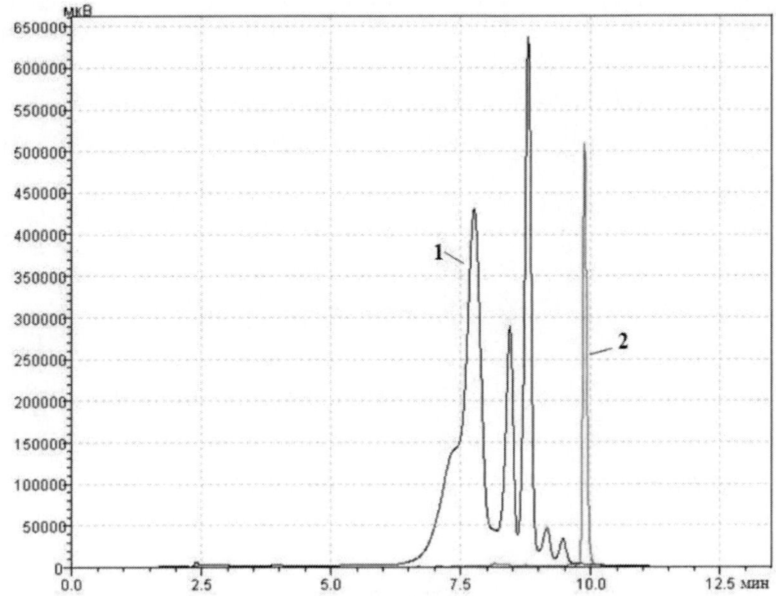

Figure 7. Chromatographic anthocyanins profile of freshly squeezed pomegranate (1) and juice «Miri Pak» (2).

From the chromatograms (Figure 6) we can see that in the juice «Packmal» there are practically no anthocyanins. The only peak by the time that corresponds to the substance, identified on the basis of the standard as the dye E 122 - Carmazin (carmoisine, azorubine). Therefore, it can be assumed that the juice was forged by using food additives: dyes, fragrances and flavors identical to pomegranate juice.

In Figure 7 there is presented the similar result as in the previous profile - anthocyanins in the commercial juice «Miri Pak», claimed by a manufacturer as pomegranate restored with the addition of sugar. The presence of anthocyanins has not been revealed, but the peak has been set, describing the presence of a large amount of dye E 122 - Carmazin.

Results and Discussion

According to results obtained by definition of synthetic dyes by high performance liquid chromatography, it can be concluded that in two samples of pomegranate juice («Packmal» and «Grunel») there has been found synthetic dye E 122 Carmazin with time output as t = 10,0 min, which is prohibited in natural juices; and the above samples of juice have also to be attributed to forged ones.

To study the impact of technology of production of reconstituted juice and temperature processing on the content of antioxidants, we have carried out research where the freshly-squeezed pomegranate juice was dried out at the boiling temperature and diluted by adding bidistilled water to the original volume; the juice was heated up during 5 hours. As a result of these experiments, it has been shown that anthocyanins while being heated are partially destroyed, but their contents reduce by no more than 50%. Therefore, in the production of reduced juices chromatographic profile of anthocyanins should remain the same for the qualitative composition of anthocyanins, while their quantitative amount can be reduced to a maximum of 50%.

Other samples of juice have been tested in the same way. It has been found that only two samples - juice of direct extraction «Grante» and recovered juice «Red Price» correspond to the requirements of quality, regulated by the Federal Law No 178 "Technical Regulations for Juice Products from Fruits and Vegetables".

Conclusion

According to the research we can conclude that the method of HPLC determination of qualitative composition of anthocyanins profile of pomegranate juice can be effectively used for the identification of their authenticity, identification of juice dilution with water and the introduction of artificial dyes in the process of juice production.

References

[1] The Federal Law of the Russian Federation of October 27, 2008. No 178 " *Technical Regulations for Juice Products from Fruits and Vegetables*";

[2] Methodical instructions 4.1/4.2.2486-09 «Methodical instructions on Eden identifying, including for the purpose of revealing of falsification, juice production of fruit and / or vegetables». Methodical instructions. - M.: *Federal Centre of hygiene and epidemiology of Rospotrebnadzor*, 2009;

[3] National standard GOST R 53137-2008 «*Juices and juice products. Identification. General provisions*» - M.: Publishers standards, 2008.

In: Quantitative Chemistry, Biochemistry and Biology
Editors: G.E. Zaikov, O.V. Stoyanov et al.
ISBN: 978-1-62948-332-0

Chapter 26

Influence of Carbon Nanofibers on the Structure and Properties of Vulcanized EPDM

M. M. Ryabenko**, *V. D. Yulovskaya and B. V. Peshnev
Moscow State University of Fine Chemical Technology
Fine Chemical Technology, Moscow, Russia

Abstract

The paper studied the effect of the structure and composition of carbon nanofibers obtained by a Co-catalyst, washed and not washed, from the metal catalyst on the kinetics and properties of vulcanized ethylene propylene diene rubber. It is shown that the fibers obtained on the Co-catalyst accelerate the crosslinking of EPDM, improve the physical and mechanical properties,and increase the molecular mobility.

Keywords: Polymer nanocomposites, carbon nanofiber, EPDM, structure, vulcanization, mechanical properties

The purpose of this research - investigation of the carbon nanofibers influence produced by Co-catalysts on the physical and mechanical properties and structure of synthetic EPDM.

Introduction

Currently, one of the promising areas is the co-building of polymer nanocomposites, which are formed by the mixing of polymers and fillers with nanoscale particles (NSP) [1].

Complex physical and mechanical properties of a substance NSP differ from the properties of the same substance in a monolithic sample, and therefore the influence of nanoscale particles (NSP) on the properties of the composite material is different. It should be

* E-mail: ryabenkom@gmail.com.

noted that it is not right to use the term "reinforcing filler in a polymer matrix" to the nanoparticles, because they act at the molecular level rather than the macro level, as in case with regular fillers. Because of this interaction nanocomposites is formed with an ordered structure. Unicity of physical and mechanical properties of nanoparticles is due to their small size, as well as a highly developed and extensive surface area. [2].

The properties of nanocomposite polymer materials are mainly due to the properties of the polymer matrix, nature and specific features, as well as the nature, shape and dispersion of nanoparticles. The interaction of the surface of nanoparticles with the elements of the matrix is very essential for these particles. The introduction of more stringent nanoparticles in elastomers and in crystalline polymers, leads to increase the initial modulus of elasticity because of increasing the number of contacts of the polymer matrix and the filler particles [3]. It is known that the mechanical properties of crystalline polymers with the introduction of nanoparticles vary more widely than amorphous ones [4]. In the polymer system to 4 ... 15% of the molecular chains go into the dense phase with a strongly reduced mobility of molecular chains, resulting in increased dimensional stability, thermal conductivity and thermal stability of the polymer composition, and reduced coefficient of thermal expansion [5].

Experimental Part

We used EPDM 505, the ratio of units of ethylene-propylene 60:40 (%), ENB content of 8% and a viscosity of 55 (ML1 +4, 1250S).

For vulcanization was used sulfuric vulcanization system: zinc oxide, sulfur, stearic acid, thiuram D, dibenztiazolildisulfid, curing mode - 10 minutes at 160°C. The mixture was prepared in laboratory mill for 10 minutes with a mixture of multiple trimming rolls.

As a filler we used a carbon nanofiber (CNF) prepared in cobalt catalyst (washed and not washed from the metal) demetallization conducted by washing the fibers with concentrated sulfuric acid, and the metal content in the sample was reduced to 38 - 40% up to 0.2 - 2% masses, and the specific adsorption surface (Ssp) increased by 2 - 2.5 times. The concentration of fibers in the rubber is 2% by weight of the rubber. The raw material was used to produce fibers on the cobalt catalyst - propane-butane fraction. The main component of synthesis is cobalt oxide. Volumetric flow rate of gas in the synthesis is 1 vol. / Hr at a temperature of 7000°C.

Comparison of diameters (Table 1) showed that as a result of demetallization (Co w /m) the outer diameter of the fibers increases, and the internal reduces.

To determine the effect of filler on the properties of rubber, the following studies were made: scanning electron microscopy, RPA, DMA, EPR.

Molecular mobility in EPDM vulcanizates before and after the introduction of CNF was determined by electron paramagnetic resonance (EPR) on the device in a temperature of 25ºC. Due to the fact that the EPDM are not systems with paramagnetic centers for their research the paramagnetic centers were introduced from outside - the spin probe. As the probe was used radical (TEMPO), because it has a high vapor pressure and easily adsorbed polymers at room temperature.

Curing kinetics was assessed on the instrument RPA2000.

The microstructure of the samples studied by scanning electron microscopy (SEM) with an electron microscope high-resolution at the University of Eastern Finland. Initially, the sample surface of polymeric materials have been processed with the help of oxygen Plasmalab 80 in the following conditions: oxygen flow 20sccm, 100W, and the working pressure 80m Torr, over 30s. Then to increasing the conductivity of the sample surface was covered with a layer of palladium in 2 nm (200Å).

Relaxation transitions in polymer EPDM samples before and after the introduction of CNF in alternating load was studied using dynamic mechanical analysis (DMA) on the instrument Netzsch DMA 242C in a temperature range from -140 to 150°C. We used the special cooler CC 200 L running on liquid nitrogen to obtain low temperatures.

Strain-strength characteristics of the materials were determined in uniaxial tension mode on a universal testing machine Instron-1122 at room temperature and constant velocity of the upper beam vε = 50 mm / min. The samples present a bilateral shoulder with the size of the working part of 4x20 mm. The thickness of the samples was measured with a micrometer with an accuracy of 0.01 mm. The curves load - elongation was measured tensile strength σ MPa and elongation at break ε%.

Table 1. Specifications nanofibres

Type of CNF	Specific surface Ssp, nm^2		Length, nm	Diameter, nm		Composition
	with calcination	without calcination		Outer(D)	Internal(d)	
Co w/m	85	78	190	30,2	8,7	-
Co	87	77	190	27,9	10,3	Co

RESULTS AND DISCUSSION

Images show that the fibers are well distributed throughout the volume of rubber and are oriented mainly in the direction of rolling (Figure 1).

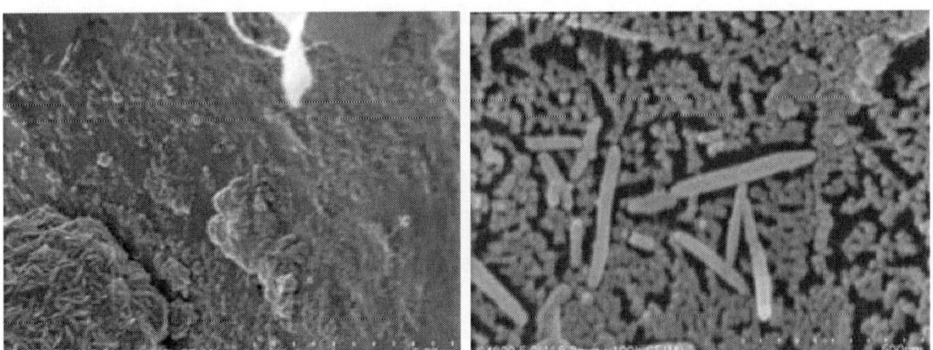

demetallized fiber fiber containing cobalt

Figure 1. CNF distribution in the matrix EPDM.

Fiber obtained at Co-catalyst has a structure of "Tree", such fibers are smooth and elongated. [6] You can also see the large agglomerates which are formed mainly with the introduction demetallized fiber, which is the main problem of getting rubbers with a uniform distribution of the fibers throughout the volume. As noted above, the nanoparticles have a high surface energy, and therefore are more likely to form aggregates ranging in size from a few to hundreds of micrometers. [7]

From the data of the kinetics of vulcanization (fiber content - 2% for all samples) we know that the addition of the fibers, which was obtained on cobalt catalyst regardless of laundered or not it reduces the induction period and the cure time, but fiber, washed from the metal, not much increase the viscosity of the technology (Table 2, Figure 2). As we can see from the data on swelling, samples containing fiber swell more than rubber without filler, it's indicative of a lower density of chemical bonds in the vulcanizates containing fiber.

Table 2. Influence CNF on the kinetics of binding EPDM (Data RPA)

Sample	Start time of vulcanization (τ_C, min)	The optimum of vulcanization (τ_{C90}, min)	Vulcanization rate (%/min)	Mooney viscosity	Qav, %
Co w/m	2,8	9	16,1	55,6	260
Co	3,9	10	16,3	51,3	210
Without CNF	5,2	12	14,6	52	150

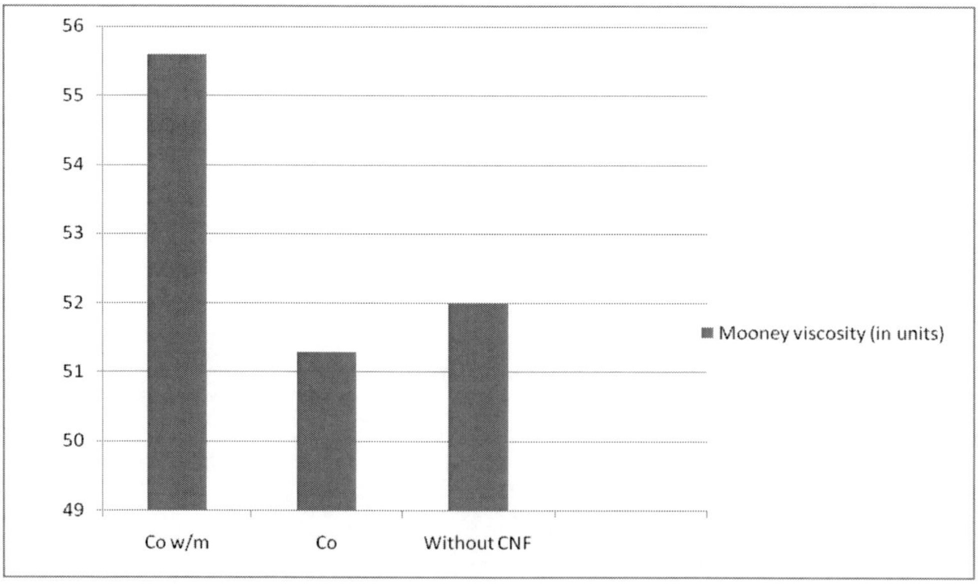

Figure 2. Mooney viscosity.

Mechanical testing shows that the addition of nanofibers obtained on Co-catalyst increases the strength of the materials, with the most units at different strains is EPDM, structured fiber, free of metal (Figures 3, 4), while the EPDM containing crude fiber breaks at a higher aspect ratio.

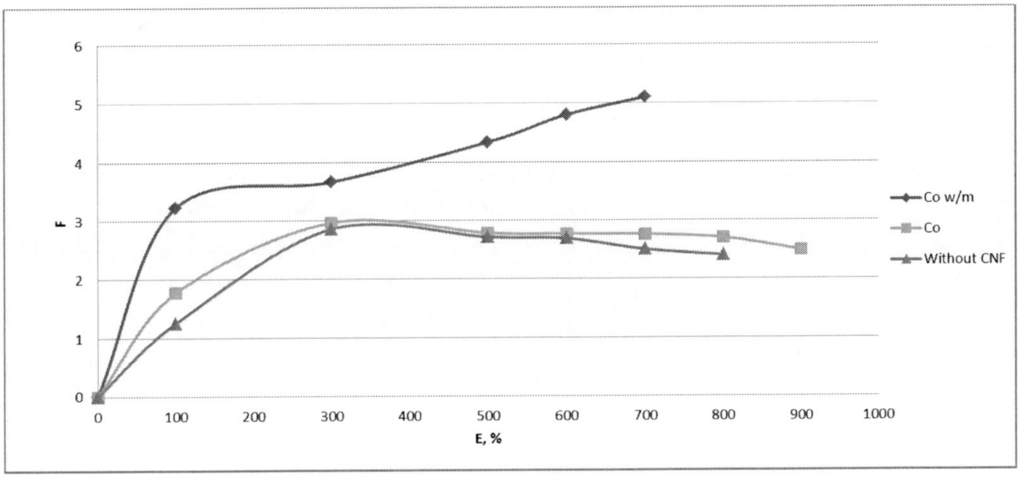

Figure 3. Influence of CNF on elastic mechanical properties of vulcanized EPDM.

Figure 4. Influence of CNF on the strength characteristics of vulcanized EPDM.

The fiber type almost no effects on the residual elongation.

From the results obtained by EPR, the fibers loosen the structure of rubber, thus increasing the mobility of the macromolecules in the whole volume, compared with the sample containing no filler. The metal content had almost no effects on the mobility (Table 3).

Table 3. The influence of composition on the glass transition temperature and the correlation time of the samples

Type of CNF	Co w/m	Co	Without CNF
The glass transition temperature, ^0C	-43,2	-45,3	-40,2
Correlation time * 10^{10}, sec.	3	2,7	4,2

The introduction of these fibers lowers the temperature of the glass-filled rubbers by 5 degrees (Table 3), which extends the operating temperature range of the material.

Conclusion

From the above, it can be concluded that, in general, the nanofibers in the future can be used as a nanofiller that affects the properties of vulcanizates. Fibers obtained by Co - catalyst not strong effect on the kinetics of binding, well-swelled samples in toluene and improve the physical and mechanical properties, molecular mobility. Thus, the fiber adjusted from the metal increases the strength of vulcanized rubber by 40%, but lowers the value of elongation at break, as opposed to a vulcanized fiber, crude from the metal. These results are the basis for further research.

References

[1] A.I. Gusev. Nanomaterials, nanostructures, and nanotechnology. *Physmathlit*. 2005. 416.

[2] I.P. Suzdalev. Nanotechnology: Physics and chemistry of nanoclusters, nanostructures and nanomaterials. M.: *KomKniga*. 2006. 596 s.

[3] Yu.S. Lipatov. Physical chemistry of filled polymers. - M.: *Chemistry*, 1977. – 304.

[4] A.D. Pomogailo, A.S. Rosenberg, A.S. Uflyand. Metal nanoparticles in polymers. *Chemistry*. 2000. 672.

[5] A.D. Pomogailo, A.S. Rosenberg, A.S. Uflyand. Metal nanoparticles in polymers. *Chemistry*. 2000. 672.

[6] R. T. K. Baker. Catalytic growth of carbon filaments / / *Carbon*. 1989. Vol. 27. P. 315,323.

[7] I.A Chmutin, N.G. Rvykina, I.L. Dubnikova etc. Control of homogeneity of distribution of nanoparticles in the polymer matrix; Abstracts of the conference "Nanotechnologies - Production - 2007." *Fryazino*, 2007 - p. 229-234.

In: Quantitative Chemistry, Biochemistry and Biology
Editors: G.E. Zaikov, O.V. Stoyanov et al.
ISBN: 978-1-62948-332-0

Chapter 27

DETERMINATION DIFFUSION COEFFICIENT OF BROWNIAN PARTICLES USING VELOCITY OR FORCE AUTOCORRELATION FUNCTION IN MOLECULAR DYNAMIC SIMULATIONS

D. A. Turchenkov*
BioSim Research Group, Moscow, Russia

ABSTRACT

This work presents the method of numerical determination of diffusion D and friction ζ coefficients of Brownian particles from velocity (VACF) and force (FACF) autocorrelation functions in molecular dynamic simulations. The electrostatic parameters of particles in the simulation were obtained using modern DFT methods of quantum chemistry. The calculations were carried out with a PCS (Patch Clamp Simulation) program package, designed for the simulation of neurotransmission processes, using Brownian and molecular dynamic methods with GPU acceleration support based on NVIDIA CUDA.

Keywords: Langevin, autocorrelation function, diffusion, friction coefficient

INTRODUCTION

The use of Brownian dynamics for biological modeling suggests involving the averaged action of a solvent on a particle. In the classical Langevin model, the averaged action is represented by a friction coefficient ζ, which depends on the approximation used. So, Stokes' law is applicable for spherical particles in continuous viscous fluid, and the theories of Adelman, Onsager [1] take into account the dielectric friction for systems with a small Knudsen number for its proper use with the Cunningham correction factor [2], etc.

* E-mail: biosim.rg@gmail.com.

Even though these approaches work well in the case of simple particles (ions of Na^+, K^+, Cl^-), for polyatomic molecules (NH_4^+, SO_4^{2-}) the situation is much more complicated. This is primarily due to the ambiguity in choice of approximation geometry and the charge distribution of particles for these theories. Therefore, for these systems method of MD simulations is one of the few guaranteed ways to obtain the diffusion D and the friction ζ coefficients.

EXPERIMENTAL PART

In this paper, the diffusion coefficient D was determined in two ways. The first used the mean square displacement (MSD) of the particle:

$$D = \lim_{t\to\infty} \frac{1}{6Nnt} \left\{ \sum_{t_0=0}^{n} \sum_{j=1}^{N} \left| \vec{r}_j(t+t_0) - \vec{r}_j(t_0) \right|^2 \right\}$$

The second used the velocity autocorrelation function (VACF) and Green-Kubo relations [3]:

$$D = \frac{1}{3Nn} \int_0^{\infty} \left\{ \sum_{t_0=0}^{n} \sum_{j=1}^{N} \vec{V}_j(t_0) \cdot \vec{V}_j(t+t_0) \right\} dt$$

The friction coefficient ζ of the Brownian particle was determined by means of the stochastic force autocorrelation function (FACF) [4]:

$$\zeta = \frac{1}{3k_B T N n} \int_0^{\infty} \left\{ \sum_{t_0=0}^{n} \sum_{j=1}^{N} \vec{F}_j(t_0) \cdot \vec{F}_j(t+t_0) \right\} dt$$

The intermolecular potential energy between atomic sites can be calculated by a sum of the Coulomb electrostatic interaction and the Lennard–Jones potential:

$$U_{tot} = \sum_{j\neq i} \left\{ \frac{1}{4\pi\varepsilon\varepsilon_0} \frac{q_i q_j}{r_{ij}} + 4\epsilon_{ij} \left[\left(\frac{\sigma_{ij}}{r_{ij}} \right)^{12} - \left(\frac{\sigma_{ij}}{r_{ij}} \right)^{6} \right] \right\}$$

Water molecules were modeled in terms of SPC approach. Electrostatic parameters of the simulated particles were calculated using the methods of quantum chemistry (DFT). Figure 1 shows the isosurface of the electron density of the glutamine molecule. The projection of the electron density onto the individual atomic sites was done according to the Hirschfeld method.

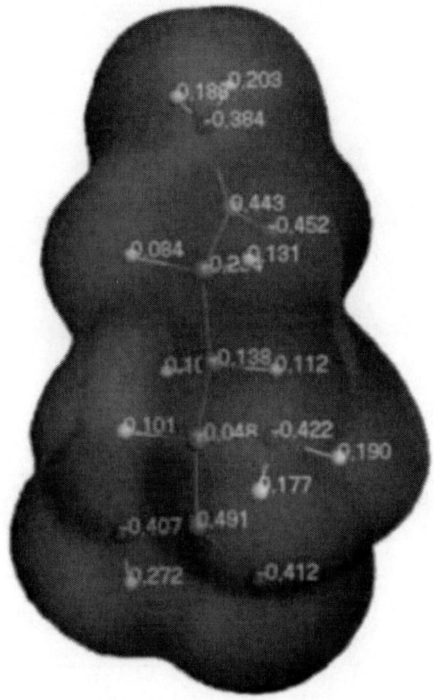

Figure 1. Isosurface of electron density of the glutamine molecule (Gln).

RESULTS AND DISCUSSION

First, we calculated the diffusion D and friction ζ coefficients for simple and complex ions.

The results of the simulation are shown in Table 1. The characteristic velocity autocorrelation function is shown in Figure 2. The results suggest that it can be used for Brownian particles in more complex models: the data is consistent through three different theoretical methods as well as with the experimental values.

Similar calculations were performed for the individual amino acids (Table 1). Some amino acids in aqueous solutions at a certain pH are present in the zwitterionic state, which affects the nature of the interaction with the solvent molecules. In this paper, the molecules of glycine (Gly) and glutamine (Gln) were examined in two states: simple and zwitterion. The geometry of the ground state is determined in the COSMO solvent model.

The comparison of the experimental data showed that, for the non-zwitterionic state, the diffusion coefficients were somewhat underestimated, indicating the need for a more accurate accounting of the charge distribution and structural features of the particles.

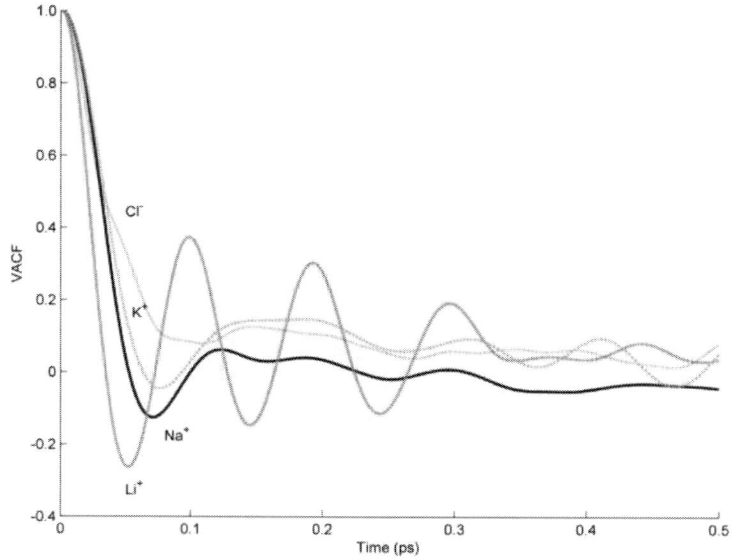

Figure 2. The normalized velocity autocorrelation functions of cations and anions in SPC water at 298 K.

Table 1. The results of simulation

Particle	ζ, 10^{-12} kg/s	D, 10^{-9} m^2/s			
		MSD	VACF	FACF	Experimental data [5,6]
Li+	4.33±0.08	0.97±0.04	1.10±0.12	0.97±0.06	1.03
Na+	3.20±0.20	1.33±0.02	1.52±0.06	1.29±0.07	1.33
K+	2.14±0.06	2.16±0.03	1.82±0.07	1.92±0.05	1.96
Cl^-	2.05±0.13	1.89±0.05	2.03±0.04	1.91±0.12	2.03
NH_4^+	2.15±0.02	1.94±0.04	2.08±0.05	1.89±0.02	1.98
NO_3^-	2.09±0.05	2.05±0.06	2.04±0.05	1.96±0.05	1.90
SO_4^{2-}	3.70±0.08	1.16±0.05	1.17±0.09	1.11±0.02	1.07
Gly	6.01±0.14	0.68±0.04	0.77±0.07	0.81±0.02	-
Gly^Z	3.91±0.01	1.01±0.07	0.96±0.05	1.05±0.03	1.06
Gln	7.48±0.04	0.39±0.04	0.51±0.09	0.45±0.08	-
Gln^Z	5.91±0.07	0.67±0.06	0.81±0.03	0.70±0.04	0.76

CONCLUSION

The proposed method extends the applicability of the Brownian dynamics for molecules of arbitrary complexity by eliminating the current shortcomings of crude methods, based on geometric approximation and the complexity of calculating the friction coefficients. To determine the electron density of the complex molecules and the final charge distribution, accurate methods of quantum chemistry were used, taking into account the effect of the solvent, which simulates the different states of amino acids in solution.

REFERENCES

[1] J.-H. Chen and S. A. Adelman. Macroscopic model for solvated ion dynamics. *The Journal of Chemical Physics* 72, 2819 (1980).

[2] E. Cunningham. On the Velocity of Steady Fall of Spherical Particles through Fluid Medium. Proceedings of the Royal Society A: *Mathematical, Physical and Engineering Sciences* 83, 357 (1910).

[3] H. Uchida and M. Matsuoka. Molecular dynamics simulation of solution structure and dynamics of aqueous sodium chloride solutions from dilute to supersaturated concentration. *Fluid Phase Equilibria* 219, 49 (2004).

[4] Pep Espanol and Ignacio Zuniga. Force autocorrelation function in brownian motion theory. *The Journal of Chemical Physics,* 98(1):574–580, 1993.

[5] Li Yuan-Hui and Sandra Gregory. Diffusion of ions in sea water and in deep-sea sediments. *Geochimica et Cosmochimica Acta*, 38(5):703 – 714, 1974.

[6] L. G. Longsworth. Diffusion measurements, at 25°C, of aqueous solutions of amino acids, peptides and sugars. *Journal of the American Chemical Society*, 75(22):5705–5709, 1953.

In: Quantitative Chemistry, Biochemistry and Biology
Editors: G.E. Zaikov, O.V. Stoyanov et al.
ISBN: 978-1-62948-332-0

Chapter 28

In Silico Simulation of Silver and Copper Ions Interacting with Fungal Cell Wall: *In Vitro* Antifungal Activity of Copper Ions and Silver Ions

M. A. Turchenkov*

BioSim Research Group, Kashirskoe shosse, Moscow, Russia

Abstract

The interaction of silver and copper ions with chitin dimers was modeled using quantum chemistry within the DFT framework. A variation of parameters of hydrogen bonds depending on the presence of different metal atoms was obtained. This variation was compared with the results of an *in vitro* study of the antifungal activity of corresponding metal ions.

Keywords: Silver, copper, chitin, DFT, fungi

Aim and Background

The aim of this work is to find the possible mechanism of the antifungal activity of silver ions using quantum chemistry simulation for large systems. The complete theory of antifungal activity of these elements is still in development [2, 4].

In computational molecular biology, systems containing metal particles are not well parameterized yet for force field calculations, so precise *ab initio* methods for that class of systems are the rational choice. However, the computational complexity of these systems suggests the use of a simplified way to choose the most probable geometry of large organic

* E-mail: biosim.rg@gmail.com.

molecules with adsorbed metal atoms in order to avoid exploring all possible geometries. The DFT codes are described in [3, 11].

The interpretation of calculated properties is also a complex problem, so we provide a comparative *in vitro* experiment to establish the possible outcome of computed values in real life.

INTRODUCTION

Silver has long been known to serve as an antibacterial and antifungal agent. Out of all the metals with similar antimicrobial properties, silver has the least toxicity to human cells. Prior to the introduction of modern antibiotics, silver was commonly used in medical treatments.

The exact mechanism of antifungal activity is still unknown and it is currently the subject of many researches. There are three suggested qualitative concepts: DNA association, cell wall damage and deactivation of enzymes. On the molecular level, it is known that silver is capable of forming a bond with S, N, and O atoms, and also catalyzes the formation of disulfide bonds.

The new computational quantum chemistry methods make it possible to directly simulate biochemical processes which involve large molecules. In this work, the interaction of silver and copper ions with chitin dimers was modeled using the DFT theory.

EXPERIMENTAL PART

The fungal cell wall is simulated as two chitin dimers. To obtain the optimized geometry of the ground state, several DFT codes were used, including PC GAMESS, SIESTA and ABINIT.

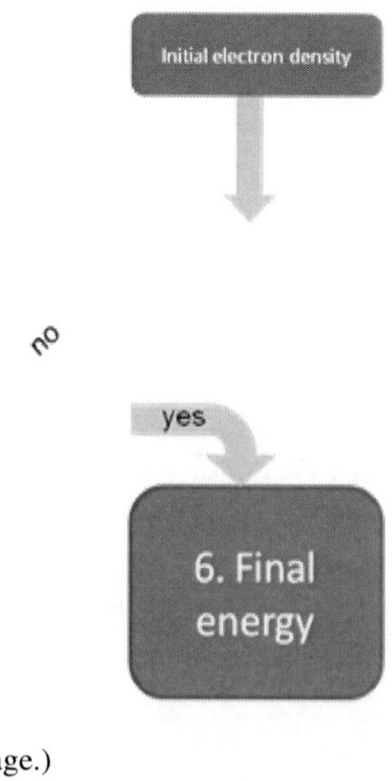

Picture 1. (Continued on next page.)

$$\nabla^2 V_{Hartree}(\mathbf{r}) = -4\pi\rho(\mathbf{r})$$

$$V_{xc}(\mathbf{r}) = \frac{\delta E_{xc}[\rho]}{\delta\rho(\mathbf{r})}$$

$$V_{eff}(\mathbf{r}) = V_{ext}(\mathbf{r}) + V_{xc}(\mathbf{r}) + V_{Hartree}(\mathbf{r})$$

$$\left[-\frac{\nabla^2}{2} + V_{eff}(\mathbf{r})\right]\psi_m(\mathbf{r}) = \varepsilon_m \psi_m(\mathbf{r})$$

$$\rho(\mathbf{r}) = \sum_m^{occup} |\psi_m(\mathbf{r})|^2$$

$$E[\rho(\mathbf{r})] = \sum_m^{occup} \varepsilon_m - \int V_{xc}(\mathbf{r})\rho(\mathbf{r})d\mathbf{r} -$$

$$-\frac{1}{2}\iint \frac{\rho(\mathbf{r}_1)\rho(\mathbf{r}_2)}{|\mathbf{r}_1 - \mathbf{r}_2|} d\mathbf{r}_1 d\mathbf{r}_2 + E_{xc}[\rho(\mathbf{r})]$$

Picture 1. Conceptual map of DFT method.

To find the most probable area of metal adsorption on a molecule, the Fukui functions scheme was used.

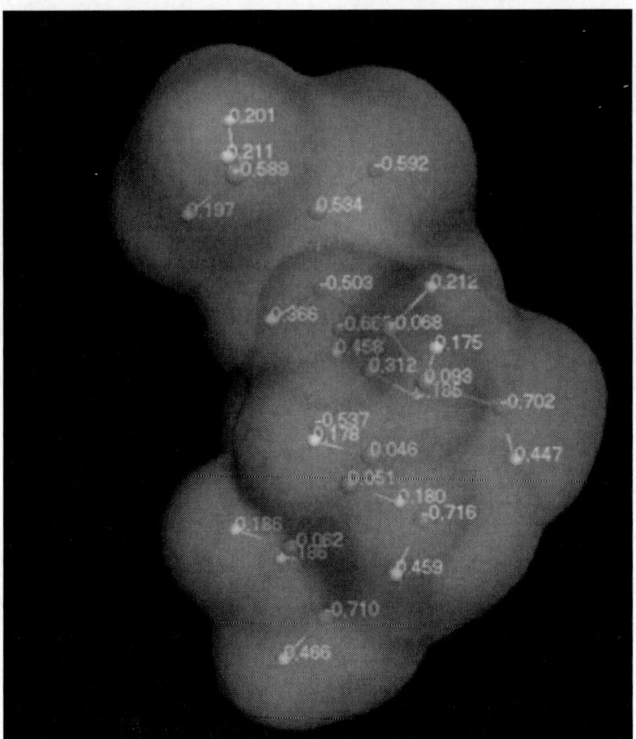

Picture 2. Visualisation of the scalar electrophilic Fukui function, mapped onto the isosurface of electron density field for detection of probable adsorbtion site.

The electron density of the ground state chitin system with adsorbed metal atoms was projected on atomic sites, responsible for hydrogen bond formation, using the Hirshfeld method.

Calculations were performed on a Linux-cluster using parallel MPI (MPICH2) and SMP (OpenMP) parallel paradigms.

In silico calculations were complemented by *in vitro* assay, in which we compared the antifungal activity of silver and copper ions in the same concentrations against *Alternaria alternata*.

RESULTS AND DISCUSSION

Chitin dimers form hydrogen bonds with each other, and there is a variability of parameters of that bond depending on the type of atom adsorbed near the bond. Adsorbed silver atom increases the symmetry of H-bond, measured by the difference of projected charges on oxygen atoms, involved in H-bond, whereas copper atom decreases that symmetry, comparing to the original system without any metal.

The *In vitro* study shows an increased antifungal activity in silver compared with that of copper in the same concentration.

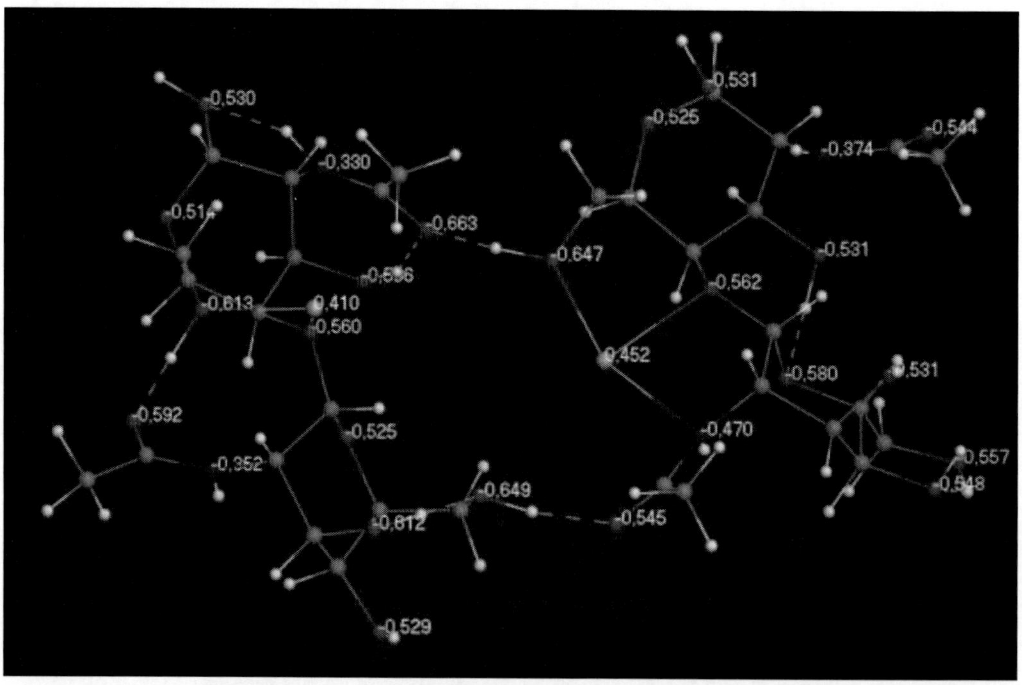

Picture 3. Ground state geometry of chitin dimers with silver atoms adsorded. Hydrogen bonds are shown by dash lines. Projected atomic charge is shown near atom.

Table 1. Calculated values of hydrogen bond parameters for original system and system with metal atoms

System	Total charge, e	Length of H-bond, Å	Charge of farthest oxygen atom, e	Charge of nearest oxygen atom, e	Charge of metal atom, e	Charge of hydrogen atom, e
Chitin	0	2.6	-0.67	-0.64		0.37
Chitin + Ag	+2	2.4	-0.66	-0.65	0.45	0.46
Chitin + Cu	+2	2.4	-0.69	-0.61	0.27	0.46

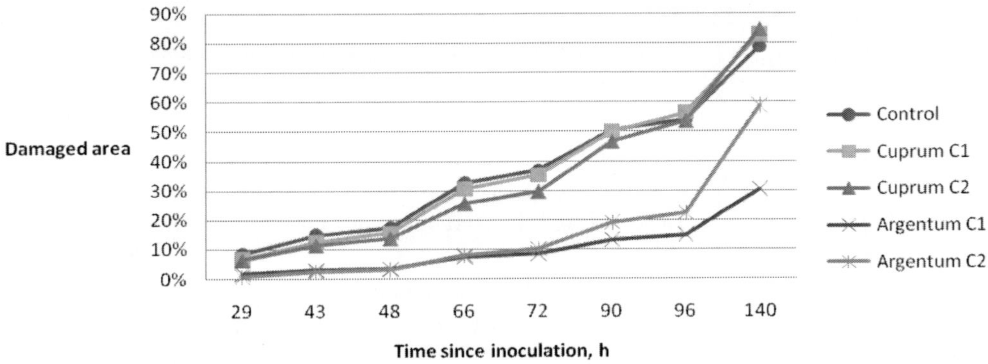

Figure 1. Growth dynamic of Alternaria alternata under two different silver and copper ions concentrations. C1 = 25 μg/ml (by metal mass) и C2 = 2.5 μg/ml (by metal mass).

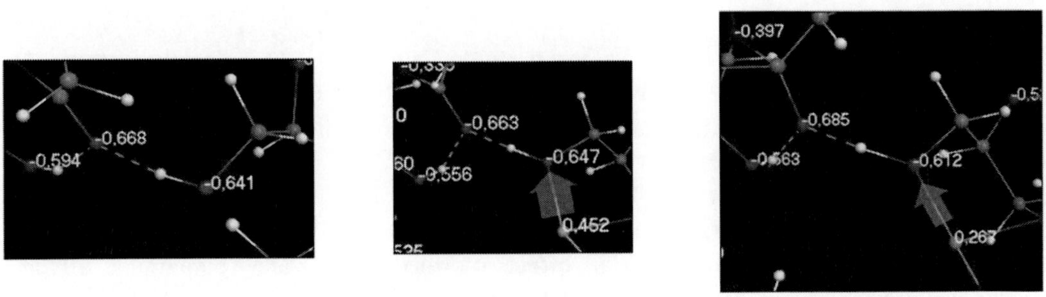

Original system Adsorbed silver Adsorbed copper

Picture 4. Difference between the effects of adsorbed copper and silver on hydrogen bonds between chitin dimers. Arrow depicts the polarity of Me-O bond.

CONCLUSION

The interaction with the cell wall and the effect on the mechanical properties of its chitin skeleton could take part in the antifungal mechanism of silver ions. It correlates with an observed low toxicity of silver for animal cells. A decrease in the symmetry of hydrogen bonds, induced by copper atom adsorption, correlates with the low antifungal activity of copper in comparison to silver atoms, which demonstrates an increase in symmetry and high antifungal activity. It could be an important parameter for mechanical properties of the chitin skeleton, and changing that parameter could play a role in the molecular mechanism of antifungal activities in silver ions.

References

[1] Damm, C.; Münstedt, H. & Rösch, A. (2008), 'The antimicrobial efficacy of polyamide 6/silver-nano- and microcomposites', *Materials Chemistry and Physics* 108(1), 61-66.

[2] Fabrega, J.; Luoma, S. N.; Tyler, C. R.; Galloway, T. S. & Lead, J. R. (2011), 'Silver nanoparticles: Behaviour and effects in the aquatic environment', *Environment International* 37(2), 517-531.

[3] Gonze, X. et al. (2009). ABINIT: First-principles approach of materials and nanosystem properties." Computer Phys. Commun. 180, 2582-2615.

[4] Jo, Y.; Kim, B. H. & Jung, G. (2009), 'Antifungal Activity of Silver Ions and Nanoparticles on Phytopathogenic Fungi', P*lant Disease* 93(10), 1037-1043.

[5] Kim, K.-J.; Sung, W.; Suh, B.; Moon, S.-K.; Choi, J.-S.; Kim, J. & Lee, D. (2009), 'Antifungal activity and mode of action of silver nano-particles on Candida albicans', *BioMetals* 22, 235-242.

[6] Luo, P. G. & Stutzenberger, F. J. (2008), Nanotechnology in the Detection and Control of Microorganisms, in Sima Sariaslani Allen I. Laskin & Geoffrey M. Gadd, ed., , *Academic Press,* , pp. 145-181.

[7] Madhumathi, K.; Sudheesh Kumar, P.; Abhilash, S.; Sreeja, V.; Tamura, H.; Manzoor, K.; Nair, S. & Jayakumar, R. (2010), 'Development of novel chitin/nanosilver composite scaffolds for wound dressing applications', *Journal of Materials Science: Materials in Medicine* 21, 807-813.

[8] Monteiro, D.; Silva, S.; Negri, M.; Gorup, L.; de Camargo, E.; Oliveira, R.; Barbosa, D. & Henriques, M. (2012), 'Silver nanoparticles: influence of stabilizing agent and diameter on antifungal activity against Candida albicans and Candida glabrata biofilms', *Letters in Applied Microbiology* 54(5), 383-391.

[9] Monteiro, D. R.; Gorup, L. F.; Takamiya, A. S.; Ruvollo-Filho, A. C.; De Camargo, E. R. & Barbosa, D. B. (2009), 'The growing importance of materials that prevent microbial adhesion: antimicrobial effect of medical devices containing silver.', *International Journal of Antimicrobial Agents* 34(2), 103-110.

[10] Morones, J. R.; Elechiguerra, J. L.; Camacho, A.; Holt, K.; Kouri, J. B.; Ramírez, J. T. & Yacaman, M. J. (2005), 'The bactericidal effect of silver nanoparticles', *Nanotechnology* 16(10), 2346.

[11] Soler, J. M., Artacho, E., Gale, J. D., García, A., Junquera, J., Ordejón, P., & Sánchez-Portal, D. (2002). The SIESTA method for ab initio order-N materials simulation. *Journal of Physics: Condensed Matter*, 14, 2745.

[12] Wales, D. S. & Sagar, B. F. (1990). Recovery of Metal Ions by Microfungal Filters. *Journal of Chemical Technology & Biotechnology* 49(4), 345-355.

In: Quantitative Chemistry, Biochemistry and Biology
Editors: G.E. Zaikov, O.V. Stoyanov et al.
ISBN: 978-1-62948-332-0

Chapter 29

THE ESTIMATION OF ANTIOXIDANTS AS NUTS QUALITY INDEX

O. V. Yurina*[*] *and L. G. Eliseeva
Moscow, Plekhanov Russian University of Economics,
Russian Federation, Moscow

ABSTRACT

A measurement of the content of antioxidants in walnuts, hazelnuts and peanuts was made by the method of coulometry. It has shown that the content of antioxidants is an indicator of the oxidative damage of the nut's fat during the process of storage.

Keywords: Antioxidants, quality, oxidative damage, nuts, coulometry

AIM

The main aim of our research is to study the dependence of oxidative damage of the nut's fat on the content of antioxidants.

BACKGROUND

Antioxidants are natural inhibitors, which slow down the processes, occurring during the storage of food products. When antioxidant's concentration is reduces the quality of products degenerates.

[*] E-mail: olga32661@mail.ru.

Introduction

Organic compounds are related to the most common objects of study and analysis. A lot of organic compounds have attracted the attention as toxicants at environmental analytical control recently. There are some problems in achieving rapidity, sensitivity and selectivity of analysis and also we have the needs of automated analysis which can be solved with methods of coulometry. For the present coulometry is only the method without using the dependence of quality from the concentration of the determined substance, which is characteristic for the analysis of others. From the point of view of the theoretical foundations and practical aspects, the galvanostatic coulometry is a very affordable method for the general use in which the main controlled parameters is the time and amperage. The devices which are used last time let us to measure these parameters with very high accuracy. All of these factors create the preconditions for the development of high accuracy methods of coulometrical definition in the different compounds.

The coulometry methods based on the physical law which sets the link between the weight of turned electricity and quantity of spent electricity. In many cases the electro generated coulometrical titrate enters in the oxidation process with organic substratum on the mechanism of reaction with the electron carriers. The most effective carriers are the variable valence metals` ions and their components: oxidants – chrome (VI), manganese (III), cobalt (III), cerium (IV), vanadium (V), copper (II); deoxidants – cobalt (II), chrome (II), vanadium (III), titanium (III), iron (II), copper (I), tin (II). The wide area of practical use of the halide ions (chloride-, bromide-, iodide-ions) highlights them apart from a number of reagents – electron carriers. Halide – ions are used at the generation of free halogens and also for the components in the positive oxidations, for example, iodine and bromide.

In the base of all chemical methods of the organic compounds` quality analysis by functional groups are the definition the substances which are formed and melted by the reaction between the sample and titrate reagent.

Such titrates electrochemical generation give us the obvious advantages: they allow us to use the unstable reagents, excluding their storages and standardization, and provide the precision and opportunity of definition the small contents of the substances. In some cases the electro generated titrates can be more reactive than their solutes. For example, this is typical for iodine (I), bromine (I), vanadium (V) and others. The advantages of the coulometry methods are precision, sensitivity, ease, expressivity and wider range of the definable contents compared to the previously known method. The confines of the definable concentrations and between 1 g/ l and 1 mkg/ml. The relative standard deviation (Sr) are between 0,05-0,002.

The necessary component of living cells is antioxidants – substances, which in the normal physical concentrations supports on the low standard level the free radical auto oxidation process, which are continuously flowing in the cells. Under the rules the processes of the spending and replenishment are balanced. In the violation of such balance the high reactive ability of the free radicals leads to the acceleration oxidation process, which breaks down the molecular basis of the living cells and causes different pathological forms.

In this connection one of the most important problems is the control of the organism`s antioxidant system state at the different pathological by way of its further therapeutic correction.

One of the most effective and widespread antioxidants are the tocopherols (vitamin E), polyphenol (catechol, Gallic acid derivatives), flavonoids (routine, quercetin), ascorbic acid. The most perspective objects of the antioxidants are the plant facilities.

EXPERIMENTAL PART

The analysis was made for 12 nuts` samples; for 3 samples it was made an analysis in the system of the accelerated storage. The accelerated storage is carried out at a temperature of 50° C in the incubator with a medium level of air circulation.

It was noted that walnuts have a higher concentration of the antioxidant than that found in the hazelnuts and peanuts. Also the unroasted hazelnuts have more antioxidants than the roasted ones. The reduction of the nuts` antioxidants connected with their oxidative damage (rancidity) that was revealed in the organoleptic analysis.

RESULTS AND DISCUSSION

So, we can state that the antioxidants` content in the nuts dramatically reduces in the storage process and that coulometry method is able to fix these changes. Also it was noted some taste and smells` letdown in the nuts after the accelerated storage. The index of the antioxidants quantity in the nuts can be used for establishing the degree of oxidative damage.

Also basing on the interim data we make up the following schedule of dependence between the antioxidants` content and the duration time of accelerated storage.

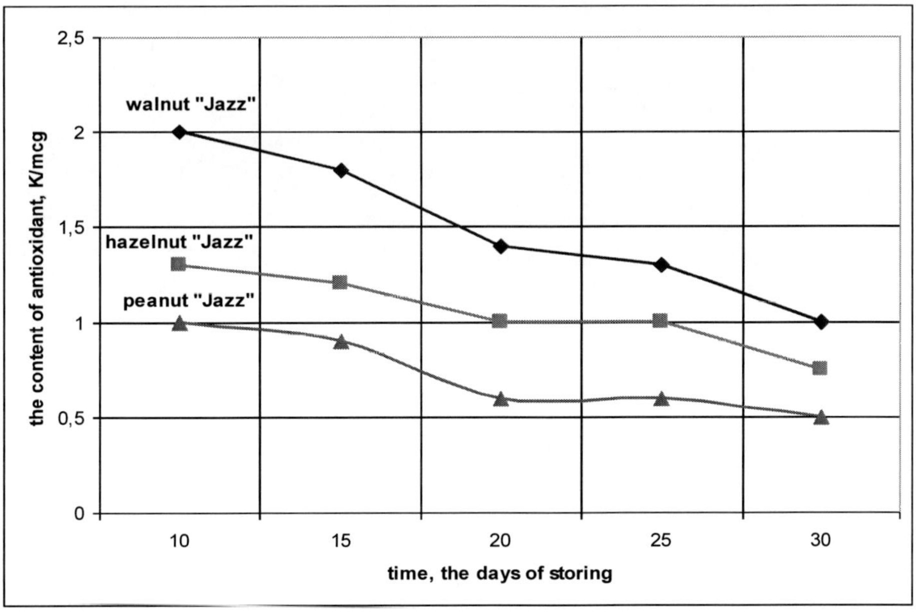

Figure 1. The dynamics of antioxidants` contents in the nuts at the accelerated storage.

We have also formed a study of the antioxidants analysis in the 7 samples of walnuts from the various producers using the amperometric method. The amperometric method is based on the measurement of electric current, which offer insight into the electrochemical oxidation of the substance or mixture of substances on the surface of working electrodes with certain potential.

The detector registers the change of current, which flows through the cells due of the changes in the concentration of antioxidants.

The sensitivity of the amperometric method is determined by the nature of the working electrode's kind and potential applying to it. The amperometric method has a number of advantages like this: the low detection limit, the high selectivity (that means it is determined only those compounds whose molecules can be oxidized), and the small cell size (0,1-5 mkl), ease of maintenance.

The results of the walnuts' antioxidants analysis are represented in figure 2. The picture shows that antioxidants reduced in the storage process. The highest antioxidants` content was noted at the first sample, which has got the highest rating according to the organoleptic indicators.

The nut's consistency in sample № 3 became crumbling and dry after 2 weeks of the accelerated storage and because of this we can't pick out the fat.

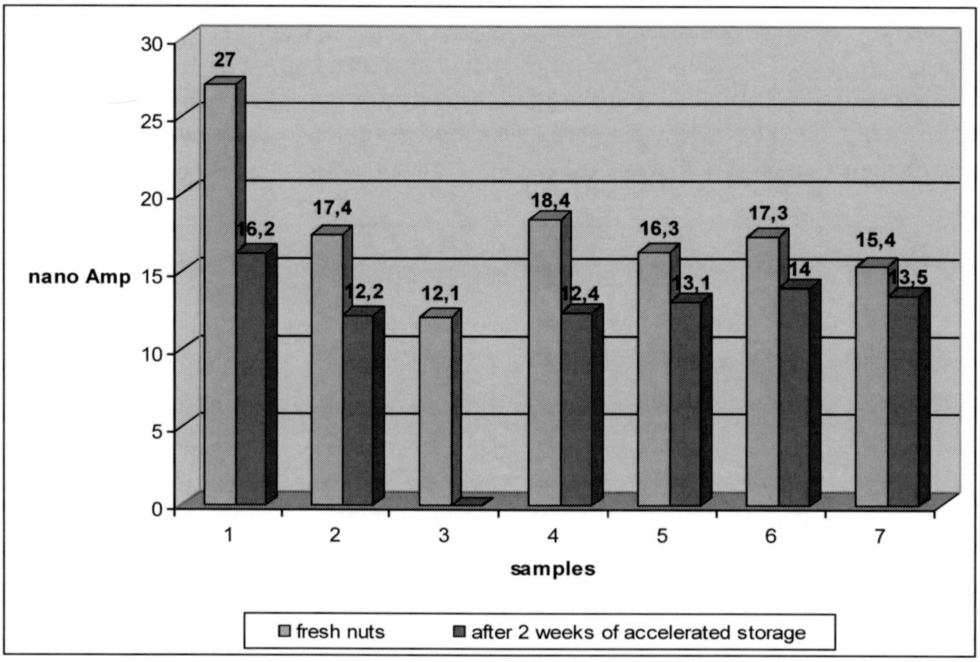

Figure 2. The dynamics of antioxidants` contents in the walnuts at the accelerated storage.

The organoleptic test was made by the tasting committee using the five-point scale taking into account the weight coefficients on the 4 main indicators: the form, color of the kernel on the breaking, taste and smell. During the experiment the statistical homogeneity of the tasters` rating was observed. The results of the organoleptic test of the walnuts in the storage process are represented in the figure 3.

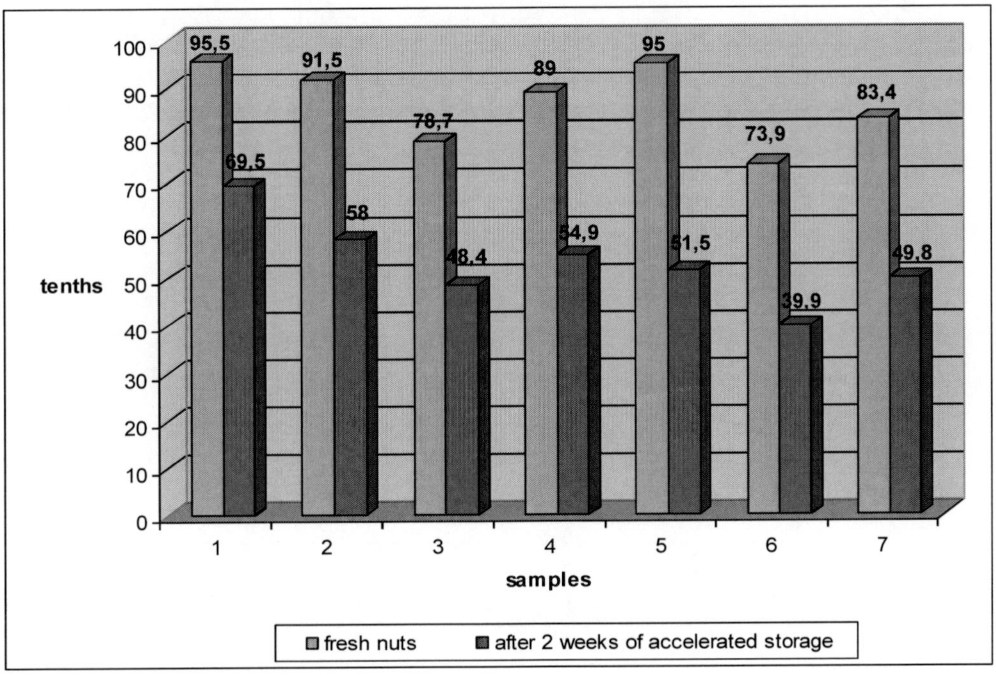

Figure 3. Comprehensive indicator of the quality of samples walnut.

Conclusion

It can be assumed that there is a certain time threshold in during which the oxidation occurs at low speed and the changes in the fat`s compounds don`t affect the organoleptic properties of the product and its biological value because the organoleptic properties` changes were fixed only in the last stage of the experiment.

It is recommended to specify the raw materials` place of origin in the marking. The transparent polypropylene packaging is not very good in preserving of nuts` quality. Because of this it is better to keep the nuts in the opaque packaging just to avoid the effecting of light on the activation of oxidative processes. It is also very important to specify the date of the raw materials` production, the date of goods` packing and the expiration date of the goods.

The study found that the oxidation`s depends on the type of the packing and the initial condition of the nuts at the packaging. Because of this it's necessary to establish the criteria, which lets us form an opinion about antioxidantion activity with the goals of forecasting the optimal period of storage depending on the type of the packing.

References

[1] Voskoboinikov V.A., Ilyina N.E. Features store kernels. Your food № 1, 2001.

[2] Komissarenkov A.A., Dmitrievich I.N., Fedorova O.V. Coulometric methods of analysis: Teaching aid, 2009.

In: Quantitative Chemistry, Biochemistry and Biology ISBN: 978-1-62948-332-0
Editors: G.E. Zaikov, O.V. Stoyanov et al.

Chapter 30

STRUCTURE OF SOLUBLE UNLINKED AND CROSS-LINKED FIBRIN OLIGOMERS

M. A. Rosenfeld*[*], *V. B. Leonova, M. I. Biryukova, E. A. Kostanova and M. V. Vasileva
Emanuel Institute of Biochemical Physics,
Russian Academy of Sciences,Moscow, Russia

ABSTRACT

Self-assembly of soluble unlinked and cross-linked fibrin oligomers formed from fibrin-desA monomer under the influence of FXIIIa was studied in the presence of non-denaturing urea concentrations. By methods of elastic and dynamic light scattering combined with analytical ultracentrifugation, fibrin-desA oligomers formed in both the presence and absence of the FXIIIa were shown to be ensembles consisting of soluble rod-like double-stranded protofibrils with diverse weight and size. Unlinked and cross-linked soluble double-stranded protofibrils can reach the length of 350-450 nm. The structure of soluble covalently-linked protofibrils is stabilized by isopeptide γ-dimers. Electrophoretic data indicate a complete absence of isopeptide bonds between α chains of fibrin-desA molecules. The molecular mechanism of formation of soluble rod-like fibrin structures and specific features of its covalent stabilization under the influence of FXIIIa are discussed.

Keywords: Soluble fibrin, double-stranded protofibrils, FXIIIa, γ-γ-dimers, α- polymers

AIM AND BACKGROUND

Fibrinogen is a soluble protein of blood plasma that is mainly responsible for generation of an insoluble fibrin gel in the presence of thrombin. A fibrinogen molecule is generally thought to be a rod-like 45 nm particle with asymmetry degree of about 5 [1–3]. Fibrinogen is

[*] E-mail: markrosenfeld@rambler.ru.

strongly hydrated and contains 5–6 g water per gram protein [4]. The fibrinogen molecule is a dimer consisting of three pairs of polypeptide chains (*A*α, *B*β, γ)2. Molecular weights of the polypeptide chains *A*α, *B*β, and γ connected by disulfide bonds are ~67.5, 55, and 47.5 kDa, respectively [2]. A fibrinogen molecule is divided into several structurally and functionally important regions: the central *E* region formed by NH2-terminal sites of all three pairs of the polypeptide chains; two peripheral *D* regions consisting of COOH-terminal sites of the *B*β and γ polypeptide chains; two α*C* regions including a significant part of the COOH-terminal sites of *A*α chains; and two *B*βN regions consisting of NH2-terminal sites of *B*β chains [5]. In turn, *E* and *D* regions of fibrinogen have multidomain structure [6–9]. Treatment with thrombin initially results in detachment from fibrinogen *A*α polypeptide chains of fibrinopeptides *A* (*A*α1-16) with production of fibrin monomer molecules (fibrin-desA) containing exposed polymerization center knobs "*A*". This polymerization center is formed by amino acid residues belonging to two different polypeptide chains of a fibrin molecule. One part includes amino acid residues of *N*-terminal site of the polypeptide α chain, α17-20GPRV, whereas the other part is located in β chain of fibrin between residues 15Gly and 42Arg [10]. The polymerization center knob "*A*" is complementary to sites of holes "*a*" that are located in the COOH-terminal site of the γ polypeptide chain in the *D* region between amino acid residues γ337 Asn-379Lys [11, 12]. Monomer fibrin molecules are connected end-to-middle with a displacement of 22.5 nm, and the *D* region of one fibrin monomer molecule interacts with the *E* region of another molecule and also contacts end-to-end with the *D* region of the third molecule, thus forming a multidomain node. This results in production of structurally rigid rod-like double-stranded fibrin protofibrils. Each strand of a protofibril consists, on average, of 7-8 fibrin monomers [13, 14]. On reaching this critical length, protofibrils display a tendency for lateral association and form fibrils, which later divide in branches and form the network for the three-dimensional structure of the fibrin gel. At the stage of protofibril formation the rate of fibrinopeptide *B* (*B*β1-14) cleavage by thrombin significantly increases with subsequent conversion of fibrin-desA into fibrin-desAB [1, 2]. This results in opening of new self-assembly regions knobs "*B*" that interact with polymerization sites holes "*b*" located in the COOH-terminal site of β polypeptide chain of the *D* region. The knob "*B*"-into-hole "*b*" interaction is thought to be insignificant for generation of protofibrils but can be manifested mainly by increasing the lateral association of protofibrils [15].

In the presence of the fibrin-stabilizing plasma transglutaminase (FXIIIa), fibrin polymers are subjected to covalent linking via formation of ε/γ glutamyl-lysine isopeptide bonds. This is associated with the covalent binding of COOH-terminal sites of γ chains, contacting *D* regions, to one another with production of intermolecular γ-dimers. These dimers are produced as a result of binding the lysine γ406 residue of the first molecule with the glutamine residue γ398 or γ399 of the other molecule and of binding γLys406 of the second molecule with γGln398 or γGln399 of the first molecule. FXIIIa also catalyzes formation of isopeptide bonds between α chains of neighboring fibrin molecules providing the interaction of the α chain of one molecule with the α chains of two other molecules with resulting production of α-polymers consisting of more than five α polypeptide fibrin chains [1-3]. Generation of equilibrium, soluble, rod-like oligomers was earlier observed in the presence of moderate urea concentrations in cases of both des*AB*- and fibrin-desA prepared from fibrinogen, respectively, under the influence of thrombin or the thrombin-like enzyme reptilase, which detaches from fibrinogen only fibrinopeptides *A* [16, 17]. Because such oligomers were prepared in the absence of FXIIIa, the monomer molecules interacted to one another only via weak noncovalent bonds that made the self-assembly process completely reversible. Thus, a reasonable question arises, whether soluble cross-linked fibrin oligomers can be generated under the influence of FXIIIa in the presence of urea. Up to now such structures have not been described in the literature. This paper presents a response to this question, which concerns molecular mechanisms of polymerization and covalent stabilization of fibrin.

EXPERIMENTAL

Isolation of fibrinogen. Fibrinogen prepared from citrated blood plasma by alcohol precipitation was used [18]. It was additionally purified from admixtures of plasminogen and fibrin-stabilizing factor by chromatography on DEAE cellulose with subsequent precipitation with ammonium sulfate [19]. The resulting fibrinogen was transferred by gel filtration with Sephadex G-25 into 0.05 M Tris-HCl buffer (pH 7.4) containing 0.15 M NaCl.

Isolation of fibrin-stabilizing factor (FXIII). FXIII was prepared from plasma by fractional precipitation with ammonium sulfate and subsequent ion-exchange chromatography on DEAE-Toy Pearl M650 (Japan) [20]. FXIII was converted into the active form XIIIa under the influence of thrombin (Roche, France) as described in [21]. The residual thrombin was inactivated with antithrombin III–heparin mixture. The completeness of thrombin inactivation was monitored with the chromogenic substrate S-2238 as described in [22]. The factor XIIIa activity determined by the Lorand method [20] was 640 standard units (S.U.) per ml (1 S.U. corresponds to the FXIIIa activity in 1 ml of donor blood serum).

Conversion of fibrin monomer molecules into fibrin. Monomeric fibrin-desAB and fibrin-desA were prepared by activation of fibrinogen with thrombin or a reptilase analog – ancistron (Tekhnologiya-Standart, Russia) using a method described in detail earlier [16]. Monomeric fibrin-desAB was dissolved in 3.5 M urea, and fibrin-desA was dissolved in 1.6 M urea. The urea solutions were prepared in 0.05 M Tris-HCl buffer (pH 7.4) containing 0.15 M NaCl. Polymerization of fibrin-desAB and fibrin-desA in the absence of FXIIIa was initiated by a decrease in the urea concentration from the initial value of 3.5 to 2.2 M in the first case and from 1.6 to 1.2 M in the second case. The urea concentration was decreased by the addition to the initial solutions of fibrin monomers of a calculated amount of Tris-HCl buffer. The urea concentration in the samples of fibrin-desAB under study was 3.5 and 2.2 M, and in the samples of fibrin-desA it was 1.6, 1.4, 1.3, and 1.2 M; the monomeric fibrin concentration in all samples was balanced with the corresponding solutions of urea and was 0.75 mg/ml. The samples were incubated at 25°C for 24 h to reach equilibration in the system. Polymers were covalently linked by adding FXIIIa to the samples of fibrin monomers in the presence of calcium ions, i.e. 1 ml of the polymer solution in urea was simultaneously supplemented with 10 μl (6.4 S.U.) of FXIIIa and 5 μl of 10% $CaCl_2$ (5 mM). The control samples (without XIIIa) were supplemented with 10 μl of Tris-HCl buffer and 5 μl of 10% $CaCl_2$. The cross-linking reaction was performed during 24 h. The ability for depolymerization of the control and experimental (XIIIa-containing) polymers of fibrin-desA prepared in 1.2 M urea was studied after increasing the urea concentration in the samples to 1.8 M.

The cross-linking of polypeptide chains was determined by electrophoresis of reduced samples in 7.5% polyacrylamide gel. The samples were treated with a mixture of 7 M urea, 2% SDS, and 1% β-mercaptoethanol.

Rayleigh light scattering. Average molecular weights M_ω of polymers and their spatial structures were assessed using Rayleigh light scattering with a 4400 spectrometer (Malvern, England) supplied with a multi-bit 64-channel K7025 correlator in cylindrical cells in the scattering angle θ range of 20-130° [17]. A helium-neon laser with wavelength of 632.8 nm was used as the light source. All protein solutions were carefully purified from dust by ultrafiltration. Incremental values of the protein refraction coefficients were determined with a Chromatix KMX-16 differential refractometer (USA).

Dynamic light scattering. *Z*-Mean coefficients of the translational diffusion D_z of samples before and after their incubation with the fibrin-stabilizing factor were determined by dynamic light scattering using a Zetasizer Nano-S spectrometer (Malvern) with the detection angle of 173° [23]. The data were processed using the Origin 7.0 program.

Analytical centrifugation. Sedimentation studies were performed using a Beckman model *E* analytical centrifuge (Austria) equipped with a photoelectric scanning absorption optical system at the rotor rotation rate of 48,000 rpm at 20°C.

Experimental results were processed with the SEDFIT computer program created by P. Schuk at the National Institutes of Health (USA), which is available free at (http://www.analyticalultracentrifugation.com/default.htm).

Results

Soluble fibrin polymers from fibrin-desAB were prepared at urea concentrations in the range 3.5-2.2 M. At the lower urea concentrations gelation occurred. Electrophoresis revealed a complete absence of the enzyme activity on the incubation of soluble fibrin polymers with FXIIIa even at the lowest possible urea concentration of 2.2 M. No accumulation of γ-dimers or α-polymers was detected even after incubation for 24 h (Figure 1). However, electrophoresis data revealed that enzymatic activity of FXIIIa appeared on decreasing urea concentration to 1.6 M. Consequently, covalently linked soluble fibrin polymers cannot be prepared under the action of thrombin on fibrinogen. However, we showed earlier [16] and confirmed in the present work that soluble fibrin polymers formed from can exist at significantly lower urea concentrations, even at 1.2 M. Such polymers can be involved in the cross-linking process catalyzed by FXIIIa. Therefore, fibrin polymers formed only from fibrin-des*A* are characterized in this work.

Hydrodynamic parameters of soluble fibrin in the absence of FXIIIa (the control sample) in the presence of 1.6 M urea had translational diffusion and sedimentation coefficients, respectively, $D_z = (2.40 \pm 0.22) \cdot 10^{-7}$ cm^2/sec and $s = (7.90 \pm 0.37)$ S, which corresponds to the physicochemical parameters of fibrin monomers [16, 24].

The decrease in the urea concentration initiates polymerization. This is confirmed by data of analytic ultracentrifugation (Figure 2) demonstrating a bimodal distribution of molecules by sedimentation rate in the urea concentration range 1.4-1.2 M. Along with the retention of monomeric fibrin-des*A*, the concentration of which decreases with a decrease in the urea concentration, a rapidly sedimenting fraction appears that represents the accumulation of fibrin oligomers. The increase in the sedimentation coefficient and enlargement of the sedimentation peak at its base with a decrease in urea concentration suggest an increase in the molecular weight and polydispersity of the self-assembly products. In FXIIIa was present the general type of the sedimentation curves is retained, but the contents of high molecular weight products increases relative to the control at correspondingly equal urea concentrations (Figure 2, the curves *6-9*).

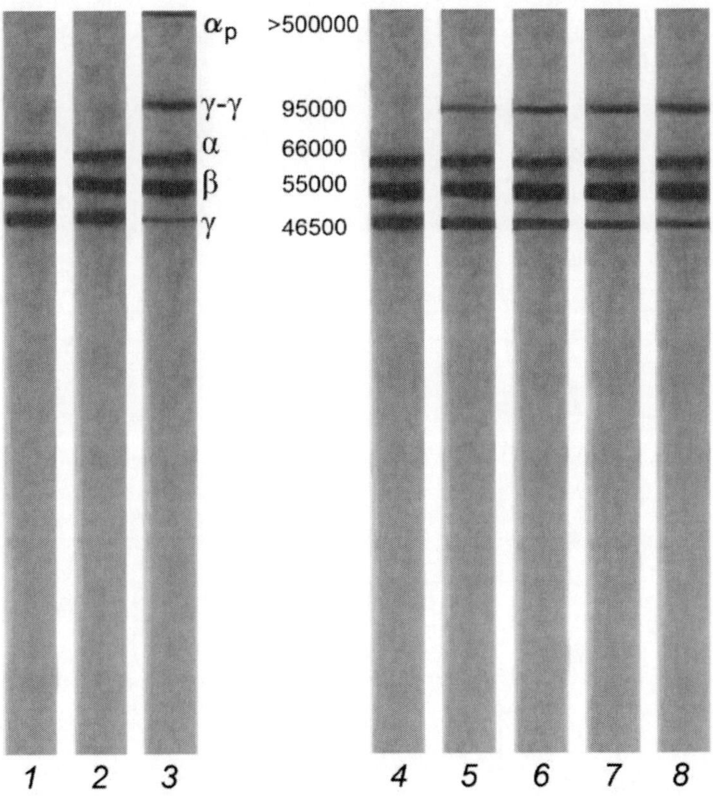

Figure 1. Electrophoresis of the polypeptide fibrin chains. *1*) Monomeric fibrin-des*AB* in 3.5 M urea; *2*) fibrin-des*AB* upon incubation for 24 h with FXIIIa at urea concentration 2.2 M; *3*) the same sample in 1.6 M urea; *4*) monomeric fibrin-des*A* in 1.6 M urea; *5-8*) cross-linked fibrin-des*A* prepared at urea concentrations of 1.6, 1.4, 1.3, and 1.2 M, respectively.

This regularity is revealed on investigation of the oligomer structure using dynamic light scattering. In the urea concentration range 1.4-1.2 M for the control samples the distribution of macromolecules by the translational diffusion coefficient D_z, as in the case of ultracentrifugation, is characterized by two peaks corresponding to the translational diffusion coefficients of fibrin monomer and oligomers (Figure 3). A decrease in the urea concentration is associated with a decrease in the translational diffusion coefficients, which suggests the polymerization of fibrin. In the presence of FXIIIa the histogram is also characterized by a bimodal distribution of translational diffusion coefficients with a relatively greater contribution (as compared to the control) of oligomers on a decrease in the urea concentration (Figure 3, curves *6-9*).

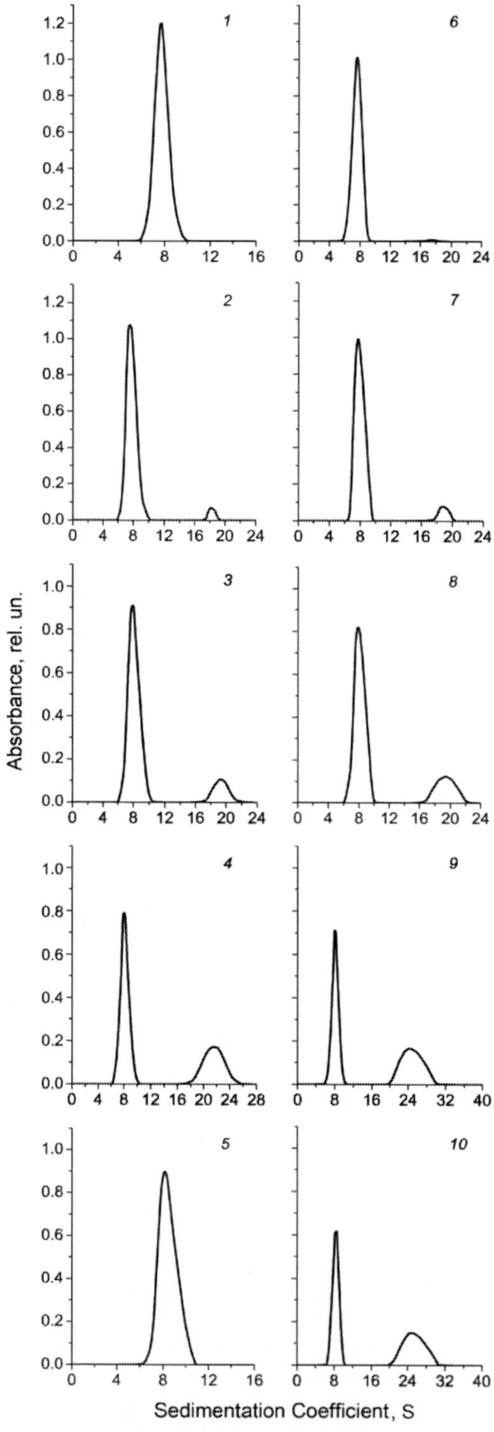

Figure 2. Results of analytical centrifugation of control (*1-5*) and experimental (*6-10*) samples of fibrin-des*A* oligomers prepared at different urea concentrations: *1, 6*) 1.6 M; *2, 7*) 1.4 M; *3, 8*) 1.3 M; *4, 9*) 1.2 M; *5, 10*) 1.8 M.

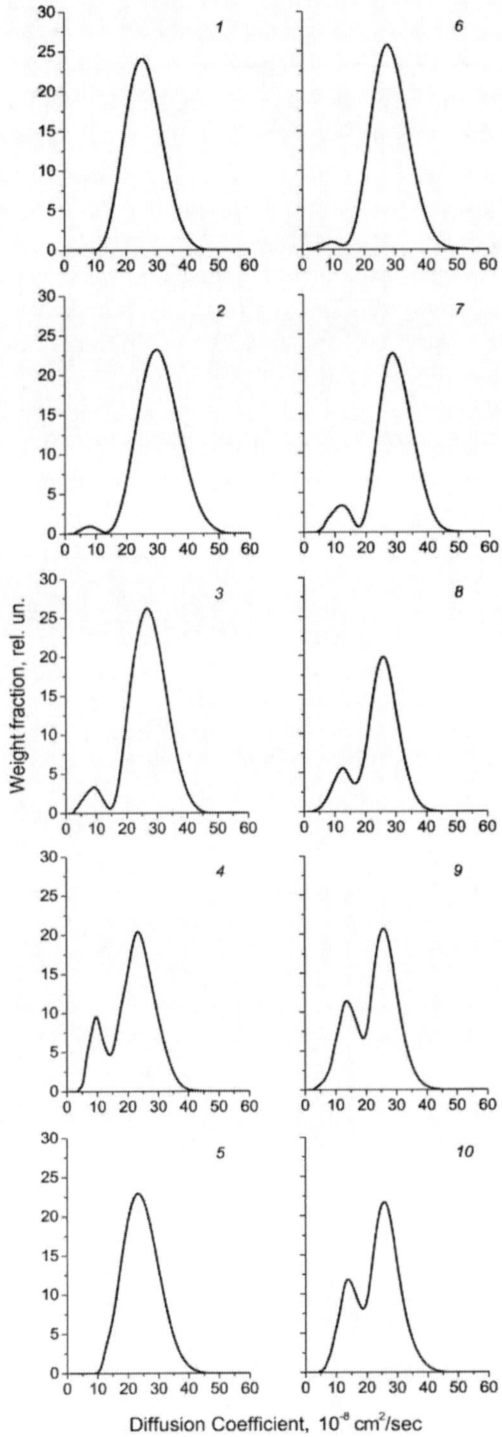

Figure 3. Distribution curves for translational diffusion coefficient D_z of control (*1-5*) and experimental (*6-10*) soluble fibrin-des*A* oligomers prepared at different urea concentrations: *1*, *6*) 1.6 M; *2*, *7*) 1.4 M; *3*, *8*) 1.3 M; *4*, *9*) 1.2 M; *5*, *10*) 1.8 M.

Overall the data of analytic ultracentrifugation and elastic light scattering allows us to characterize the molecular weight and spatial organization of polymers constituting the rapidly sedimenting fraction. Angular dependences of the light scattering plotted in Zimm coordinates allow us to determine mean values of molecular weights M of the polydisperse ensemble of macromolecules (Figure 4). Data of elastic light scattering indicate that a decrease in the urea concentration is accompanied by an increase in values of mean molecular weight M_{ω}. Because the weight contribution of oligomers is proportional to the area under the sedimentation diagram curve, and $M_{\omega} = M_i\omega_i + M_k\omega_k$, where M_i and M_k are mean molecular weights of fibrin-des*A* monomer and its oligomers, and ω_i and ω_k correspond to weight contributions of slowly and rapidly sedimenting fractions, it is possible to calculate M_k values. For the control sample the mean molecular weight of oligomers at the least possible urea concentration of 1.2 M preceding generation of fibrin gel is $(55.5 \pm 3.5)\cdot 10^5$ Da. Based on the resulting value of the molecular weight of soluble oligomers, they occur to consist of about 16 monomeric links. The covalent cross-linking of fibrin-des*A* oligomers under the influence of FXIIIa increases their critical weight to $(63.3 \pm 4.5)\cdot 10^5$ Da; thus, covalently-linked soluble oligomers can include up to 20 monomeric links.

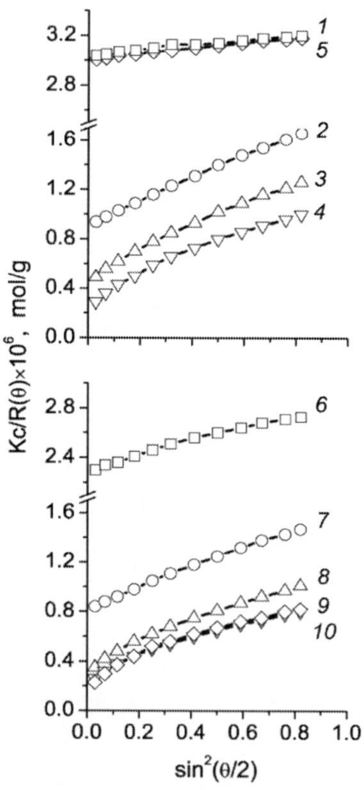

Figure 4. Angular dependences in Zimm coordinates of elastic light scattering intensity for control (*1-5*) and experimental (*6-10*) fibrin-des*A* oligomers prepared at different urea concentrations: *1, 6*) 1.6 M (3.3 ± 0.14 and 4.4 ± 0.18); *2, 7*) 1.4 M (10.9 ± 0.45 and 12.3 ± 0.6); *3, 8*) 1.3 M (21.3 ± 1.3 and 31.2 ± 1.6); *4, 9*) 1.2 M (35.7 ± 1.8 and 43.5 ± 3.5); *5, 10*) 1.8 (3.4 ± 0.17 and 46.7 ± 2.9). In parentheses M_{ω} values in 10^5 Da are presented.

Angular dependences of elastic light scattering intensity (Figure 5) plotted in Casassa coordinates gives the ratio of the mean weight to the mean length M_ω/L_ω for rather long rod-like fibrillar structures of biopolymer [25]. Similarly to the case of calculation of the mean molecular weights of fibrin oligomers, the data of elastic light scattering and analytic centrifugation together allows us to design curves $Kc_k/R_k(\theta) = f[\sin(\theta/2)]$ assigned only to the rapidly sedimenting fraction of oligomers, where c_k and $R_k(\theta)$ correspond to the concentration and the Rayleigh ratio of oligomers. The resulting experimental data show that for the whole range of urea concentration for both control and experimental samples the M_ω/L_ω values were, on average, $(1.20 \pm 0.14)\cdot 10^{11}$ g/mol·cm. Because molecular weight of fibrinogen is 340 kDa and the rod-like molecule length is 45 nm [2], the weight/length ratio for the protein is $0.63\cdot 10^{11}$ g/mol·cm. Consequently, soluble unlinked and covalently cross-linked fibrin-des*A* oligomers are double-stranded protofibrils of 350-450-nm length. Note that even at the lowest possible urea concentration no soluble fibrin fibrils were detected close to the point of the gel formation. This means that the lateral association of double-stranded protofibrils preceding the gel formation can occur only in a very narrow range of urea concentrations. Possible explanations for this will be discussed further.

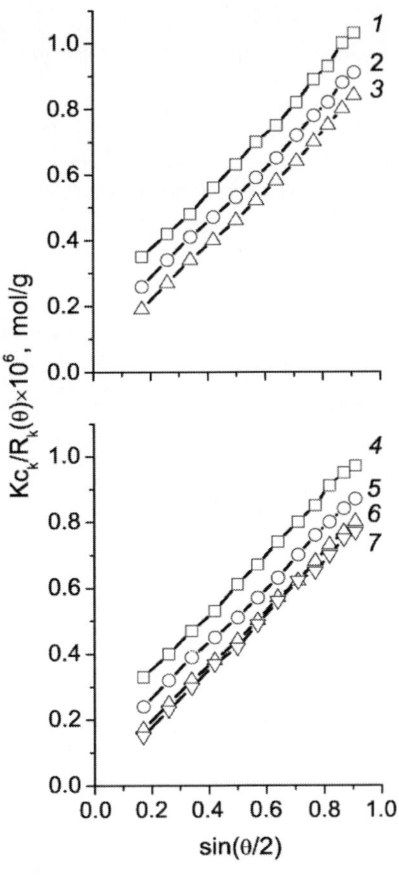

Figure 5. Angular dependences in Casassa coordinates of elastic light scattering intensity for soluble control (*1-3*) and experimental (*4-7*) fibrin-des*A* oligomers prepared at different urea concentrations: *1, 4*) 1.4 M; *2, 5*) 1.3 M; *3, 6*) 1.2 M; *7*) 1.8 M.

In the presence of FXIIIa the spatial structure of fibrin oligomers completely corresponds to unlinked structures. In other words, in the whole range of urea concentrations FXIIIa is unable to covalently cross-link soluble double-stranded protofibrils to one another with production of fibrin fibrils. But it follows from data presented in Figure 1 that the enzymatic activity of the fibrin-stabilizing factor manifests itself at the maximal urea concentration of 1.6 M. This is supported by accumulation of γ-dimers over time. Consequently, monomeric fibrin molecules not incorporated into protofibrils are covalently cross-linked, because the above-presented data show that at the urea concentration of 1.6 M fibrin-desA remains a monomer. With a decrease in the urea concentration the rate of involvement of γ-chains of fibrin monomers into covalent cross-linking increases. However, electrophoretic data indicate a complete absence of accumulation of α-polymers after 24 h of the enzymatic reaction.

The ability of unlinked and covalently-linked fibrin-desA oligomers initially formed at 1.2 M urea to be converted into monomer molecules was studied. For this purpose the oligomer structure in the control and experimental samples was analyzed on increasing the urea concentration from 1.2 to 1.8 M. The physical reason for these experiments is based on fundamentally different resistance of unlinked and covalently-linked structures to the action of urea. As was noted earlier, that fibrin formation in the absence of FXIIIa is fully reversible, whereas enzymatic cross-linking can deprive the structures of their ability for dissociation. Moreover, at the urea concentration of 1.8 M the enzymatic activity of FXIIIa is completely inhibited, and thus those covalently cross-linked structures are analyzed that were generated before the increase in the urea concentration. Data of analytic ultracentrifugation and dynamic light scattering indicate that upon increase in the urea concentration in the control sample the bimodal distribution of macromolecules is changed to a unimodal one (Figs. 2 and 3, curves *5*). Values of the sedimentation and translational diffusion coefficients equal to 8.15 S and $2.40 \cdot 10^{-7}$ cm^2/sec for the molecules suggest the complete dissociation of fibrin oligomers. However, the peak corresponding to the fraction of double-stranded protofibrils formed in the presence of FXIIIa is not noticeably changed upon the increase in the urea concentration (Figs. 2 and 3, curves *10*). Initial values of the molecular weight M_{ω} and the weight/length ratio M_{ω}/L_{ω} for the linked protofibrils are also retained (Figures 4 and 5).

DISCUSSION

The studies have shown that double-stranded fibrin-des*A* oligomers are formed in the control and experimental samples in urea-containing medium. Non-stabilized fibrin is dissolved and soluble polymers are dissociated under the action of urea before formation of initial molecules of fibrin monomer retaining all physicochemical properties inherent in the native protein. In other words, the fibrin formation is completely reversible. Data that were recently obtained on involvement of different polymerization centers in fibrin assembly allow us to analyze in detail the mechanism of formation of soluble double-stranded protofibrils. The formation of double-stranded protofibrils is mainly contributed by knobs "*A*"-into-holes-"*a*" interactions. The knob "*A*" is the so-called combined polymerization center formed by amino acid residues belonging to two different α and β polypeptide chains of fibrin molecules closely located to one another [10]. Two holes "*a*" located in the COOH-terminal part of the γ polypeptide chain of the *D* region, which are responsible for the protofibril formation on interactions with the complementary knobs "*A*", are located on one side of the fibrin molecule, whereas fibrin molecules belonging to two chains of the protofibril are located "face-to-face" [9]. In the forming double-stranded protofibrils the neighboring monomeric links of each chain are located "end-to-end", whereas molecules belonging to different chains are displaced by half of its length. X-Ray crystallography reveals an intermolecular interaction of *D* regions of fibrin molecules contacting "end-to-end" in the each chain of double-stranded protofibril. This interaction is asymmetric relative to the longitudinal axis of the two neighboring monomeric links [6]. It has been shown recently that self-assembly centers different from the knobs "*A*"-into-holes-"*a*" sites are also involved in formation of double-stranded protofibrils [26]. The so-called "*C*" site is located in the fibrin fragment *B*β12-46 and functions until the detachment of fibrinopeptide *B*. The site "*c*" complementary to site "*C*" is located in the *N*-terminal part of the γ chain of the fibrin *D* region, as differentiated from holes "*a*" located in the *C*-terminal part of the γ chain of the *D* region [26]. Upon the detachment from a fibrinogen molecule of fibrinopeptide *A*, treatment with thrombin results in detachment of fibrinopeptide *B*. However, it has been convincingly shown that the bulk of fibrinopeptide *B* is detached only after the protofibril formation and initiation of their lateral association [27]. Interactions of the polymerization center knobs "*B*" exposed as a result of the detachment of fibrinopeptides *B* with the complementary holes "*b*" located in the *C*-terminal part of the β polypeptide chain of the *D* region are mainly responsible for strengthening the lateral association of protofibrils [2, 3]. In the present work we have studied the self-assembly of fibrin oligomers on the treatment of fibrinogen with ancistron, which is incapable of detaching fibrinopeptide *B*. Consequently, the interaction between the knobs "*B*" and holes "*b*" cannot be realized. The lateral association of such protofibrils mainly occurs through binding sites located in fragments of the COOH-terminal part of the fibrin-des*A* γ polypeptide chain, γ350-360 and γ370-380 [28]. In the fibrin molecules these sites are directed differently from the longitudinal axis of the molecule, and this promotes the thickening. Thus, it seems that the assembly of double-stranded protofibrils and their subsequent lateral association are realized through interaction of the ensemble of strictly selective polymerization centers. In the present work it is shown that soluble protofibrils formed from fibrin-des*AB* monomers can exist at urea concentrations in the range 2.6-2.2 M, whereas the self-assembly of soluble protofibrils formed from fibrin-des*A* monomers occurs only at significantly lower concentrations of urea. A possible explanation is as follows. The thickness of fibrin fibrils formed under the influence of thrombin on fibrinogen is always greater than the thickness of fibrils formed as a result of polymerization of fibrin-des*AB* monomers [29]. On the self-assembly of fibrin monomers under the action of thrombin on fibrinogen, the knobs "*B*" become exposed after the protofibril formation, and this provides for the interprotofibrillar interaction of the knobs "*B*" and holes "*b*". On the polymerization of fibrin-des*AB* monomers the knobs "*B*" and holes "*b*" interact mainly inside the protofibrils [26]. Thus, the presence only of knobs "*A*" in fibrin-des*A* makes it more sensitive than fibrin-des*AB* to the action of urea.

The depolymerization activity of urea is known to be due to inhibition of formation of noncovalent hydrogen, hydrophobic, and electrostatic bonds. Just hydrogen and electrostatic bonds mainly contribute to interaction of the main assembly centers of double-stranded protofibrils – polymerization center knobs "*A*" and holes "*a*" [28]. However, urea destabilizes the protein structure mainly due to solubilization of nonpolar lateral amino acid residues. It is suggested that these influences result in the affecting by urea of complementarity of the polymerization center knobs "*A*" and holes "*a*" (possibly, also of the sites "*C*"–"*c*"). Thus, depending on the urea concentration the interaction of *D* and *E* regions of fibrin-des*A* to one another on the assembly of double-stranded protofibrils is partially or completely inhibited. It seems that this can be also assigned to intermolecular interactions of the *D* regions of fibrin molecules contacting "end-to-end" in each chain of the double-stranded protofibril with involvement of two amino acid residues. The action of urea on fibrin assembly is reversible; therefore, an increase in the inhibitor concentration is accompanied by recovery of the ability of knobs "*A*" and holes "*a*" to interact, which involves fibrin monomers into a self-assembly process of double-stranded protofibrils. However, certain thermodynamic conditions prevent a lateral association of double-stranded protofibrils, which is associated with overcoming an entropy barrier. The dimerization of structurally rigid rod-like short or long protofibrils must be accompanied by the same entropy losses [17]; therefore, protofibrils must reach a critical length. In this case the total energy of weak noncovalent bonds will compensate the unfavorable entropy effect, and the dimerization process will be spontaneous. The energy dependence on chain length causes cooperativity of the lateral association of double-stranded protofibrils. The above-said explains the causes that prevent formation of separate small aggregations of short fibrils and promote appearance of long polymeric fibrils. The present work shows (Figure 5a) that double-stranded soluble protofibrils reach the length of about 350 nm that precedes lateral association, and this is in agreement with the literature data [13, 14].

According to the electrophoresis data (Figure 1), in the presence of FXIIIa covalently cross-linked soluble fibrin oligomers are produced. All physicochemical approaches used in the present work show the greater contribution of double-stranded protofibrils as compared to the control samples at the same urea concentrations (Figs. 2-5). Double-stranded protofibrils are an ideal substrate for FXIIIa because all molecules of fibrin monomers are arranged similarly along the polymer chains of the protofibril. And the COOH-terminal sites of γ polypeptide chains of the *D* regions of the neighboring monomeric links are located immediately close to one another, and this promotes a rapid inclusion of γ chain in the enzymatic dimerization. Therefore, at the urea concentration of 1.6 M chains in fibrin-des*AB* are completely included in the dimerization (Figure 1). In this connection, it is interesting that at the urea concentration providing for retention in the control of fibrin-des*A* as monomers, γ-dimers to be generated in the presence of the fibrin-stabilizing factor (Figure 1). The molecular mechanism of cross-linking of monomeric molecules unable to interact with one another seems to be similar to the mechanism of enzymatic cross-linking of fibrinogen molecules. The *C*-terminal site of γ polypeptide chain of fibrinogen (as if of fibrin monomer) contains the binding site of FXIIIa a_2-subunits [1]. On spatial approaching of molecules and in the case of contacting COOH-terminal parts of γ polypeptide chains of the *D* regions, FXIIIa is responsible for their enzymatic cross-linking. Because not all molecules on approaching occupy the position required for the enzymatic reaction, fibrinogen molecules are cross-linked significantly less rapidly that fibrin protofibrils, and some γ polypeptide chains of the protein remain intact [2, 21]. This is the reason for the presence of a certain amount of fibrin-des*A* monomers uninvolved in the enzymatic cross-linking.

To understand the architecture of the resulting covalent bonds, we have analyzed structures of fibrin oligomers formed on the increase in urea concentration from 1.2 to 1.8 M in the experimental sample on its incubation with FXIIIa and in the control sample. As differentiated from the unlinked structures, the double-stranded protofibrils covalently cross-linked under the influence of FXIIIa lose the ability to dissociate due to formation of covalent bonds. Obviously, the dissociation of protofibrils into double-stranded fragments is impossible at any configuration of isopeptide covalent bonds resulting between the COOH-terminal parts of γ chains of the contacting *D* regions of the neighboring monomeric links. The disjoining of a double-stranded protofibril at the longitudinal position of covalent bonds into two one-stranded protofibrils could be most demonstratively recorded using analytical centrifugation. In fact, at the resulting value of the mean molecular weight of a covalently cross-linked protofibril equal to $(63.3 \pm 4.5)\cdot 10^5$ Da two rod-like one-stranded structures would be produced, each with the weight of about $3.0\cdot 10^6$ Da. According to theoretical hydrodynamic calculations [30], the sedimentation coefficient of one-stranded oligomers $s \sim s_i(M_k/M_i)^{0.2}$, where M_k is molecular weight of oligomers, s_i and M_i are, respectively, the sedimentation coefficient and molecular weight of the monomeric link. Inserting the experimental data into this expression we would obtain the s value approximately equal to 12 S. However, there is no such peak in sedimentation diagrams (Figure 2). Dissociation of a double-stranded protofibril into hydrodynamically more compact flexible-chained oligomers also does not occur because data of elastic light scattering show (Figs. 4 and 5) that M_ω and M_ω/L_ω values are not changed on the increase in the urea molarity to 1.8 M. Consequently, our findings have shown the inability of double-stranded protofibrils to dissociate and to produce oligomeric structures. Do these findings allow us to consider the geometry of isopeptide γ-dimers?

The available data on the architecture of isopeptide γ-dimers are very contradictory. According to one standpoint, isopeptide bonds are formed between γ polypeptide chains of two fibrin molecules contacting end-to-end, or longitudinally, within the same chain of a protofibril [31-34]; according to the other standpoint these bonds are formed between γ polypeptide chains of the neighbor monomeric links from the different chains of a protofibril, i.e. located side-to-side, or transversally [35]. The lateral cross-linking of two chains of a protofibril seemed to explain the ability of a cross-linked protofibril to completely recover its initial form after the nearly twofold elongating, as differentiated from an unlinked protofibril [1, 35]. However, together the X-ray crystallography and electron microscopy data about production of one-stranded fragments as a result of plasmin hydrolysis of fibrin stabilized by the FXIIIa convincingly confirm the production of the end-to-end formed γ-dimers [31-34].

Unfortunately, our findings that soluble covalently cross-linked double-stranded protofibrils are unable to dissociate with an increase in the urea concentration do not present information that would sufficient to suppose the geometry of isopeptide γ-dimers. The resistance of protofibril structures to the action of urea seemed to suggest the lateral orientation of covalent bonds because in this case isopeptide bonds resulting between γ polypeptide chains of the neighboring monomeric links belonging to the different chains of the protofibril would prevent their dissociation. However, even at the end-to-end geometry of γ-dimers no dissociation of protofibrils can be observed. Protofibrils possess a helical conformation [36, 37]. Isopeptide bonds between γ polypeptide chains within the same chain of a protofibril stabilize coils of the helix and thus prevent their untwisting and subsequent separation. Thus, at any possible geometry of γ-dimers protofibrils can be resistant to the action of urea.

According to data of electrophoresis, α polypeptide chains of soluble structures of fibrin oligomers are not involved in formation of α-polymers. Now it is well known that formation of isopeptide bonds between α chains of neighbor fibrin molecules occurs an order magnitude more slowly than dimerization of γ chains [2]. In our experiments this can be the reason for the absence of accumulation in the urea-containing medium of a noticeable amount of α-polymer. However, this can also be due to the ability of FXIIIa to covalently cross-link γ chains of molecules only at the stage of protofibril joining into fibrils. There are no data in the literature about the possibility of the covalent stabilization of protofibrils due to cross-linking of γ chains; therefore, at present this idea is purely hypothetical. We think that glutamine and lysine residues of fibrin α polypeptide chains involved in α-polymerization are sterically available to one another only in the structure of a fibril. This hypothesis is supported by our findings on the presence of α-polymers in fibrin-des*AB* produced in the presence of 1.6 M urea. Moreover, there are also direct confirmations that the production of α-polymers correlates in time with the rate of gel formation [38]. Thus, it is supposed that γ-dimers with any orientation should stabilize double-stranded protofibrils due to presence of isopeptide bonds, whereas α-polymers appear between protofibrils and stabilize fibrils. In general, this is responsible for formation in fibrin of a strong carcass of covalent bonds.

The authors are grateful to researchers of the Belozersky Institute of Physico-Chemical Biology, Moscow State University, P. V. Kalmykov and N. N. Magretova, for their help in performing analytic centrifugation experiments.

REFERENCES

[1] Mosesson, M. W. (2005) *J. Thromb. Haemost.*, 3, 1894-1904.

[2] Blomback, B. (1996) *Thromb. Res.*, 83, 1-75.

[3] Weisel, J. W. (2005) *Adv. Protein. Chem.*, 70, 247-299.

[4] Marguerie, G., and Stuhrmann, H. B. (1976) *J. Mol. Biol.*, 102, 143-156.

[5] Medved, L., and Weisel, W. (2009) *J. Thromb. Haemost.*, 7, 355-359.

[6] Spraggon, G., Everse, S. J., and Doolittle, R. F. (1997) *Nature*, 389, 455-462.

[7] Madrazo, J., Brown, J. H., Livinovich, S., Dominguez, R., Yakovlev, S., Medved, L., and Cohen, C. (2001) *Proc. Natl. Acad. Sci. USA*, 98, 11967-11972.

[8] Brown, J. H., Volkman, N., Jun, G., Henschen-Edman, A. H., and Cohen, C. (2001) *Proc. Natl. Acad. Sci. USA*, 97, 85-90.

[9] Yang, Z., Kollman, J. K., Pandi, L., and Doolittle, R. F. (2001) *Biochemistry*, 40, 12515-12523.

[10] Budzynski, A. Z., Olexa, S. A., and Pandya, B. V. (1983) *Ann. N. Y. Acad. Sci.*, 408, 301-314.

[11] Yee, V. C., Pratt, K. P., Cote, H., Trong, I. L., Chung, D. W., Davie, E. W., Stencamp, R. E., and Teller, D. C. (1997) *Structure*, 5, 125-138.

[12] Pratt, K. P., Cote, H., Chung, D. W., Stencamp, R. E., and Davie, E. W. (1997) *Proc. Natl. Acad. Sci. USA*, 94, 71-81.

[13] Doolittle, R. F. (1984) *Annu. Rev. Biochem.*, 53, 195-229.

[14] Blomback, B., Hessel, B., Hogg, D., and Therkildsen, L. (1978) *Nature*, 257, 501-505.

[15] Mihalyi, E. (1988) *Biochemistry*, 27, 967-976.

[16] Rosenfeld, M. A., Gershkovich, K. B., Kuznetsov, D. V., Meshkov, B. B., and Gontar, I. D. (1986) *Mol. Biol.*, 20, 1098-1110.

[17] Rosenfeld, M. A., Gershkovich, K. B., and Kuznetsov, D. V. (1988) *Mol. Biol.*, 22, 86-93.
[18] Blomback, B., and Laurent, T. C. (1958) *Ark. Kemi.*, 2, 137-146.
[19] Lawrie, J. S., Ross, J., and Kemp, G. G. (1979) *Biochem. Soc. Trans.*, 7, 693-694.
[20] Lorand, L., Gredo, R. B., and Janus, T. J. (1981) *Methods Enzymol.*, 809, 333-341.
[21] Rosenfeld, M., and Vasileva, M. V. (1991) *Biomed. Sci.*, 2, 155-161.
[22] Svedsen, L., Blomback, B., Blomback, M., and Olsson, P. J. (1972) *Tromb. Res.*, 3, 267-278.
[23] Rosenfeld, M. A., Leonova, V. B., Shchegolikhin, A. N., Razumovskii, S. D., Konstantinova, M. L., Bychkova, A. V., and Kovarskii, A. L. (2010) *Biochemistry (Moscow)*, 75, 1285-1293.
[24] Shulman, S. (1953) *J. Am. Chem. Soc.*, 75, 5846-5852.
[25] Casassa, E. F. (1955) *J. Chem. Phys.*, 23, 596-597.
[26] Lugovskoi, E. V., Gritsenko, P. G., and Komissarenko, S. V. (2009) *Bioorg. Khim.*, 35, 437-456.
[27] Hogg, D. H., and Blomback, B. (1978) *Thromb. Res.*, 12, 953-964.
[28] Yang, Z., Mochalkin, I., and Doolittle, R. F. (2000) *Proc. Natl. Acad. Sci. USA*, 97, 14156-14161.
[29] Weisel, J. W., Veclich, Y., and Gorkun, O. (1993) *J. Mol. Biol.*, 232, 285-297.
[30] Tsvetkov, V. N., Eskin, V. E., and Frenkel, S. Ya. (1964) *Structure of Macromolecules in Solutions* [in Russian], Nauka, Moscow, pp. 93-480.
[31] Weisel, J. W., Francis, C. W., Nagaswani, C., and Marder, V. J. (1993) *J. Biol. Chem.*, 268, 26618-26624.
[32] Veklich, Y., Ang, E. K., Lorand, L., and Weisel, J. W. (1998) *Proc. Natl. Acad. Sci. USA*, 95, 1438-1442.
[33] Doolittle, R. F. (2003) *J. Thromb. Haemost.*, 1, 1559-1565.
[34] Weisel, J. W. (2004) *J. Thromb. Haemost.*, 2, 1467-1469.
[35] Mosesson, M. W. (2004) *J. Thromb. Haemost.*, 2, 388-393.
[36] Krakow, Enders, G. F., Siegel, B. M., and Scheraga, H. A. (1972) *J. Mol. Biol.*, 71, 95-103.
[37] Williams, R. C. (1983) *Ann. N. Y. Acad. Sci.*, 408, 180-193.
[38] Gron, B., Filion-Myklebust, C., Bennick, A., Nieuwenhuizen, W., Matsueda, G. R., and Brosstad, F. (1992) *Blood Coagul. Fibrinol.*, 3, 731-736.

In: Quantitative Chemistry, Biochemistry and Biology
Editors: G.E. Zaikov, O.V. Stoyanov et al.
ISBN: 978-1-62948-332-0

Chapter 31

KINETICS AND MECHANISM OF THE OZONE REACTION WITH ALCOHOLS, KETONES, ETHERS AND HYDROXYBENZENES

Slavcho Kirillov Rakovsky[1*], Metody Anachkov[1], Mikhail M. Belitskii[2] and G.E. Zaikov[2]

[1]Institute of Catalysis, Bulgarian Academy of Sciences, Sofia Bulgaria
[2]N.M. Emanuel Institute of Biochemical Physics, Russian Academy of Sciences, Moscow, Russia

ABSTRACT

The review, based on 92 references, is focused on degradation of organics by ozonation and it comprises various classes of oxygen-containing organic compounds – alcohols, ketones, ethers and hydroxybenzenes. The mechanisms of a multitude of ozone reactions with these compounds in organic solvents are discussed in details, presenting the respective reaction schemes and the corresponding kinetic parameters are given and some thermodynamic parameters are also listed. The dependences of the kinetics and the mechanism of the ozonation reactions on the structure of the compounds, on the medium and on the reaction conditions are revealed. The various possible applications of ozonolysis are specified and discussed. All these reactions have practical importance for the protection of the environment.

Keywords: Ozonation, alcohols, ketones, ethers, hydroxybenzenes, kinetics, mechanisms

INTRODUCTION

The ozonolysis of oxygen-containing compounds is a promising process that takes place under mild conditions and yields compounds of a higher oxidation state than that of the starting

* Corresponding author: E-mail address: rakovsky@ic.bas.bg.

compounds. It may find various applications in chemical and pharmaceuticals industries, fine organic synthesis, etc. [1-2]. The widest application, in this respect, has been found for the oxidation of primary and secondary alcohols respectively into their corresponding aldehydes and ketones.

For example in the cases of the oxidation of open-chain (simple) and cyclic secondary alcohols the yield of ketones is within the range 57% up to 83% [3]. Moreover, the considered interactions are extremely important from an ecological point of view for the utilization and purification of industrial wastewaters, originating from hydroxybenzene production, through their partial or complete oxidation [4-24].

The importance of this process for theory and practice gave us an impetus to carry out systematic investigations [25-35]. The aim of the present paper is the precise determination of the rate constants of ozonolysis of some more widely occurring representatives of the studied classes of organic compounds.

Applying the activated complex method (ACT) [36] and collision theory (CT) [36] some theoretical investigations have also been carried out elucidating the structure of the transition state formed in the course of the reaction.

On the basis of the correlation between the results from the experimental and theoretical studies some peculiarities in the mechanism of ozone reactions with the considered classes of oxygen containing compounds have been established.

EXPERIMENTAL

Ozone has been obtained from dry oxygen by means of a silent discharge of 5-8 kV at an oxygen flow rate of 0.1 L/min. The ozone concentration, 10^{-5} -10^{-3} M, has been measured spectrophotometrically in the wavelength region of 254-300 nm in a 5 cm quartz gas cell.

METHODS

The UV, IR, ESR spectra were registered on standard equipments, as well as HPLC, and GC analyses.

KINETIC MEASUREMENTS

Static Method

Pure reagent or reagent solution was injected into thermostatic 1 cm quartz cuvette, containing a solution of ozone in CCl_4, the time of mixing being less than 0.2 second. Ozone concentration was monitored spectro-photometrically in the region of 270-290 nm. At $[RH]_o$/ $[O_3]_{lo}$>100, the ozone pseudomonomolecular constant $k`=k[RH]_o$ was determined on the basis of the equation $lg([O_3]_{lo}/[O_3]_{lt})=k`t$ where $[O_3]_{lo}$ and $[O_3]_{lt}$ are the initial and current concentrations of ozone in solution, respectively (27).

Dynamic Method

The ozone was bubbled through a cylindrical glass reactor with inner diameter ϕ=1.7-3.7 cm and height 7-15 cm, supplied with porous glass grit-G2 at its bottom. The accuracy of maintaining constant temperature was ±0.1°C. Conventionally, gas flow rate was v=0.1 L/min; the solutions volume was V=10 ml; the ozone concentrations at the reactor inlet ($[O_3]_o$) varied from 10^{-6} to 10^{-3} M; the solvent was CCl_4; $[RH]_o=10^{-4}-10^{1}$ M. The inlet and outlet ozone concentrations were measured in the gas phase in the 254-300 nm wavelength range. The determination of rate constants is based on the approach, which connects the balance of consumed ozone with the rate of the chemical reaction - Eq. (1):

$$\omega([O_3]_o-[O_3]_g) = k[O_3]_l[RH] \quad (1)$$

where ω is the relative flow rate of ozone–oxygen gas mixture (in litres per L of solution per sec); $[O_3]_o$ and $[O_3]_g$ are the ozone concentrations at the reactor inlet and outlet, respectively, $[O_3]_l$ is ozone concentration in the solution; [RH] is concentration of the reagent. This model is valid in all cases, when the rate of ozone absorption is considerably greater than the rate of the chemical reactions. If in the case of a bimolecular reaction, in accordance with Henry's Law, $[O_3]_l$ is substituted with $\alpha[O_3]_g$, where α is Henry's coefficient, Eq. (1) can be transformed into Eq. (2) (1):

$$k = \omega.\Delta[O_3]/([RH].\ \alpha[O_3]_g) \quad (2)$$

One of the widely applied criteria with respect to the conditions of validity (applicability) of Henry's Law is the expression:

$$D_{O3}.k_I'/k_L^2 << 1$$

where D_{O3} is the diffusion coefficient of ozone in the solution; $k_L=D_{O3}/\delta$ is the coefficient of mass transfer in the liquid phase, and δ is thickness of the boundary layer in the hydrodynamic model of renovation surface; or $k_L=(D_{O3}.s)^{1/2}$, where s is the time interval of renovation. In the case of applying the bubbling method with small bubbles (diameters up to 2.5-3 mm) then $k_L=0.31x(g\nu)^{1/3}x(D_{O3}/\nu)^{2/3}$, where $\nu=\eta/\rho$, is the kinematic viscosity of the solvent, η is the viscosity of the solvent, ρ is the solvent density, g is the earth acceleration. Usually the k_L values are of the order of 0.1-0.05 cm/s. As k_I`$=k[RH]_o$ in most of the cases it is possible to select such values for $[RH]_o$, at which the criterion for applicability of Henry's Law is fulfilled [34].

In order to minimize the influence of the so-called "effect of delay in the response function"- $[O_3]g=f(\tau)$ upon calculating the values of k such sections of the kinetic curves are selected, which appear to be practically parallel or only slightly inclined with respect to the abscissa: k_I`$\alpha[O_3]_g >> d[O_3]_g/d\tau$. The advantages and limitations of this method have been discussed in detail in (37, 38). Despite some contradictory observations, the significant part of rate constants of ozone with organic compounds and polymers are obtained on the basis of Eq. (2) [1, 34].

Alcohols

A number of authors have tried to elucidate the kinetics and mechanism of the ozonolysis of alcohols and their application to the selective preparation of ketones and aldehydes in high yields under mild conditions [3, 39-46]. The basic concepts on such reactions are discussed in several references [1, 2, 25, 34, 35, 39]. It has been established that the rate of ethanol ozonolysis does not change when the hydroxyl group is deuterated [43] and its value is 4.17 times higher than that when the methylene group is deuterated. On the basis of this fact, together with the data from the analysis of the product composition and the kinetics of their formation the authors have suggested that the H-atom abstraction by the ozone molecule is the rate-determining step of the reaction. This conclusion is confirmed by the data in Table 1 as reported by various authors [41-43]. The ratio between the relative reactivities of tertiary: secondary: primary alcohols is 1:12:241 according to reference [41], or 1:156:817 according to reference [7]. On the basis of the studies on the reaction of ozone with methyl, ethyl and 2-propyl alcohol, the following reaction mechanism is proposed (Scheme 1) [40]:

$$CH_3OH + O_3 \Leftrightarrow [HOC^{\bullet}H_2...HO_3^{\bullet}]^{\#} \rightarrow HOCH_2OOOH,$$

and in the presence of a base:

$$CH_3OH + O_3 + (B) \rightarrow HOCHO + {}^{-}O_2H + BH^{+}$$
$$CH_3CH_2OH + O_3 \Leftrightarrow [CH_3HC^{\bullet}(OH)...HO_3^{\bullet}]^{\#} \rightarrow CH_3HC(OOOH)OH \rightarrow$$
$$CH_3OOH + H_2O_2 \text{ and/or } CH_3CHO + O_2 + H_2O$$
$$(CH_3)_2C(OH)H + O_3 \Leftrightarrow [(CH_3)_2C^{\bullet}(OH)...HO_3^{\bullet}]^{\#} \rightarrow (CH_3)_2C(OH)OOOH$$
$$\downarrow -O_2$$
$$CH_3C(OH){=}CH_2 + H_2O + O_2 \quad (CH_3)_2CO + H_2O$$
$$\downarrow +O_3$$
$$CH_3COOH + HCHO$$

Scheme 1

It is supposed that an intermediate ion or pair of radicals is formed, whose recombination in the kinetic cage yields α-hydroxy-hydrotrioxide. The latter leaves the cage and passes into the volume of solution.

Table 1. Kinetic parameters of the ozone reaction with aliphatic alcohols at 25°C

Alcohols	k, $M^{-1}.s^{-1}$	lg A	E, kcal/mol	Reference
t.-Butanol	0.01	4.6	9.0	[32]
t.-Butanol	0.05	-	-	[31]
Ethanol	0.35	6.7	9.8	[32]
Ethanol	0.25 (22 °C]	6.7	9.8	[33]
n-Butanol	0.54	7.3	10.3	[32]
n-Butanol	0.39			[31]
iso-Propanol	0.89	7.3	10.0	[32]
Cyclopentanol	1.35			[31]

The authors of reference [40] found that the product composition of 2-propanol ozonolysis depends on the method of propanol purification. Taking into account the above given consideration about the reaction pathway, in our opinion, the mechanism still remains unclear and new data should be provided for its elucidation. In this connection we have carried out intensive experimental studies on the kinetics of ozonolysis of MeOH, *t*-BuOH, EtOH, *n*-PrOH, *n*-BuOH, *i*-PrOH, *s*-BuOH, *c*-HexOH both by the static and barbotage methods, the results of which are summarized in Figures 1-3 and Tables 2-4.

Figure 1 demonstrates the kinetic data of ozone consumption in solutions of methanol, ethanol and iso-propanol by means of the static method with time interval of mixing less than 0.2 seconds.

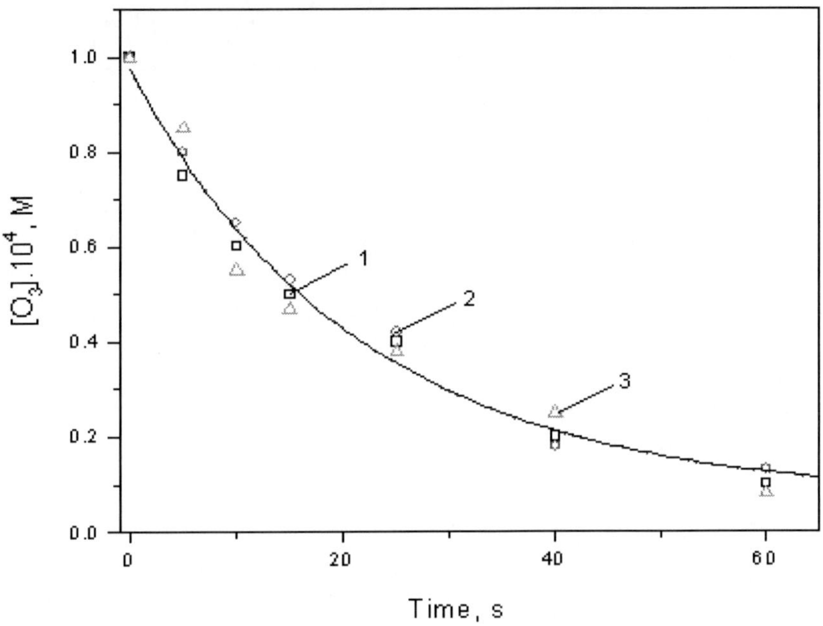

Figure 1. Kinetics of ozone consumption in the reaction of ozone with: 1 - methanol, 22 °C, 0.74 M; 2 - ethanol, 3 °C, 0.24 M; and 3 - isopropanol, 3.5 °C, 0.037 M.

It is seen that regardless of the type of alcohol being ozonized, the kinetic curves coincide with first-order rate law. With a view to more precise evaluation and comparison of the respective rate constants such concentrations of the respective alcohols have been selected, at which the rates of ozone consumption have close values. The kinetic curves of ozone reactions with MeOH, EtOH and i-PrOH gave the following values of the rate constants: 0.057, 0.17 and 1.13 $M^{-1}.s^{-1}$, respectively.

The same values were also obtained by carrying out the reactions in a bubble reactor. The values of *k* are calculated on the basis of Equation 2 – the data are given in Table 2 and they are represented graphically in Figures 2 and 3.

Processing the data from Table 2, columns 2, 3, and 5, based on Equation 2, yields the linear dependencies shown in Figure 2a and 2b. The dependence of the rate W on ΔO_3 has also been found out to be a linear one.

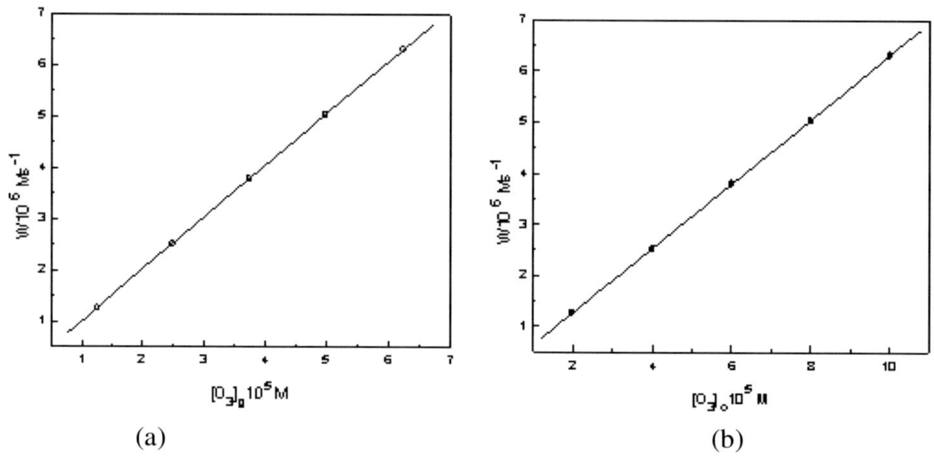

Figure 2. Dependence of the rate of ozone consumption on the ozone concentrations- (a) inlet $[O_3]_o$ and (b) outlet $[O_3]_g$, at MeOH concentration of 0.74 M.

The dependence of the rate W on the concentration of alcohol [ROH] (column 1) is a curve, which approaches a limit value at $W = 1.67 \times 10^{-5}$ $M.s^{-1}$. The nonlinear character of this relationship is in accordance with the complexity of Equation 2 (see also Figure 3). In the case when $[ROH] \rightarrow \infty$, then we have $[O_3]_g \rightarrow 0$ and $\Delta O_3 \rightarrow [O_3]_o$. The values of *k,* calculated based on Equation 2 at each one of the points on the curve (Figure 3), are equal to 0.057.

Table 2. Kinetic parameters of ozone reaction with MeOH in carbon tetrachloride (CCl_4) and pure MeOH solutions: 22 °C, $\omega = 0.167$ s^{-1}; $v = 1.67 \times 10^{-3}$ L/s, maximum rate of ozone inlet - 1.67×10^5 $M.s^{-1}$

[MeOH], M	$[O_3]_0 \times 10^5$, M	$[O_3]_g \times 10^5$, M	$\Delta[O_3] \times 10^5$, M	$W \times 10^6$, $M.s^{-1}$	k, $M^{-1}.s^{-1}$
1	2	3	4	5	6
0	10	0	10	0	-
0.247	10	8.32	1.68	2.80	0.057
0.439	10	7.35	2.65	4.43	0.056
0.618	10	6.64	3.36	5.61	0.058
0.740	10	6.22	3.78	6.31	0.055
0.987	10	5.53	4.47	7.46	0.057
1.0	10	5.49	4.51	7.53	0.056
2.0	10	3.79	6.21	10.37	0.055
3.0	10	2.89	7.11	11.87	0.057
4.0	10	2.34	7.66	12.79	0.058
5.0	10	1.96	8.04	13.42	0.054
10.0	10	1.09	8.91	14.88	0.056
24.7	10	0.47	9.53	15.91	0.058
0.740	8	4.98	3.02	5.04	0.057
0.740	6	3.73	2.27	3.79	0.055
0.740	4	2.49	1.51	2.52	0.056
0.740	2	1.24	0.76	1.27	0.058

Table 3. Dependence of *k* on the temperature in °C for ozone reaction with three types of alcohols

k, $M^{-1}.s^{-1}$	0 °C	10 °C	20 °C	25 °C	30 °C
MeOH	0.008	0.021	0.049	0.072	0.108
t-BuOH	0.005	0.013	0.029	0.045	0.064
EtOH	0.14	0.28	0.54	0.74	1.10
n-PrOH	0.19	0.36	0.67	0.89	1.18
n-BuOH	0.15	0.30	0.56	0.76	1.10
i-PrOH	0.93	1.61	2.71	3.46	4.39
s-BuOH	0.88	1.54	2.58	3.29	4.18
c-HexOH	0.92	1.59	2.65	3.37	4.27

Therefore the experimentally obtained linear dependences of W on $[O_3]_o$, $[O_3]_g$ and on $\Delta[O_3]$ as well as the profile of the curve in Figure 3 correspond to the mathematical description of ozonolysis in a barbotage reactor - Equation 2.

Judging from the analysis of the data obtained (Tables 3 and 4) it follows that the rate constant and the activation energies are strongly dependent on the alcohol structure. The interaction of ozone with MeOH possessing primary α-H atoms and with *tert*-BuOH having only primary C-H bonds has been found to be the slowest, and the value of *k* at 20 °C per one α-H atom in the first case is 1.62×10^{-2} $M^{-1}.s^{-1}$, and in the second case, related to one primary H atom, is 3.22×10^{-3} $M^{-1}.s^{-1}$. The difference in the values of these constants is due to the fact that while the OH group in MeOH directly affects the α-H atom, in the case of the *tert*-BuOH, which does not possess any α-H atoms, the effect of the OH groups is transferred through one σ-bond and so it is considerably weaker.

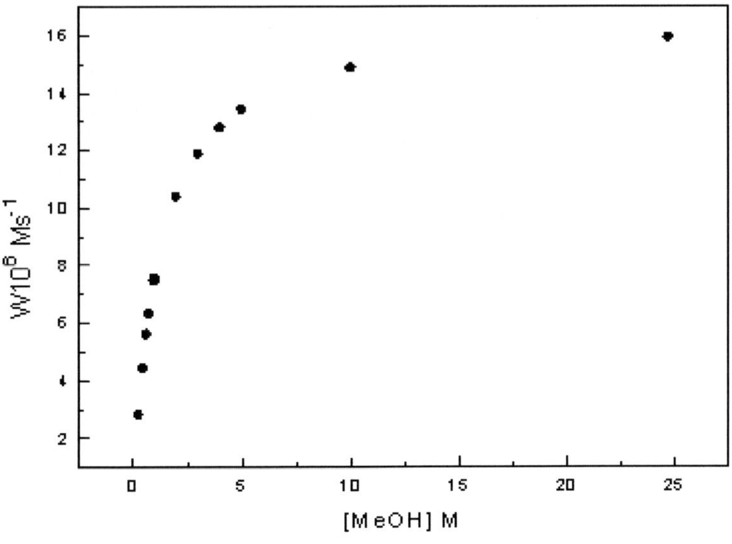

Figure 3. Dependence of the rate of ozone consumption on methanol concentration at $[O_3]_o = 1.10^{-4}$ M (according to Equation 1).

As a result of this, the reactivity of *tert*-BuOH becomes similar to that of methane and neopentane. The interactions of ozone with EtOH, *n*-PrOH, *n*-BuOH take place at higher rates and the values of *k* per one α-H atom amount to: 0.27, 0.34 and 0.28, respectively. The higher rates of the ozone reactions with these alcohols are associated with the presence of secondary α-H atoms in their molecules, which have lower bond energies than the primary ones. The enhanced reactivity of *n*-PrOH compared with that of EtOH could be assigned to the donor effect of the second CH_3 group while the donor effect of the C_2H_5 group in *n*-BuOH is weaker than that of the methyl group and therefore *k* is lower. It has been found that the reactions of ozone with *i*-PrOH, *s*-BuOH and *s*-HexOH alcohols possessing *tert*-H atoms with the lowest bond energies is the fastest with *k* equal to 2.71, 2.58 and 2.65, respectively, i.e., in fact they are almost the same. The ratio between the reduced values of *k* at 25 °C for methanol: ethanol: *sec*.butanol are 1:15:137. Simultaneously, the values of E_a decrease with the decrease of the α-H atoms bond energy (D) and their ratio is 1:0.78:0.61 [12, 15]. This fact can be regarded as important evidence for the mechanism of α-C-H-atom abstraction by ozone.

The theoretical estimates of A were performed considering the two possible structures of the activated complex (AC): linear (LC) and cyclic (CC) (Figure 4). The good agreement between the experimental data and theoretical estimates for A allows the determination of the AC structure and on this basis enables the selection of the most probable mechanism. Simultaneously the value of the pre-exponential factor was calculated by the collision method and by comparing its value with the one calculated based on the AC theory the steric factor -p was determined (Equation 3) [48-52]:

Table 4. Kinetic parameters of ozone reaction with some alcohols at 25°C

Parameters	MeOH	*t*-BuOH	EtOH	*n*-PrOH	*n*-BuOH	*i*-PrOH	*s*-BuOH	*c*-HexOH
$k \times 10^2$, $M^{-1}.s^{-1}$	7.2	4.5	74	89	76	346	329	337
n, α-C-H	3	9 (β-C-H)	2	2	2	1	1	1
$k \times 10^2$/n, $M^{-1}.s^{-1}$	2.4	0.5	37	44	38	346	329	337
E_a, kcal/mol	13.9	13.7	10.9	10.1	10.4	8.5	8.5	8.4
$A \times 10^{-7}$, $M^{-1}.s^{-1}$	41	5.8	3.9	2.4	1.7	0.62	0.59	0.51

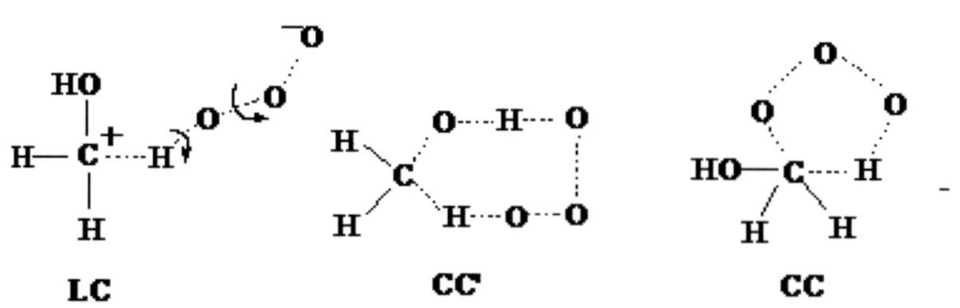

Figure 4. Structure of the probable activated complexes in the reaction of ozone with aliphatic alcohols: LC - linear with free fragment rotation and CC′ and CC - cyclic complexes without free rotation.

$$k = p.Z_0.\exp(-E_a/RT), \quad (3)$$

where: $Z_0 = \pi.(r_A+r_B)^2 \times (k_B\ T/\pi m^*)^{1/2}$; Z_o is the collision factor; r_A and r_B are the van der Waals radii of the reagents; k_B is Boltzmann's constant; T is the absolute temperature; and m* is the reduced mass.

The pre-exponential factors for the reactions of ozone with alcohols, calculated according to the activated complex method (ACT) and collision theory (CT) are represented in Table 5.

An interesting fact is the good agreement between the values of A^{LC} and A^{CC}. The values of A^{CC} with the five- and seven-member cyclic form of AC are practically similar as they differ by not more than 1%.

However the free rotation produces an increase in $A^{LC}{}_{cal}$ by three orders of magnitude compared with that of A^{CC}.

The steric factor has a value, which is in agreement with a liquid phase reaction. The ratio between the calculated and experimental values of A^{LC}, is increasing in the sequence primary: secondary: tertiary alcohols is 1:3.4:8.8, i.e., the coincidence in this sequence is decreasing. This is connected with the fact that E_{fr} is not zero and it is increasing with the changes in the alcohol structure in the same sequence.

It has been found that the values of $A^{LC}{}_{calc}$ and A_{obs} coincide when the values of rotation energy are those given in line 10 in Table 5. We have obtained the same values by means of the *MOPAC 6* software package and therefore the data on the free rotation energy (E_{fr}) in line 10 can also be considered as theoretically evaluated.

The good agreement between $A^{LC}{}_{calc}$ and A_{obs} suggests that the rate-determining step of the ozone reaction with alcohols involves the formation of linear AC and abstraction of the α-H atom from the alcohol molecule.

The experimental and theoretical results conform well to Scheme 2.

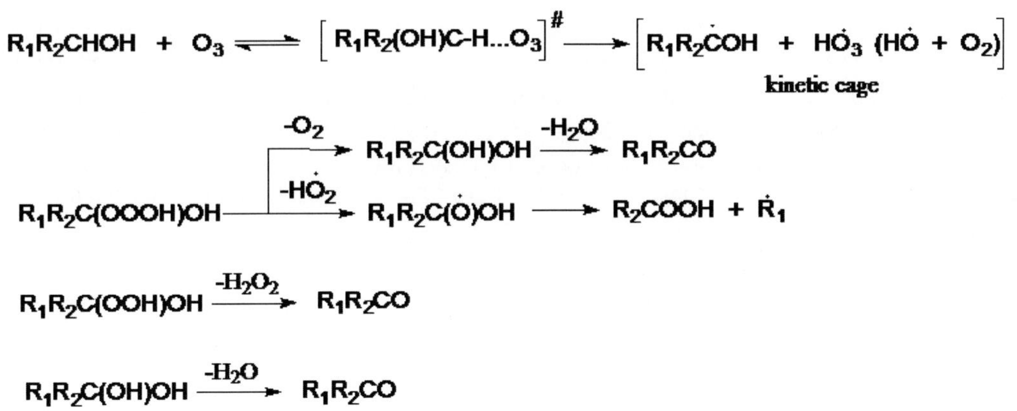

Scheme 2.

Alinger method (*PCMOD4* program) at the radius of the ozone molecule of 2A; E_{fr} - the calculated values of the rotation energy at ratio of $A^{LC}{}_{cal}/A_{obs} = 1$.

Ozone attacks the α-H atom, forming a LC, which further undergoes decomposition into a radical (or ion) pair in one kinetic cage.

The α-hydroxy alcohol, α-hydroxyperoxy alcohol and α-hydroxytrioxy alcohol being unstable leave the cage and decompose rapidly to the corresponding aldehyde or ketone liberating water, hydroperoxide and oxygen or lead to the formation of hydroperoxy and alkoxy radicals. The latter species can further undergo monomolecular decomposition.

Table 5. The values of A calculated by ACT with LC and CC and by CT, the sums (R_{fr}) and energy (E_{fr}) of inner rotation, steric factors (p), VDW radii of the molecules (r) and the ratio between the calculated and experimental values of pre-exponentials $A^{LC}{}_{cal}/A_{obs}$

Parameters	MeOH	EtOH	*n*-PrOH	*i*-PrOH	*n*-BuOH	*s*-BuOH	*t*-BuOH	*c*-HexOH
$A^{CC} \times 10^{-4}$, $M^{-1}.s^{-1}$	52	8.3	3.7	3.5	2.4	2.2	6.3	1.6
$A^{LC} \times 10^{-4}$, $M^{-1}.s^{-1}$	56	15	6.7	6.3	4.4	4.0	4.9	2.9
$F_{fr} \times 10^{-2}$	7.1	9.8	12	11	13.8	12.6	12.9	12.6
$A^{LC}{}_{cal} \times 10^{-7}$, $M^{-1}.s^{-1}$	40	15	8.1	6.8	6.1	5.0	6.3	3.6
r, Å	2.24	2.57	2.91	2.88	3.13	3.23	3.13	3.23
$A^{CT} \times 10^{-11}$, $M^{-1}.s^{-1}$	2.6	2.6	2.7	2.7	2.8	2.8	2.8	2.7
$p \times 10^{5}$	22	5.8	2.5	2.3	1.6	1.4	1.8	4.1
$A^{LC}{}_{cal}/A_{obs}$	1.0	3.8	3.4	11	3.1	8.3	1.1	7.2
$E_{fr,}$ cal/mol	0	730	670	1368	701	1172	52	1081

Note: ALC is the pre-exponential factor calculated with LC without free fragment rotation (it is absent at $E_{fr} \geq 4$ kcal); A_{CC} - calculated with CC; ALC_{calc} = at free fragment rotation, i.e., when exp(-E_{fr}/RT) = 1 or E_{fr} = 0 kcal; A_{ct} - calculated according to the collision theory at 300 K and r - calculated by the

Ketones

The investigation of the reactions of ozone with ketones is of interest from the viewpoint of the theory of reactivity, the ozone chemistry, preparation of oxygen-containing compounds, and the degradation and stabilization of organic materials. In particular, the ozonation of methylethylketone (MEK) [53-55] yielded acetic acid (AcAc), diacetyl and hydroperoxide as the main reaction products in CCl_4. It has been assumed that the process is not a oxidation chain-radical reaction. Similar products are also formed when the reaction takes place in water.

However, in this case ozone interacts with the enol form of MEK (53). The addition of nitric acid accelerates enolization, which is evidenced by the increase of the oxidation rate [55]. The decrease of the oxygen content in the ozone-oxygen mixture results in an increase of the oxidation rate, which is due to a change of the radical leading to oxidation, from $RO_2^{\bullet}$ to $RO^{\bullet}$. The ozonation of isomeric decanones indicates that the reactivity of ketones depends on the length and structure of the alkyl chain. In this case the main products are α-keto-hydroperoxides, monocarboxylic acids, keto alcohols and alkylbutyric lactones. However, it should be noted that the conclusions in these studies were based on product analysis at very high degrees of conversion whereby the secondary reactions play an important role in the product formation.

This gives us reasons to doubt whether the proposed mechanism describes adequately the initial steps of the reaction. The rate constant and the activation energy of interaction between cyclohexanone and ozone are 1.6×10^{-2} $M^{-1}.s^{-1}$ and 11,2 kcal/mol, respectively.

Table 6. Kinetic parameters of the reaction of ozone with some aliphatic ketones at 21 °C

Ketone	$k_0 \times 10^3$, $M^{-1}.s^{-1}$	$k_{st} \times 10^3$, $M^{-1}.s^{-1}$	k_{rel}	lg A	E_a, kcal/mol
CH_3COCH_3	3.5	7.7	0.2	10.823	18
$CH_3COC_2H_5$	12	24	0.7	10.63	17
$CH_3COC_5H_{11}$	23	67	1	10.54	16.5
$CH_3COC_6H_{13}$	45	58	2.4	10.46	16
$CH_3COC_7H_{15}$	448	54	2.3	10.31	15.8
$CH_3COC_{10}H_{21}$	42	113	2.2	10.58	16.2
$C_2H_5COC_5H_{11}$	20	74	1	11.48	15.5
$CH_3COCH_2COCH_3$	29000	-	-		-
c-$C_7H_{12}O$	87	80	4.1	9.00	17
CH_3CO-2-Naph	22000	-	-	5.77	6
4-CH_3-*c*-C_6H_9O	117	101	6.2	8.97	12
4-*t*-Bu-*c*-C_6H_9O	166	123	8.1	7.93	11.8
Ketone	$k_0 \times 10^3$, $M^{-1}.s^{-1}$	$k_{st} \times 10^3$, $M^{-1}.s^{-1}$	k_{rel}	lg A	E_a, kcal/mol

Note: k_o is calculated according to Equation 2; k_{rel} is the relative rate constant determined by gas chromatography relative to ethylpentylketone.

The reactivity of ozonized methyl derivatives of cyclohexanone is considerably higher than that of the H form.

An increase in the rate of interaction between the above derivatives and ozone is also observed depending on the location of substitution and it follows the sequence (56-58): 3-methylcyclohexanone <4-methylcyclohexanone <2-methylcyclohexanone.

It is evident from the foregoing that so far no systematic investigations on the ozonolysis of ketones have been carried out. In our studies on the ozonolysis of ketones we combined the kinetic approach and product analysis with theoretical methods. The results from the ozonolysis of some aliphatic ketones are shown in Table 6.

The following six types of ketones have been the subject of our investigations:

1. Acetone containing only α-C-H bonds,
2. Ketones with the general formula $CH_3CO(CH_2)_nCH_3$, where n = 1, 4, 5, 6 and 9, which contain secondary α, β, γ C-H bonds,
3. Ketones with the general formula $CH_3(CH_2)_m$ $CO(CH_2)_nCH_3$, where m = 1, n = 4, which contain α, α'-secondary C-H bonds and cycloheptanone, which contains α, β and γ secondary C-H bonds, 4-4-methyl or 4-*tert*-butyl cyclohexanone with γ-*tert*-C-H bond,
4. 2-Naphthylmethylketone with various substituents, and
5. Acetylacetone containing α-C-H bond activated by two keto groups.

The rate constant of keto-enol tautomerism for the latter compound is considerably higher than those of the other ketones studied, because of appearance of the C+C-double bonds.

$$CH_3\text{-}CO\text{-}CH_2\text{-}CO\text{-}CH_3 \Leftrightarrow CH_3\text{-}CO\text{-}CH{=}C(OH)\text{-}CH_3$$

The values of *k* vary, depending on the ketone structure, particularly on the content of primary, secondary or tertiary α-C-H-bonds, in a wide range. The ratio of their reactivities is 1 : 3.4÷24.8 : 33.4÷ 47.4. In contrast to paraffins these values are more similar, which are associated to the activation of α-C-H-bonds by the keto group. Ozone is sensitive not only to the activation of the α-C-H bonds but also to the keto-enol equilibrium. The rate of its reaction with acetylacetone is by three orders of magnitude higher than that with monoketones. Ketones with different substituents, for example, 2-naphthylmethylketones also show higher reactivity than the aliphatic ketones. It is concluded that the reaction centre in this case is the naphthyl ring and not the α-C-H bond in the methyl group.

Another interesting observation is the higher values of E_a compared with those of paraffins (with some 2-3 kcal/mol). As ozone is an electrophilic reagent and the activation of the α-C-H bonds from the keto group favours the nucleophilic attack, it is reasonable to expect an increase in E_a.

Figure 5 shows a typical kinetic curve illustrating the change of ozone concentration at the reactor outlet during ozonolysis of ketones in solution.

The horizontal line describes the inlet ozone concentration and the curve corresponds to the change in its concentration at the reactor outlet. It is seen that the addition of ketone results in a sharp decrease in the ozone concentration, followed by a section with only slight changes of $[O_3]_{g1}$.

After some time the ozone concentration rises up again to reach a region, where the respective curve is practically parallel to the abscissa ($[O_3]_{g2}$).

The shape of the kinetic curve of ozone consumption (Figure 5) is complicated and this implies that ozone reacts with more than one compound [1, 26, 27]. The two horizontal sections of the curve are the result of the interaction of ozone with two compounds, the rate constants of which differ considerably. The areas over the curve correspond to the amount of absorbed ozone. Knowing the concentrations of the reacted compounds the stoichiometry coefficients and other kinetic parameters can be calculated.

This type of kinetic curve for the ketones is associated with the keto-enol equilibrium and with the ability of ozone to react with the C=C bonds 106 times faster than with the C-H bonds with almost zero activation energy (1).

The processes corresponding to the ozone-gas curve (Figure 5) are the following:

1. The ozone starts its reaction with ketone (probably with the enolic form) and an abrupt fall of its concentration is registered,
2. When the rate of ozone supply becomes equal to the rate of the chemical reaction, the first step is formed,
3. The end of the step is connected with the consumption of the enolic form and the ozone concentration begins to rise up,
4. At this moment the ozone begins to react with the keto form and the second step is formed, whereby the rate of ozone supply becomes equal to the rate of the second reaction. The sharp transition between the two steady-state regions gives evidence that the rate of restoration of the keto-enolic equilibrium is considerably lower than the rate of ozone interaction with the keto form.

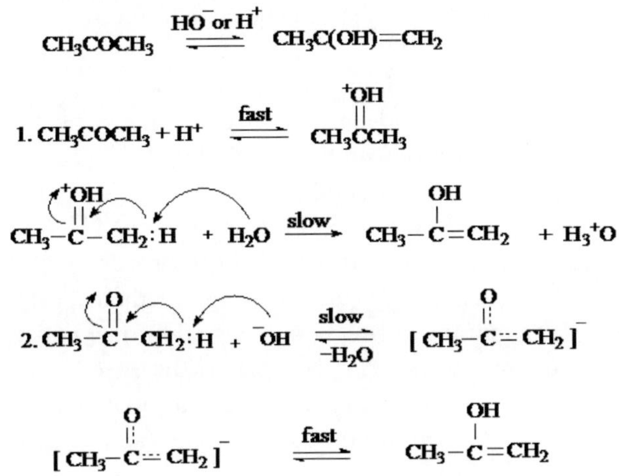

The concentration of the enolic form can be estimated from the area enclosed below the kinetic curve in Figure 5 and then the equilibrium constant of enolization could be measured:

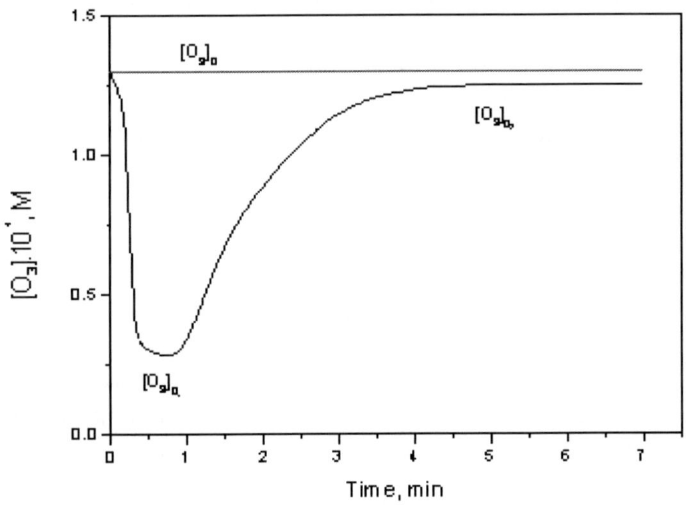

Figure 5. Ozonolysis of 0.712 mmol cycloheptylketone in 10 ml CCl4, 21 oC, v - 0.1 L/min.

It is known that the stoichiometry of the reaction of ozone with C=C is 1:1 and thus the amount of the enolic form can be judged based on the amount of absorbed ozone, i.e., in this particular case the concentration of the enolic form is 0.134 mM. As the initial concentration of ketone is known, the equilibrium quantity will be the difference between the initial concentration and that of the enolic form. Then the equilibrium constant can be easily calculated using the formula:

K_e=[Enol]$_e$/[Keton]$_e$,

and it amounts to 1.9×10^{-2}.

Based on the curve in Figure 5 one can also determine the rate constants of the ozone interaction with the two tautomeric forms. Taking into account the parameters of the first section of curve ($[O_3]_{g1}$) and by using Equation 2, the value of the rate constant with the enolic form is found to be $k_E = 2.9 \times 10^2\ M^{-1}.s^{-1}$ and therefore the second rate constant with the keto form is $k_K = 0.048\ M^{-1}.s^{-1}$. In this way the possibility of application of the method of ozone titration to the measurement of the equilibrium constant and the rate constants of ozone interaction with the two tautomeric forms in one kinetic experiment has been demonstrated.

With a view to evaluate the possible influence of the processes of mass transfer on the values of the rate constants, determined by the barbotage method, beside the observation of the criteria, described in the experimental section, some additional experimentation has been carried out, connected with the changes in the size of the reactor, in the volume of the solution, and in the concentration of the ozone at the reactor inlet. The results, obtained upon varying the above-listed parameters within the range up to about 30-50%, did not show any significant deviations in the respective values of k, and therefore there is noticeable effect of the conditions of mass transfer on these values.

Table 7. Equilibrium constants of keto-enol tautomerism of some ketones in CCl_4 solution at 21 °C determined by titration with ozone

Ketone	CH_3COCH_3	$CH_3COC_2H_5$	$CH_3COC_5H_{11}$	$CH_3COC_6H_{13}$	$CH_3COC_7H_{15}$	$CH_3COC_{10}H_{21}$	$c\text{-}C_7H_{12}O$
$K_e \times 10^4$	1.2	1.2	46	29	96	130	190

It is evident that the rate constants of ozone interaction with ketone, measured by the static method (Figure 6) through mixing of ozone and ketone solutions at $[K]_o >> [O_3]_o$ are higher than those found by the bubbling method (Table 6). This is not difficult to explain because in the former case the obtained values represent the total effective constant of interaction of both the enol and the keto form.

The results from the calculation of the equilibrium constants of keto-enol tautomerism for some aliphatic ketones in CCl_4 are given in Table 7.

The data for acetylacetone and 2-naphthylmethylketone are not presented in the table because in the former case the rate of reaching the equilibrium is commensurable with the rate of ozone consumption and in the latter case the ozone reacts with the double bonds in the naphthyl ring.

The equilibrium constants do not differ from those found within the temperature range of 21 °C to 3 °C and agree with data from the literature (59-62).

The kinetics of ozone reactions with ketones is also determined using gas chromatographic analyses. The relative rate constants shown in Table 6, column 4 demonstrate that only acetone and methylketone possess lower reactivity than the standard. It is seen that the rate constants, calculated from the relative values and the value of the standard constant correspond to those found by the bubbling method. The main products of ozone interaction with methylethylketone are 2-hydroxymethylethylketone, diacetyl, peroxides - alkyl and hydro, acetaldehyde and AcAc.

On the basis of the kinetic results obtained and the product analyses we suggest Scheme 3 for the proceeding of ketone ozonolysis:

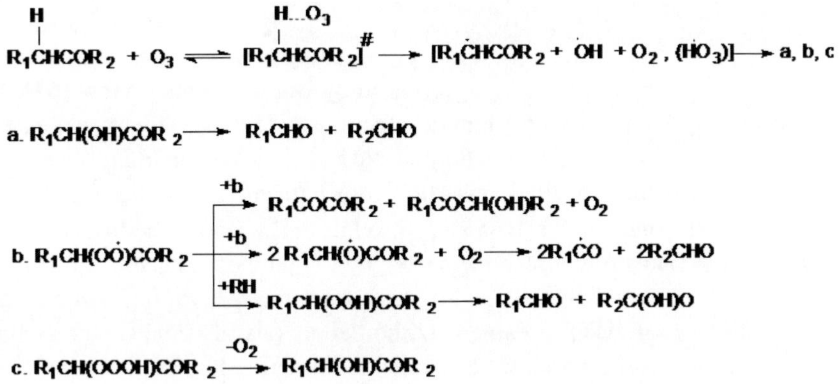

Scheme 3.

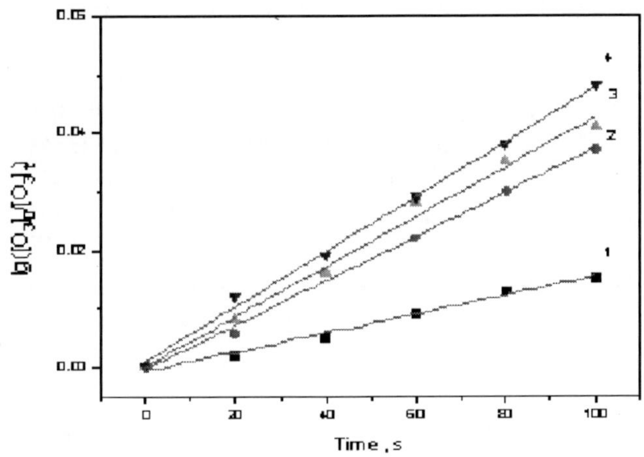

Figure 6. Semilogarithmic anamorphosis of the kinetic curve of ozonolysis of: 1 - methylethylketone; 2 - methylpentylketone; 3 – ethylpentylketone; and 4 – cycloheptylketone; 21 oC, ketone concentration 28 mM at [O3]lo=1.10-3 M.

Ozone attacks the α-H atom forming a LC, which further undergoes decomposition into a radical (or ion) pair in one kinetic cage, or leads to the formation of hydroperoxy and alkoxy radicals. The latter can further undergo monomolecular decomposition.

The intermediate formation of LC is assumed in the first stage, followed by breaking of α-C-H bond leading to the formation of a radical (or ion) pair. Then α-hydroxyketone, α-peroxykctone radicals and α-hydroxytrioxyketone appear as a result of C-C bond breaking can decompose into two aldehydes.

The α-peroxyketone radical, reacting with the initial ketone, can be transformed into an α-hydroxyperoxyketone, which is decomposed through breaking of the C-C bond to aldehyde and acid or the two radicals recombine giving rise to diketone and α-hydroxyketone or two α-oxyketone radicals. The latter react with the initial ketone and are transformed into α-hydroxyketone or through a monomolecular decomposition and breaking of the C-C bond to an alkoxy radical and aldehyde. α-Hydrotrioxyketone is converted into α-hydroxyketone and it is transferred in its turn into two aldehydes.

Ethers

Investigations of ozone reactions with ethers began as early as the last century [63-67]. Results of these studies are summarized and reported by Bailley in his review [67]. The main products formed during ozonolysis of aliphatic ethers are alcohols, aldehydes, esters, acids, hydrotrioxides (HTO), hydrogen and organic peroxides, singlet oxygen and water. Price and co-workers [68] proposed the so-called ‘insertion’ mechanism, according to which ozone is inserted into the α-C-H bonds at the first step through a 1,3-dipolar addition, thus forming unstable HTO. This mechanism was also supported by the data of Erickson and Bailey [2, 67, 69] and Murray and co-workers (70), who measured the activation parameters of the decomposition of a series of HTO, obtained by the ozonolysis of some ethers. Giamalva and co-workers [71] summarized the possible mechanisms known today, i.e., interaction with the ether oxygen atom, 1,3-dipolar insertion of ozone into the α-C-H bonds, homolytic abstraction of the α-H atom and heterolytic abstraction with carboanion and carbocation formation. These authors clearly indicate the predominance of the one-step mechanisms with transfer of a hydrogen atom, hydride anion or cation.

As it was mentioned above, the low-temperature ozonolysis of ethers yields HTO, which however are stable at very low temperatures. At normal temperatures the composition of the products is found to be different from that after the decomposition of HTO. These two reasons gave us impetus to carry out extended studies on the ozone reaction with ethers in order to establish the real mechanism of these reactions at normal temperatures.

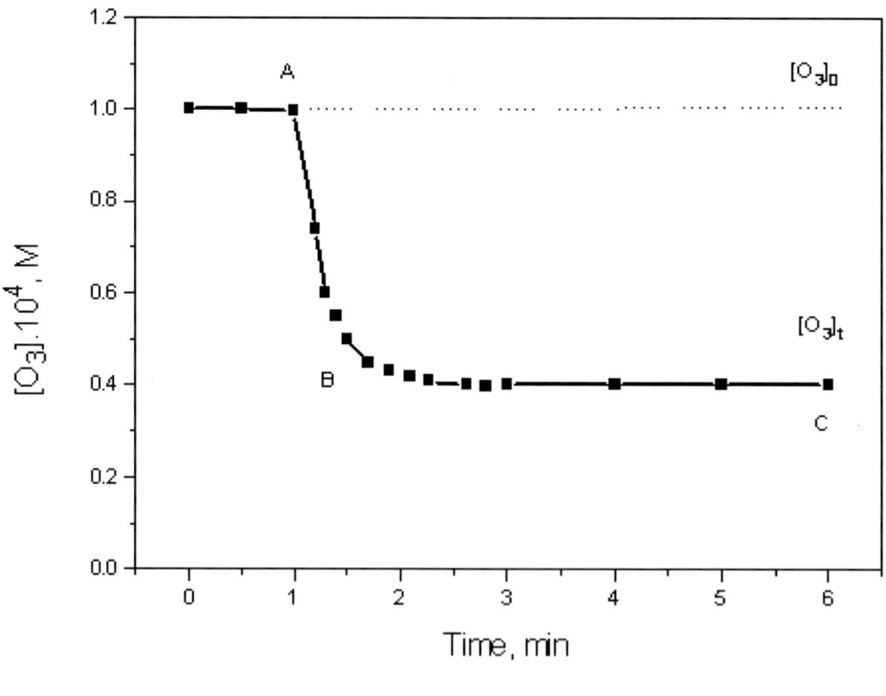

Figure 7. Kinetic curve of ozone concentration at the reactor outlet: 10 ml, 92 mM DEE, at –5 °C.

It was estimated, based on kinetic curve of ozone consumption (Figure 7) and eq (2), that the kinetic constant of the ozonation of diethyl ether (DEE) at t° = –5 °C is 0.9 $M^{-1}.s^{-1}$. Taking into account the area, comprised between the straight line $y=[O_3]_o$ and the kinetic curves of DEE, similar to that in Figure 5, which have been obtained at the lower concentrations of the reactant, the amount of ozone, consumed in the reaction has been determined. Juxtaposing this amount with the initial quantity of DEE in the reactor the stoichiometric coefficient of DEE ozonolysis has been estimated to be 1.

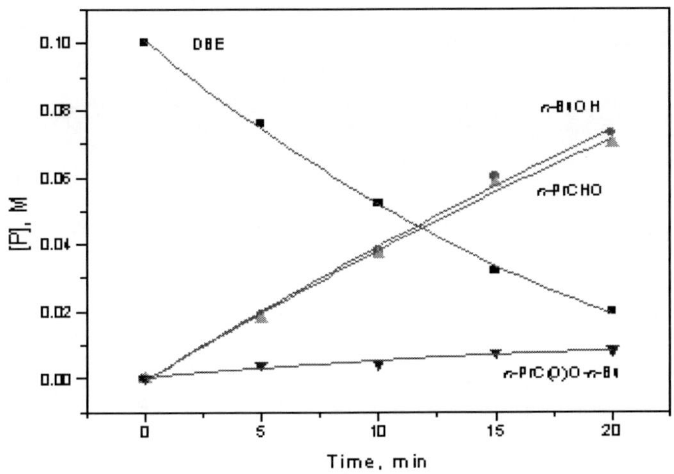

Figure 8. Kinetics of product formation and consumption of DBE during ozonolysis at ambient temperature, $[O_3]_o = 2.34 \times 10^{-4}$ M.

The calculated stoichiometric coefficient of the ozonation reaction with other ethers was also unity. The kinetics of ozonolysis was studied with the example of *n*-dibutylether (DBE). The kinetic curves of the reaction product formation and the initial ether consumption are shown in Figure 8.

The following products were identified by gas chromatography-mass spectrometry in the reaction mixture after 10 minutes (mass spectra of the obtained compounds are given below):

130 * 57 41 87 56 55 101 39 130 43 45 * *n*-Dibutylether,
74 * 56 41 43 42 55 39 57 45 40 41 * *n*-Butanol,
72 * 44 43 41 72 57 42 38 37 40 71 * Butanal,
88 * 60 73 41 42 43 45 39 55 61 88 * Butiryc acid;
144 * 71 89 56 43 41 57 60 73 55 42 * Butyl butyrate and
164 *& 247 * - Chloro-containing compounds like:
$CH_3(CH_2)_2CH(Cl)O(CH_2)_3CH_3$ (M_w 164.4) and
$CH_3(CH_2)_2CH(CCl_3)O(CH_2)_3CH_3$ (M_w 247)

The rate of DBE decomposition (7.4×10^{-5} $M.s^{-1}$ determined on the basis of Figure 8) is almost equal to the sum of butanal (6.5×10^{-5} $M.s^{-1}$), butanol (6.71×10^{-5} $M.s^{-1}$) and butyl butyrate (1.0×10^{-5} $M.s^{-1}$) formation rates. The kinetic curves of butanol and butanal formation have almost the same slope and they start without any induction period. This could mean that they are formed in parallel reactions from a common precursor. The rate of butylbutyrate accumulation, as it is demonstrated by its kinetic curve, is approximately 7 times lower. The latter indicates that their formation is occurring in a parallel reaction to butanol (butanal) formation, most probably from the same precursor. The ratio of product amount to initial DBE amount after 10 minutes was found to be 1:1. This means that the share of the auto-oxidation process is negligibly small, although we have identified some butyric acid by its IR spectrum (1765 cm^{-1} (monomer) and ε = 1470 $M^{-1}.cm^{-1}$) resulting from the butanal oxidation.

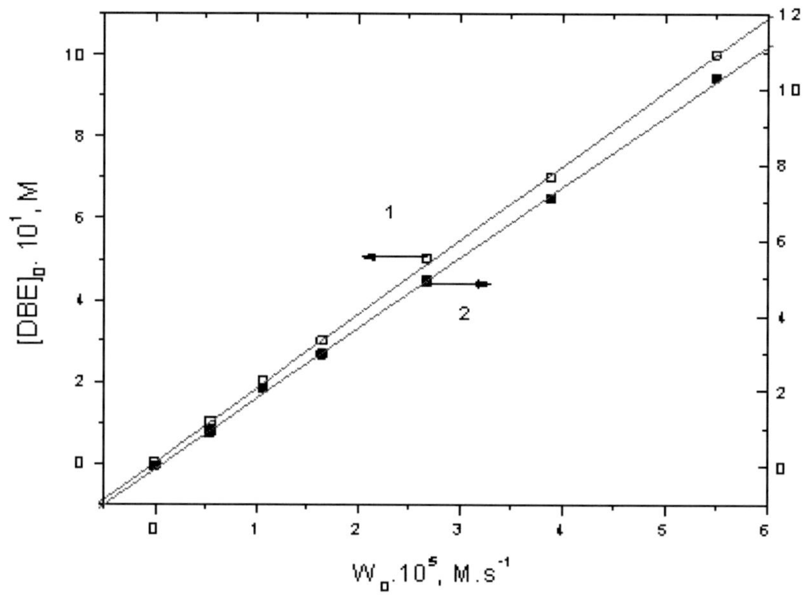

Figure 9. Dependence of the reaction rate of DBE ozonation (21 °C) on the concentration of ozone (1) at [DBE] = 10 mM and DBE (2) at $[O_3]$ = 0.1 mM.

The possible existence of a common precursor raises the question about its nature. Such a precursor could be either α-hydroxyether (EOH) or α-hydrotrioxyether (EOOOH), which can produce further through intermolecular disproportionation simultaneously aldehyde and alcohol:

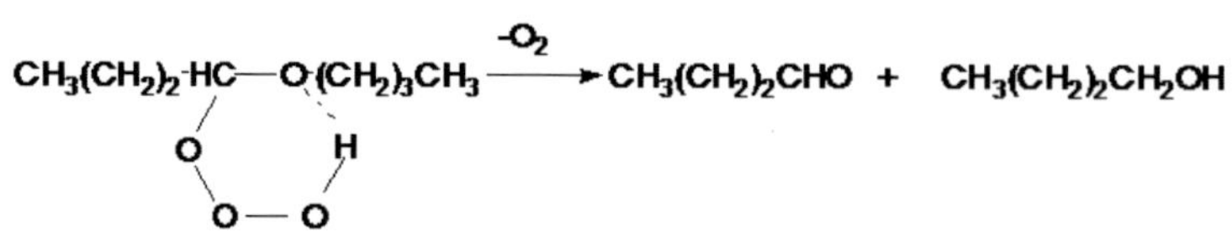

(2)

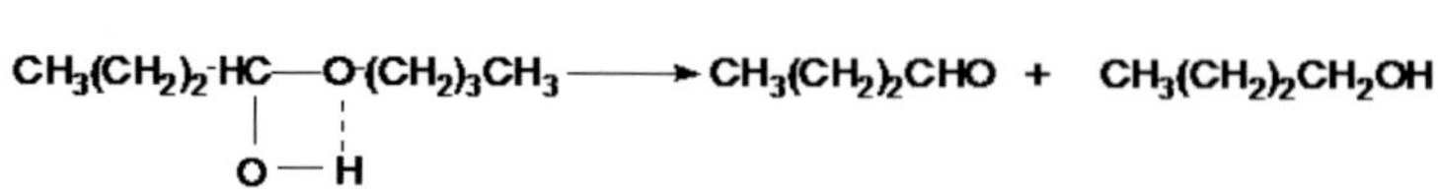

(2')

Probably, the intermolecular reaction of EOOOH disproportionation is preferable, because of the six-member-ring transition state, while EOH disproportionation occurs via a strained four-member-ring transition state, which is energetically unfavorable.

If EOH is assumed to be the precursor, then the butylbutyrate should be a major reaction product as a result of the rapid oxidation of EOH by ozone:

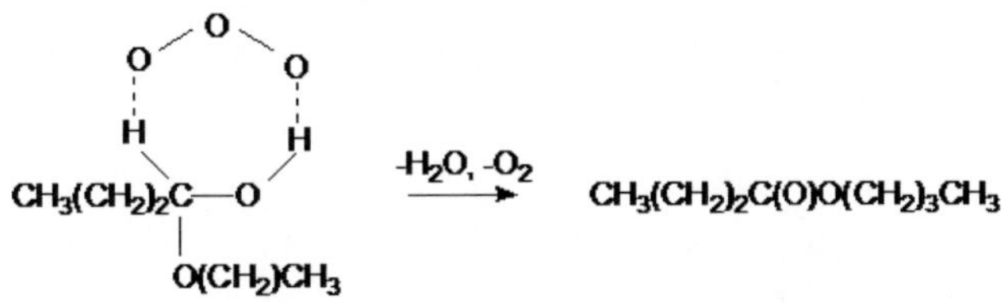

(3)

which has not been observed experimentally.

When EOH is assumed to be the precursor, it is difficult to imagine its one-step transformation into an ester without any additional assumptions. Also the solvent, CCl_4, could not affect at all this transformation, bearing in mind its weak oxidizing properties and its non-specificity.

The formation of the ester from EOOOH in a single step can be easily presented through a four-member-ring transition state:

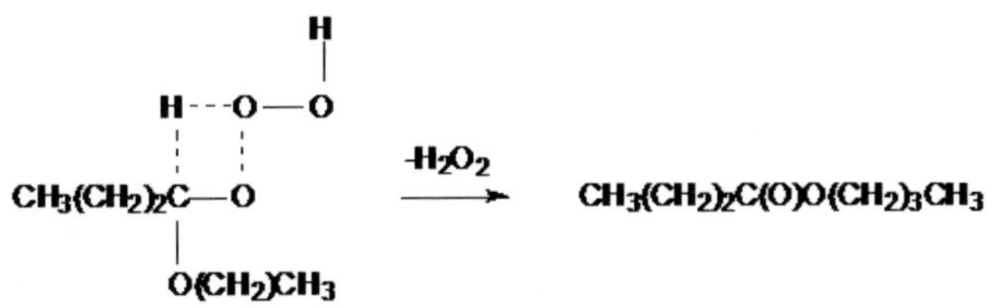

(3')

The formation of four-member transition state will be more unfavourable than the six-member transition state. This conclusion is in agreement with the kinetic data, namely, the rate of ester formation is approximately 7 times lower than that of alcohol and aldehyde formation. Direct evidence for EOOOH formation was found only after prolonged ozonation for 24 h at –78 °C [72-75]. The NMR spectrum of the oxidate has a signal at δ = 13.52 ppm, which is attributed to OOOH.

Table 8. Rate constants of ethers ozonations at various temperatures (°C)

Ether	–15 °C	–6 °C	–5 °C	3.5 °C	4 °C	10 °C	13 °C	21 °C	24 °C
EtE	0.5	0.9	1.0	1.2	1.6	2.2	2.4	3.1/3.0	3.6
DClEtE	-	-	0.009	0.016	-	-	0.028	0.035/0.04	-
i-PrE	-	-	1.6	1.9	-	-	3.4	5.4/5.1	-
n-ButE	-	-	1.9	2.1	-	-	4.3	6.3/6.3	-
i-AmE	-	-	1.9	2.5	-	-	4.1	7.8/7.0	-
n-AmE	-	-	2.2	2.8	-	-	4.2	7.8/7.5	-

Note: In column 9, after the slash the values of the constants determined by stop-flow method are given.

Figure 9 represents the dependences of the rate of ozonation of DBE on the concentrations of the ozone and that of DBE. Judging from the linear character of the dependences, in conformity with equation 2, it follows that the order of the reaction is unity with respect to each one of the two reactants and therefore rate law can be written in the form W = $k \times$ [DBE][O_3].

Based on the results discussed above, a scheme of the ozone reaction with aliphatic ethers is proposed in Scheme 5.

Ether + $O_3 \rightarrow$ EOOOH (k_1)
EOOOH $\rightarrow$ Alcohol + Aldehyde + O_2 (k_2)
EOOOH $\rightarrow$ Ester + H_2O_2 (k_3)

Scheme 5

The formation of chlorine-containing compounds (1-2%) can be explained by the presence of radical intermediates in the reaction mixture. This means that EOOOH will also be decomposed via radical route:

$EOOOH \rightarrow EO^{\bullet} + {}^{\bullet}O_2H$ (k_4)
$EO^{\bullet} + EH \rightarrow EOH + E^{\bullet}$ (k_5)
$E^{\bullet} + O_2 \rightarrow EO_2^{\bullet}$ ($k_{p'}$)
$E^{\bullet} + CCl_4 \rightarrow ECl + CCl_3^{\bullet}$ ($k_{p'}$)
$EO_2^{\bullet} + EH \rightarrow EOOH + E^{\bullet}$ (k_p)
$E^{\bullet} + CCl_3^{\bullet} \rightarrow ECCl_3$ ($k_{t'}$)
$2EO_2^{\bullet} \rightarrow$ non-radical products (k_t)

The validity of the mechanism, indicated above, was confirmed by the good agreement between the experimental points on the DBE decomposition curve and the product accumulation and by the theoretical curves, calculated according to the scheme and given in Figure 8. For example, the curve describing the DBE consumption is obtained from Equation 4:

$$[DBE]_t = [DBE]_0.\exp(-k't) \quad (4)$$

where k' is the constant, measured by the stop-flow method - 1.26 $\times 10^{-3}$ s^{-1}, while the corresponding bimolecular constant is 6.3 $M^{-1}.s^{-1}$.

A comparison of the experimental values and calculated estimates of A, assuming linear (LC) and cyclic (CC) forms of the activated complex (AC) has been made. The kinetic constants of ozone reaction with diethylether (EtE), dichlorodiethylether (DClEtE), di-iso-propylether (*i*-PrE), di-*n*-butylether (*n*-ButE), di-iso-amylether (i-AmE) and di-*n*-amylether (*n*-AmE) in CCl_4 are given in Table 8.

The rate constants for EtE, *n*-ButE, *i*-AmE and *n*-AmE have similar values. All these ethers have close α-C-H bond energies and similar electronic environment. The value of the rate constant for *i*-PrE ozonation is also close to those mentioned above, while the constant for DClEtE has 100 times lower value. In DIPE the presence of *tert*-C-H bonds, should contribute to their higher reactivity because of the lower energy of these bonds [34]. The presence of oxygen, however, and the more difficult stabilization of the transition state due to steric factors, makes this interaction slower. The very low rate constant of DClEtE ozonation can be attributed to the strong electron-accepting properties of the chlorine atom.

Arrhenius parameters, calculated on the basis of the data in Table 8, are summarized in Table 9. The comparison of the results, obtained with reference data, shows a good coincidence in regard to the values of the rate constants at 21 °C. Only a slight difference in the activation energies was observed. Perhaps, it is the heat of ozone dissolution in CCl_4 that needs to be taken into account as the reason for these differences. All reaction schemes, found in the current literature, describe the ozonation of ethers proceeding through two geometric forms of the activated complex - LC and CC:

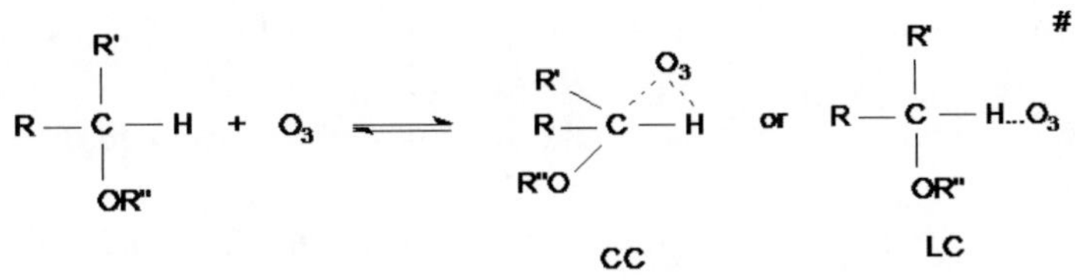

The CC has a more compact structure, without any possibility for free rotation. It can be a transition state for the following reaction:

Table 9. Arrhenius parameters of ozone reaction with some aliphatic ethers

Ether	k, $M^{-1}.s^{-1}$	lg A	E_a, kcal/mol
EtE	3.1	6.4 (7.3)	7.9 (9.9)
DClEtE	0.035	4.9 (5.5)	8.5 (10.1)
i-PrE	5.4	6.3 (7.9)	7.6 (10.0)
n-ButE	6.3 (6.0)	6.6 (7.4)	7.8 (9.7)
i-AmE	6.8	6.6	7.8
n-AmE	7.8	6.5	7.6
THydP	(1.3)	(6.8)	(9.7)
THydF	(12.3)	(6.6)	(8.2)

Note: The values in brackets are taken from reference (43); THydP and THydF are tetrahydropyrane and tetrahydrofurane, respectively.

Table 10. Heats of formation (ΔH) in kcal of initial, intermediate and final products, van der Waals radii of the ethers and free rotation energy (E_{fr}) around H-O bond

Ether	r, [Å]	ΔH, EH	ΔH, EOOOH	ΔH, $E^{\bullet}$	ΔH, E^{+}	E_a, kcal/mol
EtE	3.18	–68	–66	–32	135	1.3
dClEtE	3.51	–74	–73	–41	132	2.2
i-PrE	3.48	–76	–74	–42	125	1.4
n-ButE	4.03	–81	–79	–47	120	1.3
i-AmE	4.31	–87	–86	–53	115	1.3
n-AmE	4.45	–86	–85	–51	116	1.3

$$EH + O_3 \rightarrow EOOOH \rightarrow \text{products} \qquad (a)$$

where EOOOH is being formed in one step via 1,3-dipolar insertion.

The LC structure is an open one, allowing free rotation around H-O and O-O bonds. LC can be a transition state for the reactions:

$$EH + O_3 \rightarrow E^{\bullet} + HO_3^{\bullet} \rightarrow EOOOH \rightarrow \text{products} \qquad (b)$$
$$EH + O_3 \rightarrow E^{+} + HO_3^{-} \rightarrow EOOOH \rightarrow \text{products} \qquad (c),$$

in which an H atom or hydride anion abstraction occurs and EOOOH is formed during the second step.

The necessary parameters for estimating the A values are represented in Table 10.

The calculated values of A are shown in Table 11. The good agreement between the calculated estimates and experimental values for A confirms the conclusion that the geometry of the transition state is a linear (LC) and abstraction mechanisms (a) and (c) seem more probable.

In order to evaluate which one of the mechanisms is occurring we used the thermodynamic parameters shown in Table 10. The calculated heats of formation of the individual compounds fit well to the experimentally measured values [48, 49]. Using different expressions for the Hamiltonians, we have obtained the following values for ΔH: O_3 –33.3 kcal against the literature data of (34.1); HOOOH –17.8 (–17.1); HOOO –9.2 (–17.8); HOOO –20.3(–24.9); HO –5.2(–25.8).

$$\Delta H = -\Delta H\,(EH) - \Delta H\,(O_3) + \Delta H\,(EOOOH) \qquad (a)$$
$$\Delta H = -\Delta H\,(EH) - \Delta H\,(O_3) + \Delta H\,(E^{\bullet}) + \Delta H\,(HO_3^{\bullet}) \qquad (b)$$
$$\Delta H = -\Delta H\,(EH) - \Delta H\,(O_3) + \Delta H\,(\text{Å}^{+}) + \Delta H\,(HO_3^{-}) - e^2/r_{ip} - E_s \qquad (c)$$

where both types of values: 1) for energy of interaction between two charged particles (e^2/r_{ip} = 5.4 eV = 2.06x10^{-16} kcal), and 2) for energy of solubility in tetrachloromethane solution (E_s = 17 kcal) were taken from reference [76]. The calculated values for ΔH are depicted in Table 12. From thermodynamic point of view, (a) is the preferable mechanism and out of (b) and (c), (b) is more plausible.

Hydroxybenzenes

The reactions of ozone with mono- and dihydroxybenzenes have provoked so far a particular interest [1, 2, 34, 73- 75, 77-79], namely, because of their great importance for environment protection, chemical stabilization and the theory of reactivity. The ozonation of phenol, pyrocatechol, resorcinol and hydroquinone has been studied in different solvents - aqueous and organic, aimed at the deriving of the kinetic parameters and product composition [21-24, 80–86]. The rate constants of phenol and resorcinol ozonation in water at room temperature are 1.3×10^3 $M^{-1}.s^{-1}$ and >3×10^5 $M^{-1}.s^{-1}$, respectively, whereas the rate constants of benzene, toluene and anisole ozonation in organic media are 2, 14 and 2.9×10^2 $M^{-1}.s^{-1}$ (72-75).

Table 11. The calculated pre-exponents (A) and steric factors (p)

Ether	lg A, CT	lg A, CC	lg A, LC	p × 10^6, CC	p × 10^6, LC	A_{calc}^{LC}/A_{exp}
EtE	11.66	4.02	6.40	0.02	5.5	1
dClEtE	11.66	3.62	4.99	0.009	0.2	1.23
i-PrE	11.67	4.28	6.35	0.04	4.8	0.91
n-ButE	11.72	3.99	6.09	0.02	2.3	0.31
i-AmE	11.75	3.90	6.00	0.01	1.8	0.25
n-AmE	11.74	3.77	5.87	0.01	1.3	0.23

Note: the values of A in columns 3 and 4 are per one equivalent α-C-H atom.

Gurol and co-workers [77] found that the relative rates of pyrocatechol/phenol and resorcinol/phenol ozonation in water medium are 220 and 70, respectively. Provided that the rate constant of phenol ozonation is known [75], the calculated values of the rate constants of pyrocatechol and resorcinol ozonation are 2.86×10^5 $M^{-1}.s^{-1}$ and 9.1×10^4 $M^{-1}.s^{-1}$, respectively. However, in the case when the reaction is carried out in organic solvents the values are quite different. For example, in CCl_4 and at room temperature the following values have been obtained for: benzene - 0.06, ethylbenzene - 0.2, anisole - 1.1, phenol - 230 and pyrocatechol - 3.2×10^3 $M^{-1}.s^{-1}$ [70-73]. One of the possible explanations for the different values of the rate constants of hydroxyphenols ozonation obtained by the various researchers could be the great influence of the water on this reaction, for example for phenol they vary within the range 100-180 $M^{-1}.s^{-1}$, 500 $M^{-1}.s^{-1}$ for pyrocatechol and 300 $M^{-1}.s^{-1}$ for 3,6-di-*tert*-butylpyrocatechol. Pryor and co-workers [72] have reported that the rate constants of ozone reaction with α-tocopherol in CCl_4 and water are 5.5×10^3 $M^{-1}.s^{-1}$ and 1×10^6 $M^{-1}.s^{-1}$, with α-tocopherol acetate - 1.45×10^2 $M^{-1}.s^{-1}$ and for the reaction with α-tocopherolquinone it is 1.15 $\times 10^4$ $M^{-1}.s^{-1}$.

Table 12. Calculated heats of investigated reactions according to mechanisms a, b and c

Ether	a, kcal/mol	b, kcal/mol	c, kcal/mol
EtE	–32	–16	0
dClEtE	–33	–19	6
i-PrE	–32	–18	1
n-ButE	–32	–18	1
i-AmE	–32	–18	2
n-AmE	–33	–17	2

Ozone is unstable when dissolved in water. The mechanism and the kinetics of the elementary reactions, involved in the ozone decomposition, have been investigated in numerous studies [87, 88, and references therein]. The stability of the ozone depends to a great extent on the water matrix, especially on its pH value. The pH of the water is important because the hydroxide ions initiate ozone decomposition, which involves the following reactions [87]:

$O_3 + OH^- \rightarrow HO_2^- + O_2$ $k = 70\ M^{-1}\ s^{-1}$ (1)
$O_3 + HO_2^- \rightarrow {}^{\bullet}OH + O_2^{\bullet -} + O_2$ $k = 2.8 \times 10^6\ M^{-1}\ s^{-1}$ (2)
$O_3 + O_2^{\bullet -} \rightarrow O_3^{\bullet -} + O_2$ $k = 1.6 \times 10^9\ M^{-1}\ s^{-1}$ (3)

In aqueous solution the ozonide radical decomposes quickly into an OH radical, O_2 and OH anion [87, 88].

Additionally in the case of ozone reaction with electron-rich compounds, such as phenols (phenolates) and akoxylated aromatics in aqueous phase again OH radicals are formed, most probably via the mechanism of formation of the respective ozone adduct [88, 89]. For example it has been established that as a result of the electron transfer from phenol to ozone (reaction 4) at neutral pH the radical anion $O_3^{\cdot -}$ is obtained with an yield of 22% [90].

$$PhO^- + O_3 \rightarrow PhO^{\bullet} + O_3^{\bullet -} \quad (4)$$

As a result the rate constant of one and the same compound (of phenols or alkoxylated aromatics type) in water, determined on the basis of its ozone consumption, varies within a wide range of values depending on the pH values, and on the presence of hydroxyl radical scavengers minimizing the interfering radical chain reactions. This rate constant differs from that one, determined on the basis of the compound consumption [91]. However, this aspect requires special studies and discussion, which are beyond the scope of the present investigation.

Product analysis of pyrocatechol ozonation in aqueous medium shows that 3 moles of ozone are readily absorbed to give CO_2 (24.8%), CO (6%), formic acid (32.5%) and glyoxal (4.2%). As CO_2 and formic acid are the main reaction products, it seems very likely that most of the pyrocatechol undergoes anomalous ozonolysis [82]. Radical formation has been observed during 2,6-di-*tert*-butylphenol ozonolysis. When the ratio of absorbed ozone to phenol is 0.8, 50% viscous yellow oil and 3,5, 2,6-di-*tert*-butyl-*o*-quinone have been identified in the products [85]. In 3,6-di-*tert*-butyl-pyrocatechol ozonation the corresponding quinone was found to be the main product after complete consumption of the initial substrate [86]. Side products such as 2,5-di-*tert*-butyl-, 3-hydroxy-*p*-quinone and 3,6-di-*tert*-butyl-1,2-phenylacetal-5-hydroxy-3,6-di-*tert*-butyl-*p*-benzophenone have also been found. The rate constant of this reaction amounts to $3 \times 10^2\ M^{-1}.s^{-1}$. The same constant with the corresponding pyrocatechol with O-hydroxy-acetylated groups has been estimated to have a two orders of magnitude lower value.

A mechanism has been supposed to involve the formation of either a 1,3-cyclic activated complex between two hydrogen atoms from two OH groups and one ozone molecule or π- or σ-complexes, formed with the benzene ring.

The different values of the literature constants and the various mechanisms proposed for this reaction impose the necessity of further research of its kinetics and reaction pathway. In this connection we have studied the ozonolysis of the following hydroxybenzenes (Scheme 6).

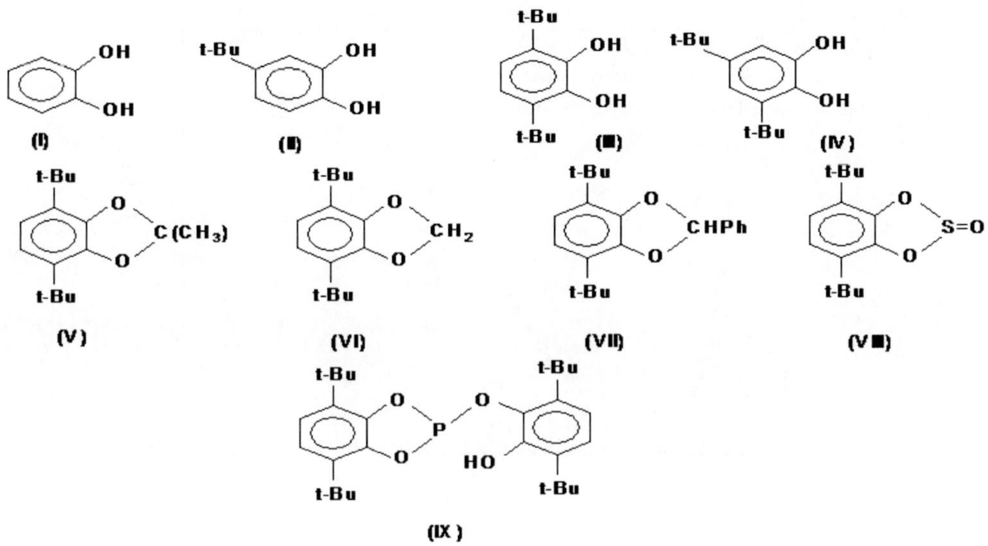

Scheme 6.

The probable mechanisms of ozone interaction with dihydroxybenzenes are represented in Scheme 7.

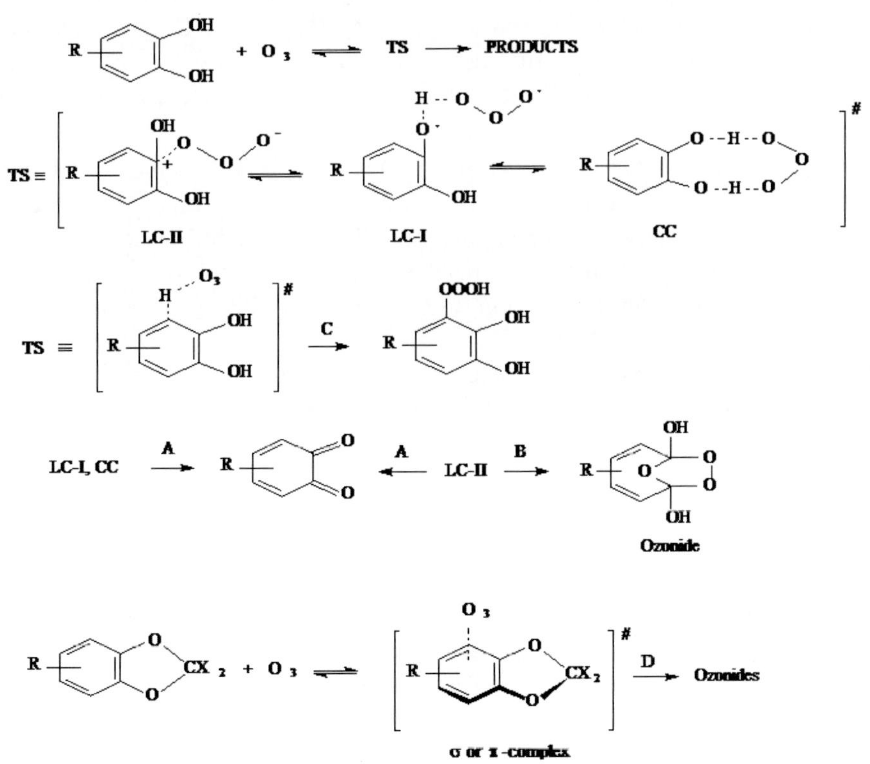

Scheme 7.

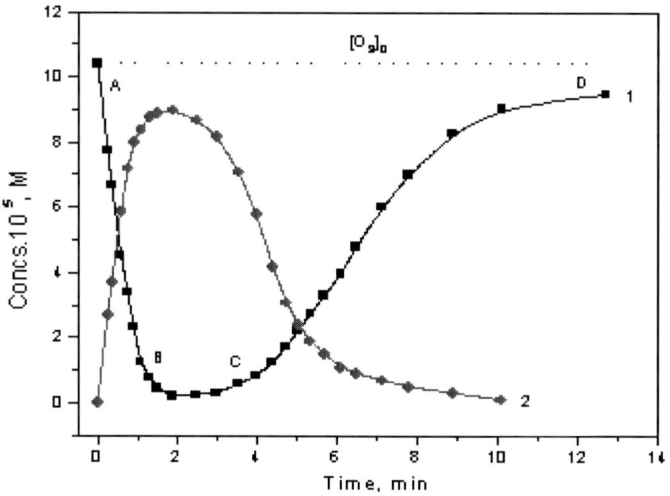

Figure 10. Kinetic curves of ozone absorption at reactor outlet (1) and *o*-quinone accumulation (2) in the course of pyrocatechol ozonation at ambient temperature, [PC] = 0.227 mM in CCl_4 (10 ml) during bubbling of ozone with 0.1 L/min flow rate.

The mechanism A with a cyclic complex (CC) formation in the transition state was supposed by Razumovskii and co-workers [86]. The mechanism B with linear complex LC-II in the transition state was put forward and discussed by Bailey [2]. The interaction of ozone with C-H bonds with the formation of trioxide [71, 73] as a possible parallel reaction is indicated in mechanism C. The acetylated forms of dihydroxybenzenes can react only via attack on the benzene ring according to mechanism D or C. Formally, mechanism D can be regarded as an extended version of mechanism B, involving the formation of TS similar to π- or σ-complexes. We propose a new mechanism, an extended version of the Razumovskii mechanism, whereby the transition state is linear with LC-I structure.

Upon ozonation of any of the investigated catechols directly inside the electron spin resonance cell or after freezing the reaction products in liquid nitrogen no signals have been detected.

The kinetic curve of the changes in the ozone concentration at the bubbling reactor outlet (Figure 10, 1) is characterized by three different regions: AB - fast ozone consumption after the addition of pyrocatechol, BC - steady-state part, when the rate of the chemical reaction becomes equal to the rate of ozone supply, and CD – the ozone concentration begins to rise up due to the pyrocatechol consumption. The BC part of the curve allows calculation of the rate constant, and based on the area below the curve ABCD – evaluation of the stoichiometry of the reaction. The straight line designated $[O_3]_0$ is the ozone concentration at the reactor inlet. Curve 2 presents the *o*-quinone formation in the course of the reaction time. Its profile suggests the intermediate formation of *o*-quinone.

The kinetic curves of the product formation in pyrocatechol ozonolysis, its consumption and *o*-quinone consumption during its ozonation in a separate experiment are given in Figure 11. The rate constants of pyrocatechol and *o*-quinone consumption, calculated based on the kinetic curves in Figure 11, were 3.2×10^4 $M^{-1}.s^{-1}$ and 7.1×10^4 $M^{-1}.s^{-1}$, respectively. The initial rate of the *o*-quinone formation had almost the same value as that of pyrocatechol consumption. The small variation of the constants is due to the participation of pyrocatechol in parallel reactions. Actually, during the reaction small amounts of open-chain products have been identified.

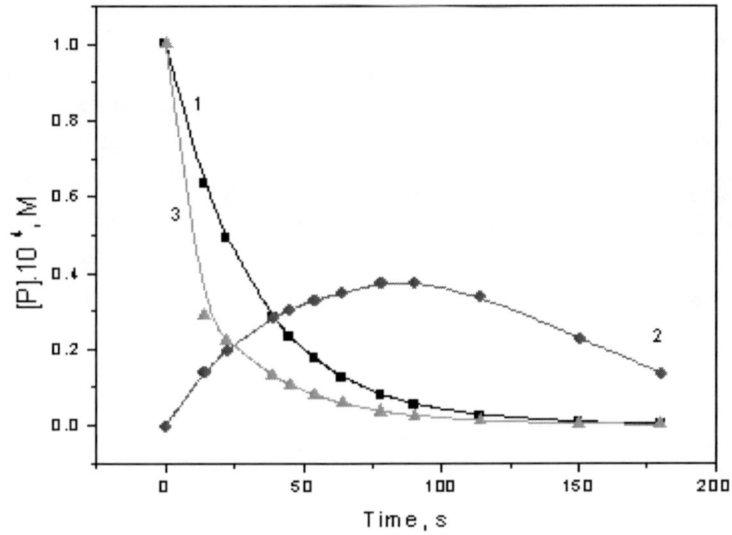

Figure 11. Kinetics of pyrocatechol consumption (1), *o*-quinone accumulation (2) and *o*-quinine consumption (3), at ambient temperature, $[O_3] = 1 \times 10^{-5}$ M.

Pyrocatechol ozonation at 15% conversion degree gave the following yields: *o*-quinone - 85%, pyrogallol - 3%, ozonide - 10%, muconic acid -2%, maleic acid and fumaric acids and the polymeric products - 1%. The ratio between the amount of absorbed ozone and the consumed pyrocatechol was calculated to be 6. Similar ratio values have also been obtained for the other hydroxybenzenes with free hydroxy groups.

It was found that with the increase of the conversion degree from 0 to 100%, the amount of open-chain products is continuously increasing, while that of the remaining products passes through a maximum. The individual compounds identified by ^{13}C-NMR are: muconic semialdehyde and acid - 20%, maleic and fumaric semialdehyde and acid - 40%, glyoxal, formic acid, oxalic acid, carbon dioxide and polymeric products - 40%. The reaction rate of di-hydroxybenzene ozonation follows first-order kinetics in relation to each reagent (Figures 12 A and 12 B).

Table 13 represents the obtained experimental kinetic data on the ozonation of the investigated dihydroxybenzenes at various temperatures.

A drastic difference between the values of the kinetic parameters of catechols I-IV and V-IX, in which the OH groups are acetylated, has been observed. The rate constants of catechols I-IV manifested 4-28 times higher values than those of catechols of V-IX types. On the other hand the pre-exponential factors demonstrated with about two orders of magnitude lower values with the former compounds. The activation energies are 2.05 ± 0.5 and 5.9 ± 0.5 kcal/mol, respectively, or the acetylation of the HO groups leads to an increase in E_a by about 4 kcal/mol.

The ozone probably reacts predominantly with the hydrogen atoms of the HO groups and, only to a very small extent, with the benzene ring. The ratio of these ozone interactions varies from 94:4 to 80:20 depending on the hydroxybenzene nature.

The lower activation energies of catechols I-IV with respect to those of V-IX are more consistent with the formation of an activated complex of contact type in the transition state, i.e., with structure LC-I (Scheme 7).

This assumption could also be confirmed by the analysis of the kinetic parameters of the ozonation of anizol, benzenes, phenol and toluene. The reaction of ozone with benzene proceeds at a low specific rate of 0.06 $M^{-1}.s^{-1}$ and a relatively high activation energy equal to 12.8 kcal/mol.

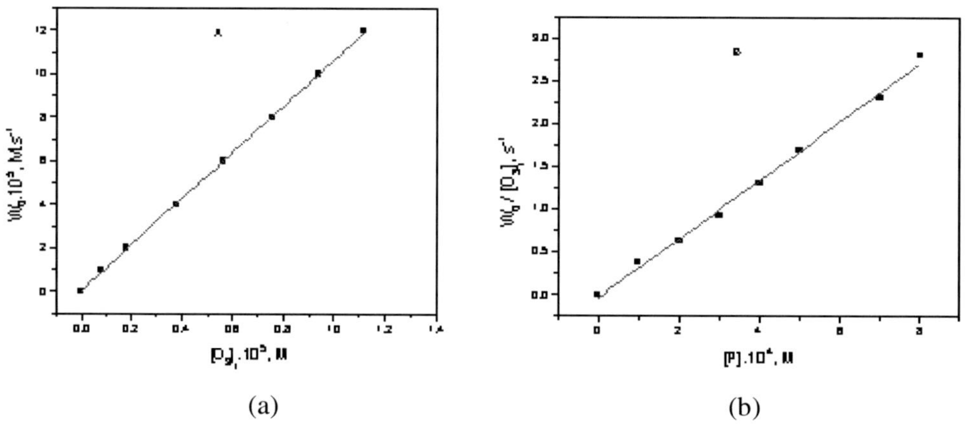

Figure 12. Dependence of the rate of pyrocatechol ozonation on the concentration of ozone (a) and on the concentration of pyrocatechol (b).

The methyl derivatives of benzene, i.e., toluene, reacts 6.7 times faster and the E_a is decreased by 2.5 kcal/mol, and the rate of anizol ozonation is even higher - 10 $M^{-1}.s^{-1}$. The exchange of ketyl group by a hydroxyl one leads to significant acceleration of the ozonation reaction rate and it approaches values of 160 $M^{-1}.s^{-1}$, and the activation energy is reduced by 5.8 kcal/mol. Probably, in this case the mechanism of the reaction is changed and ozone interacts predominantly with the hydrogen atoms of the OH groups. The decrease in the reactivity of anizol compared with that of phenol (~16 times) supports this assumption.

The analysis of the kinetic parameters of catechols in acetylated form and those of benzene, toluene, anizol and phenol shows that the rate constants increase and the activation energies go down upon increasing the number of the electron-donating substituents (Table 13). The high values of the pre-exponential factors could be associated with the high values of the energy of free rotation in the activated complex as a result of the steric hindrances caused by the presence of the bulky *tert*-butyl groups. This means that in the case when the hydroxyl groups are acetylated, the activated complex should have an LC-II like structure and the ozonation would lead only to the formation of ozonides and open ring products.

The kinetic parameters of some dihydroxybenzenes have been determined in the case of ozonation in aqueous medium (for those ones, which are water soluble - I-IV, VIII and IX). However, we have found that because of the diffusion limitations in the bubbling reactor we have not been able to measure constants higher than 1×10^4 $M^{-1}.s^{-1}$. In fact, the literature values for some of these constants are of this order, but they have been measured by a modified stop-flow technique [32].

Table 13. Kinetic data of dihydroxybenzene ozonolysis in CCl_4, 20 °C

Substrate	k, $M^{-1}.s^{-1}$	lg A	E_a, kcal/mol
I	3230	5.35	2.5
II	3100	4.91	1.9
III	3100	5.07	2.1
IV	3200	4.80	1.7
V	500	7.14	6.0
VI	749	7.17	5.8
VII	828	7.06	5.6
VIII	598	6.84	5.5
IX	111	6.81	6.4
Benzene	0.06	8.32	12.8
Toluene	0.40	7.05	10.3
Anisole	10	-	-
Phenol	160	7.31	7.0

Table 14. Symmetry numbers (σ), VdW radii (r), heats of formation (ΔH) of dihydroxybenzenes and the corresponding quinones and ozonides

Substrate	σ	r, Å	ΔH, kcal/mol	$\Delta H^{quinone}$, kcal/mol	$\Delta H^{ozonide}$, kcal/mol
I	2	3.02	–62 (–85)	–39 (–43)	–73
II	1	3.82	–88	–66	–99
III	2	4.46	–109	–86	–125
IV	1	4.39	–110	–90	–123
V	2	5.07	–116		–125
VI	2	4.56	–90		–103
VII	1	5.37	–63		–86
VIII	1	4.77	–73		–77
IX	1	5.30	–135		–140
Benzene	6	2.86	19 (13)		21
Toluene	1	3.37	12 (3)		11
Anisole	1	3.45	–16 (–26)		–20
Phenol	1	2.96	–21 (–38)		–24

Note: The experimental values are in parentheses (92).

The analyses made so far show that the reaction path of the ozone reactions with dihydroxybenzenes depends strongly on their nature and they precede via transition states with activated complex - LC-I or LC-II. The parameters needed for the calculations are represented in Table 14.

In addition, we have calculated the heats of formation of pyrogallol - –117.6 kcal/mol (–129 kcal/ mol [[92]]), 3-trihydroxy pyrocatechol - –59 kcal/mol, H_2O_3 - –21.2 kcal/mol (–17.7 kcal/mol [[92]]) and ozone - 34.1 kcal/mol.

Thus, the heats of the ozonation reaction according to the different mechanisms have been calculated and they amount to the following values: A - 33.1; B and D - 46.1 and C - 32.1 kcal/mol. All the mechanisms are exothermic and, therefore thermodynamically favorable. In this case only the magnitude of the activation energy and the entropy benefits will determine the reactivities of these compounds and the reaction pathway.

Table 15. Calculated pre-exponential values for ozone reaction with dihydroxybenzenes

Substrate	lg A TC	lg A, CC	lg A, LC	lg A, calc.	A_{calc}/ A_{obs}	lg p
1	2	3	4	5	6	7
I	11.318	4.346	7.69	5.39	1.10	5.928
II	11.422	3.905	7.25	4.95	1.10	6.472
III	11.500	4.070	7.42	5.12	1.12	6.380
IV	11.490	3.780	7.12	4.82	1.05	6.670
V	11.572	4.017	7.36	7.36	1.66	4.212
VI	11.511	4.056	7.40	7.40	1.70	4.102
VII	11.801	3.934	7.28	7.28	1.66	4.521
VIII	11.533	3.717	7.06	7.06	1.66	4.473
IX	11.575	3.689	6.88	6.88	1.10	4.695
Benzene	11.319	5.124	8.47	8.47-	1.41	2.849
Phenol	11.322	4.196	7.54	7.54	1.70	3.782

Note: p is the collision factor.

Calculated pre-exponential factors (Table 15) were compared with those obtained experimentally (Table 13) and it is seen that A have lower values (~10 times) for the cyclic form of the activated complex, if compared with the experimental values. At the same time the values of A for LC are about 200 times higher. The values of A for CC are the highest as predicted by the theory and they are lower than the experimental values. This supposes that the reaction takes place via a cyclic complex. The values of A calculated for LC in Table 15 have been obtained without taking into account the energy of free rotation. The latter was calculated as a sum of the rotation around H-O and O-O axes (by *mopac6*) and it amounts to 3.1 kcal. This means that the real values in column 4 are about 200 times smaller. The comparison between the A values corrected in this way and the experimental data for the compounds I-IV reveals complete agreement.

The reaction pathway of the ozonation reactions of compounds V-IX is quite different as the transition state includes the formation of a σ- or π-complex. For these compounds the formation of AC in the transition state is impeded due to the breaking of their aromatic character but simultaneously the free rotation is facilitated, whose energy may be even zero, due to the action of the principle of the lowest energy. If the energy of free rotation is assumed to be very low then the values of A calculated for LC-II would coincide with the experimentally found ones. Such a coincidence is observed for compound V in column 5 (Table 15).

The basic conclusion from the analysis of the results obtained for this reaction is that the kinetics and mechanism of the ozonation reaction of dihydroxybenzenes depend strongly on their structure and the type of the reaction medium.

REFERENCES

[1] Razumovskii, S.D.; Rakovsky, S.K.; Shopov, D.M.; Zaikov, G.E. Publishing House of the Bulgarian Academy of Sciences: Sofia, Bulgaria, 1983.

[2] Bailey, P.S., In *Ozonation in Organic Chemistry, Volume 2 - Nonolefinic Compounds*; Academic Press: New York, NY, USA, 1982, p 255.

[3] Waters, W.L.; Rollin, A.J.; Bardwell, C.M.; Schneider, J.A.; Aanerud, T.W. *J. Org. Chem.* 1976, *41*, 889.

[4] Otal, E.; Mantzavinos, D.; Delga, M.V.; Hellenbrand, R.; Lebrato, J.; Metcalfe, I.S.; Livingston, A.G. *Journal of Chemical Technology and Biotechnology* 1997, *70*(2), 147.

[5] Benitez, F.J.; Beltran-Heredia, J.; Acero, J.L.; Pinilla, A. *Journal of Chemical Technology and Biotechnology* 1997, *70*(3), 253.

[6] Amalric, L.; Guillard, C.; Pichat, P. *Research on Chemical Intermediates* 1995, *21*(1), 33.

[7] Heinzle, E.; Stockinger, H.; Stern, M.; Fahmy, M.; Kut, O.M. *Journal of Chemical Technology and Biotechnology* 1995, *62*(3), 241.

[8] Beltran, F.J.; Alvarez, P.M.; Legube, B.; Allemane, H. *Journal of Chemical Technology and Biotechnology* 1995, *62*(3), 272.

[9] Lee, M.J.; Arai, H.; Miyata, T. *Chemistry Letters*, 1994, *6*, 1069.

[10] Benko, E.M. ; Kovaleva, V.V.; Mamleeva, N.A.; Mitrofanova, A.N.; Voblikova, V.A.; Pryakhin, A.N.; Lunin, V.V. *Zhurnal Fisicheskoi Khimii* 1994, *68*(11), 1964.

[11] Artemev, Y.M.; Artemeva, M.A.; Vinogradov, M.G.; Ilika, T.I. *Russian Journal of Applied Chemistry* 1994, *67*(9) (Part 2), 1354.

[12] Scheuer, C.; Wimmer, B.; Bischof, H.; Nguyen, L.; Maguhn, J.; Spitzauer, P.; Kettrup, A.; Wabner, D. *Journal of Chromatography A*, 1995, *706*(1-2), 253.

[13] Hoger, B.; Gilbert, E.; Eberle, S.H. *Vom Wasser*, 1993, *80*, 187.

[14] Ruck, W.; *Vom Wasser* 1993, *80*, 253.

[15] Gab, S.; Turner, W.V.; Wolff, S.; Becker, K.H.; Ruppert, L.; Brockmann, K.J. *Atmospheric Environment* 1995, *29*(18), 2401.

[16] Takeuchi, K.; Fushimi, C.; Nakamura, K.; Ibusuki, T. *Bunseki Kagaku* 1993, *42*(10), 625.

[17] Pan, G.Y.; Chen, C.L.; Gratzl, J.S.; Chang, H.M. *Research on Chemical Intermediates* 1995, *21*(3-5), 205.

[18] Wada, H.; Naoi, T.; Kuroda, Y. *Nippon Kagaku Kaishi* 1995, *4*, 306.

[19] Marco, A.; Chamarro, E.; Esplugas, S. *Afinidad* 1994, *51*(452), 265.

[20] Starek, J.; Zukal, A.; Rathousky, J. *Carbon* 1994, *32*(2), 207.

[21] Mvula, E.; von Sonntag, C. *Organic & Biomolecular Chemistry* 2003, *1*(10), 1749.

[22] Ben'ko, E.M.; Bokova, M.N.; Pryakhin, A.N.; Lunin, V.V. *Russian Journal of Physical Chemistry* 2003, *77*(5), 739.

[23] Ramseier, M.K.; von Gunten, U. *Ozone-Science & Engineering* 2009, *31*(3), 201.

[24] Valsania, M.C.; Fasano, F.; Richardson, S.D.; Vincenti, M. *Water research* 2012, *46*, 2795.
[25] Rakovsky, S.K.; Cherneva, D.R. *International Journal of Chemical Kinetics* 1990, *22*(4), 321.
[26] Rakovsky, S.K.; Cherneva, D.R.; Shopov, D.M.; Razumovskiy, S.D. *Communications of the Department of Chemistry, Bulgarian Academy of Sciences* 1976, *9*(4), 711.
[27] Rakovsky, S.K.; Cherneva, D.R. *Oxidation Communications* 1989, *12*(3), 108.
[28] Rakovsky, S.K.; Kulak, L.G.; Kuramshin, E.M.; Zlotsky, S.S.; Rakhmankulov, D.L. *Communications of the Department of Chemistry, Bulgarian Academy of Sciences* 1989, *22*(3), 722.
[29] Rakhmankulov, D.L.; Zlotsky, S.S.; Rudnik, L.Z.; Teregulova, G.T.; Rakovsky, S.K. *Communications of the Department of Chemistry, Bulgarian Academy of Sciences* 1989, *22*(3), 652.
[30] Anachkov, M.P.; Rakovsky, S.K.; Stoyanov, A.K.; Fotty, R.K., *Thermochemica Acta* 1994, *237*, 213.
[31] Rakovsky, S.K.; Cherneva, D.R.; Deneva, M., *International Journal of Chemical Kinetics* 1995, *27*(2), 153.
[32] Rakovsky, S.K.; Cherneva, D.R.; Deneva, M.; Ershov, V.V. *Oxidation Communications* 1997, *20*(2), 169.
[33] Rakovsky, S.K.; Sheldon, R.A.; Rantwijk, F.V. *Oxidation Communications* 1996, *19*(4), 482.
[34] Zaikov, G.E.; Rakovsky, S.K., *Ozonation of Organic & Polymer Compounds*, *i*Smithers, Smithers Rapra, Shawbury, Shreswsbury, Shropshire, SY4 4NR, UK, 2009.
[35] Popova, D.; Rakovsky, S.; Anachkov, M. *Oxidation Communications* 2007, *30*(3), 529.
[36] Eyring, H.; Lin S.H.; Lin S.M. *BASIC Chemical Kinetics*, Moskow, Mir, 1983 (In Russian).
[37] Rozovskii, A.Ja. *Heterogeneous Chemical Reactions (Kinetics and Macrokinetics)*, Nauka: Moscow, 1980, (In Russian).
[38] Rozenberg, M.M.; Brun, E.B. *Teor. Osn. Khim. Teknol.* (In Russian), 1990, *24*(2), 198.
[39] Denisov, E.T.; Mitskevich, N.I.; Agabekov, V.E. *Mechanism of Liquid-Phase Oxidation of Oxygen Containing Compounds*, Science and Technique Publishing House, Minsk, 1975.
[40] Whiting, M.; Bolt, A.; Parish, J. *Oxidation of Organic Compounds - III, Advances in Chemistry Series*, No. 77, ACS, Washington, DC, USA, 1968, p 4.
[41] Williamson, D.; Cvetanovich, R.J. *Journal of American Chemical Society* 1970, *92*, 2949.
[42] Gerchikov, A.Ya.; Kuznetzov, E.R.; Denisov, E.T. *Kinetika i Kataliz* 1970, *15*, 109.
[43] Shereshovetz, V.V.; Shafikov, N.; Komissarov, V.D. *Kinetika i Kataliz* 1980, *21*, 1596.
[44] Shafikov, N.Ya.; Gusmanov, A.A.; Zimin, Yu.S.; Komissarov, V.D. *Kinetics and Catalysis* 2002, *43*(6), 799.
[45] Gerchikov, A.Ya.; Zimin, Yu.S.; Trukhanova, N.V.; Evgrafov, V.N. *React.Kinet. Catal.Lett.* 1999, *68*(2), 257.
[46] Borodin, A.A.; Razumovskii, S.D. *Kinetics and Catalysis* 2009, *50*, 3385.
[47] Rakovsky, S.K. Kinetics and mechanism of the ozone reactions with paraffins in liquid phase, *Moscow, Institute of Chemical Physics, Academy of Science of the USSR*, 1975. [PhD Thesis].

[48] Kondratiev, V.N. (Chief Editor), *Bond Energy, Ionization Potential and Electron Affinity*, [in Russian], Nauka Publishers: Moscow, 1974, p 83.
[49] Gordon, A.J.; Ford, R.A. *The Chemists Companion*, Mir Publishing House: Moscow, 1976.
[50] Eiring, G.; Lin, S.G.; Lin, S.M. *The Principles of Chemical Kinetics*, Mir Publishing House: Moscow, 1983.
[51] Karadakov, K.; Rakovsky, S.K. *Calculation of Pre-exponents ver.1.0*, (Software) Copyright, 1987, based on the J. Howbert, Software: The Molecular Animator, vers. 1-3, Copyright, 1983.
[52] Thiel, W.; Software, IBM, Program MNDOC, QCPE ver., Fachbereich Physikalische Chemie der Philipps-Universitat, auf den Zahbergen, D-3550, Marburg, Germany, based on M.Y.S. Dewar and W. Thiel, *Journal of the American Chemical Society* 1977, *99*, 4899.
[53] Gerchikov, A.Ya.; Komissarov, V.D.; Denisov, E.T.; Kochemassov, G.B., *Kinetika i Kataliz* 1972, *13*(1126); *Kinetika Kataliz* 1974, *15*, 509.
[54] Gerchikov, A.Ya.; Komissarov, V.D.; Denisov, E.T.; *Kinetika i Kataliz* 1974, *15*, 230.
[55] Gerchikov, A.Ya.; Komissarov, V.D.; Galimova, L.G.; Denisov, E.T. *Doklady Akademii Nauk SSSR*, 1973, *213*, 881.
[56] Korotkova, N.P.; Syroezhko, A.M.; Proskuryakov, V.A. *Zhurnal Prikladnoi Khimii* 1981, *54*, 885.
[57] Komissarov, V.D.; Galimova, L.G.; Denisov, E.T. *Kinetika i Kataliz* 1974, *15*, 1063.
[58] Syroezhko, A.M.; Korotkova, N.P.; Vihorev, A.A.; Proskuryakov, V.A. *Zhurnal Prikladnoi Khimii* 1978, *15*, 2562.
[59] Barton, D.; Li, V.D., Eds., *General Organic Chemistry*, Volume 2, p 577, Khimiya Publishers: Moscow, 1982.
[60] House, H.O.; Benjamin, W.A., Eds., *Modern Synthetic Reactions*; 2nd Edition, Menlo Park, CA, USA, 1972.
[61] Zabicky, J., Ed., *The Chemistry of the Carbonyl Group*, (2 Volumes) Interscience Publishers, London, UK, 1970.
[62] Alcais, P.; Brouillard, R. *Journal of the Chemical Society - Perkin II* 1972, *1214*.
[63] Von Balo, L., *Annalen* 1866, *140*, 348.
[64] Berthlot, M. *Comptes Rendus* 1881, *92*, 895.
[65] Harries, V. *Analen* 1905, *343*, 311.
[66] Fischer, F.G. *Analen der. Chemie* 1929, *476*, 233.
[67] Bailey, P.S. *Chemical Reviews* 1958, *58*, 925.
[68] Price, C.C.; Tumolo, A.L. *Journal of the American Chemical Society* 1964, *86*, 4691.
[69] Ericson, R.E.; Bakalik, D.; Richards, C.; Seanlon, M. *Journal of Organic Chemistry* 1966, *31*, 461.
[70] Stary, F.E.; Emge, D.E.; Murray, R.W. *Journal of American Chemical Society* 1976, *98*, 1880.
[71] Giamalva, D.H.; Church, D.F.; Pryor, W.A. *Journal of American Chemical Society* 1986, *108*, 7678.
[72] Pryor, W.A.; Ohto, N.; Church, D.F. *Journal of American Chemical Society* 1983, *105*, 3614.
[73] Hoigne, J.; Bader, H. *Water Research* 1983, *17*(2), 173.
[74] Hoigne, J.; Bader, H. *Water Research* 1983, *17*(2), 185.

[75] Hoigne, J.; Bader, H.; Haag, W.R.; Staehlin, J. *Water Research* 1985, *19*(8), 993.
[76] Kuramshin, E.M.; Ivashov, U.B.; Zlotskii S.S.; Rahmankulov D.L. *Izvestia Visshikh Uchebnih Zavedenii, Khimia i Khimicheskaya Tekhnologia* 1984, *27*, 13.
[77] Gurol, M.D.; Nekouinaini, S. *Industrial and Engineering Chemistry Fundamentals* 1984, *23*(1), 54.
[78] Bernatek, E.; Frangen, C. *Acta Chemica Scandanavica* 1961, *15*, 471.
[79] Bernatek, E.; Moskeland, J.; Valen, K. *Acta Chemica Scandanavica* 1961, *15*, 1454.
[80] Bernatek, E.; Frangen, C. *Acta Chemica Scandanavica* 1962, *16*, 2421.
[81] Bernatek, E.; Vincze, A. *Acta Chemica Scandanavica* 1965, *19*, 2007.
[82] Jarret, M.; Bermond, A.; Ducauze, C. *Analysis*, 1983, *14*, 185.
[83] Razumovskii, S.D.; Nikiforov, G.A.; Globenko, G.M.; Kefely, A.A.; Gurvich, Ya.A.; Karelin, N.A.; Zaikov, G.E. *Neftekhimia* 1972, *12*, 376.
[84] Konstantinova, M.L.; Razumovskii, S.D.; Zaikov, G.E. *Izvestiya Akademii Nauk SSSR, Seriya Khimicheskaya;* 1991, p 324.
[85] Razumovskii, S.D.; Konstantinova, M.L.; Zaikov, G.E. *Izvestiya Akademii Nauk SSSR, Seriya Khimicheskaya*; 1992, p 1203.
[86] Konstantinova, M.L.; Vol'eva, V.B.; Razumovskii, S.D., *Izvestiya Akademii Nauk SSSR, Seriya Khimicheskaya*; 1992, p 1443.
[87] Hoigne´, J. In: *The Handbookof Environmental Chemistry*; Hrubec, J., Ed.; Springer-Verlag: New York, 1998, Vol. 5, p 83.
[88] von Gunten U. *Water Research* 2003, *37*, 1443.
[89] Pi, Yu.; Schumacher, Jo.; Jekel, M. *Water Research* 2005, *39*, 83.
[90] Buffle, M-O, von Gunten U. *Environ. Sci. Technol.* 2006, *40*, 3057.
[91] Yao, C.C.D. Haag, W.R. *Water Research* 1991, *25*(7), 761.
[92] *Handbook of Chemistry and Physics, 66th Edition*, p.D278, CPC Press, Boca Raton; FL, USA, 1985-1986.

In: Quantitative Chemistry, Biochemistry and Biology
Editors: G.E. Zaikov, O.V. Stoyanov et al.
ISBN: 978-1-62948-332-0

Chapter 32

THE MECHANISM OF THE SELECTIVE ETHYLBENZENE OXIDATION, CATALYZED WITH TRIPLE BINUCLEAR METALLOCOMPLEX CATALYSTS, INCLUDING Ni AND REDOX-INACTIVE METAL Na (OR Li): NANOSTRUCTURES $\{Ni^{II}(ACAC)_2 \cdot NaSt\ (OR\ LiSt) \cdot (Ph_{OH})\}_N$ FORMATION DUE TO INTERMOLECULAR H-BONDS

L. I. Matienko*[*]*, V. I. Binyukov, L. A. Mosolova and G. E. Zaikov

N M Emanuel Institute of Biochemical Physics, Russian Academy of Sciences, Moscow, Russian Federation

ABSTRACT

The role of the intra- and intermolecular H-bonds in mechanisms of catalysis with triple heterobinuclear hetero ligand complexes $Ni^{II}(acac)_2 \cdot NaSt$(or LiSt)·PhOH, including nickel and redox-inactive metal Na (Li), in the selective ethylbenzene oxidation by dioxygen into α-phenyl ethyl hydroperoxide is discussed. The AFM method has been used for research of possibility of the stable supramolecular nanostructures $\{Ni^{II}(acac)_2 \cdot NaSt$(or LiSt)·PhOH}n formation, due to intermolecular H-bonds.

Keywords: Nanostructures, triple binuclear complexes $Ni^{II}(acac)_2 \cdot NaSt$(or LiSt)·PhOH, H-bonds, catalysis, ethylbenzene, oxidation, dioxygen, α-phenyl ethyl hydroperoxide

[*] Fax (7-495) 137 41 01, tel. (7-495) 939 71 40, e-mail: matienko@sky.chph.ras.ru (L.I. Matienko).

I. Aim and Background

The problem of selective oxidation of alkylarenes to hydroperoxides is economically sound. Hydroperoxides are used as intermediates in the large-scale production of important monomers. For instance, propylene oxide and styrene are synthesized from α-phenyl ethyl hydroperoxide, and cumyl hydroperoxide is the precursor in the synthesis of phenol and acetone [1]. The method of modifying the Ni^{II} and $Fe^{II,III}$ complexes used in the selective oxidation of alkylarenes (ethylbenzene and cumene) with molecular oxygen to afford the corresponding hydroperoxides aimed at increasing their selectivities has been first proposed by Matienko [2,3]. This method consists of introducing additional mono- or multidentate modifying ligands into catalytic metal complexes. The mechanism of action of such modifying ligands was elucidated. New efficient catalysts of selective oxidation of ethylbenzene to α-phenyl ethyl hydroperoxide were developed [2,3].

The phenomenon of a substantial increase in the selectivity (*S*) and conversion (*C*) of the ethylbenzene oxidation to the to α-phenyl ethyl hydroperoxide upon addition of PhOH together with alkali metal stearate M'St (M' = Li, Na) as ligands to metal complexes $Ni^{II}(acac)_2$ was discovered in our works [1,2,4].

The observed values of *C* [*C* >35% at $(S_{PEH})_{max}$ = 85%–87%], $[ROOH]_{max}$ (1.6–1.8 mol $litre^{-1}$ far exceeded those obtained with other ternary catalytic systems $\{Ni^{II}(acac)_2 + \mathbf{L^2} + PhOH\}$ (L^2 is N-metyl-2-pyrrolidone (MP), hexamethylphosphorotriamide (HMPA),) and the majority of active binary systems. These results by Matienko and Mosolova are protected by the Russian Federation patent (2004).

The advantage of these ternary systems is the long-term activity of the in situ formed complexes $Ni^{II}(acac)_2 \cdot L^2 \cdot PhOH$. The best results were obtained in the case of catalysis with system $\{Ni^{II}(acac)_2 + NaSt + PhOH\}$ (Figure 1).

The high efficiency of three-component systems $\{Ni^{II}(acac)_2 + MSt + PhOH\}$ (M = Na, Li) in the reaction of selective oxidation of ethylbenzene to α-phenyl ethyl hydroperoxide was associated with the formation of extremely stable binuclear heteroligand complexes $Ni(acac)_2$ MSt PhOH. The stability of complexes $Ni^{II}(acac)_2 \cdot$ MSt · PhOH can be associated with the formation of intermolecular H-bonds [5].

Nanostructure science and supramolecular chemistry are fast evolving fields that are concerned with manipulation of materials that have important structural features of nanometer size (1 nm to 1 μm) [6,7]. The self-assembled systems and self-organized structures mediated by transition metals are considered in connection with increasing research interest in chemical transformations with use of these systems [8].

Earlier we have applied AFM technique for evidence of the possibility of stable supramolecular structures formation in the process of ethylbenzene oxidation to the to α-phenyl ethyl hydroperoxide in the presence of systems $\{Ni^{II}(acac)_2 + L^2\}$ [9].

In this paper we have applied AFM technique for research the possibility of supramolecular structures formation in the process of ethylbenzene oxidation to the to α-phenyl ethyl hydroperoxide at catalysis with three-component systems $\{Ni^{II}(acac)_2$+NaSt(or LiSt)+PhOH$\}$:

$$\{Ni^{II}(acac)_2 + NaSt(or\ LiSt) + PhOH\} \rightarrow Ni^{II}(acac)_2 \cdot NaSt(or\ LiSt) \cdot PhOH \rightarrow$$
$$\rightarrow \{Ni^{II}(acac)_2 \cdot NaSt(or\ LiSt) \cdot PhOH\}_n$$

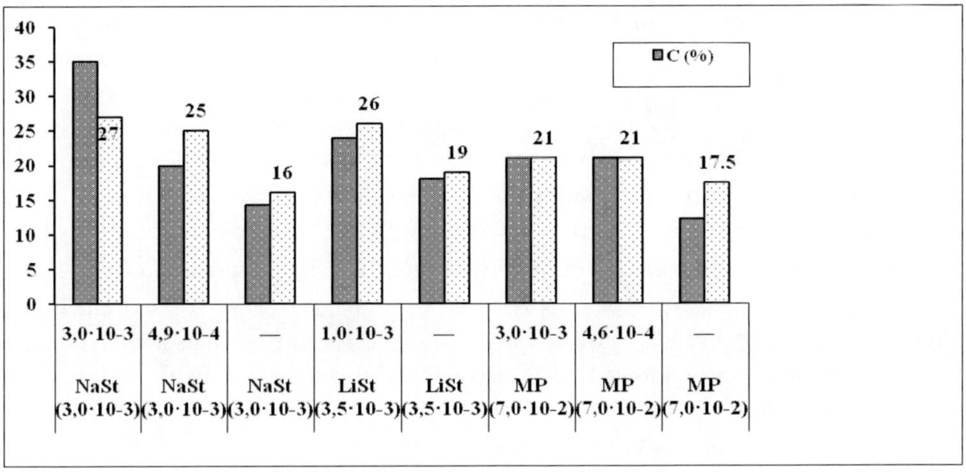

Figure 1. Values of conversion **C** (%) (I row), maximum values of hydroperoxide concentrations $[PEH]_{max}$ (mass.%) (II row) in reactions of ethylbenzene oxidation in the presence of triple catalytic systems {Ni(II)$(acac)_2$+L^2+PhOH} (L^2 = NaSt, LiSt, MP: [PhOH] (mol/l) - on an axis of abscises (the top number), $[L^2]$ (mol/l)- on an axis of abscises (the bottom number)). $[Ni^{II}(acac)_2]=3.0\cdot 10^{-3}$ mol/l, 120°C.

II. Introduction

Often metals of constant valency compounds are used in combination with redox-active transition-metal complexes to promote a variety of reactions involving the transfer of electrons [10]. This effect is typified in metalloproteins such as the copper zinc superoxide dismutase, in which both metal ions have been proposed to be functionally active [11].

The adducts of salts of metals of constant valency with β - diketonates and N,N'-ethylene bis (salycylideniminates) of Co^{II}, Ni^{II}, Cu^{II} are known [12, 13]. Thus the coordination of salts of metals of constant valency M′L (M′ – metal of constant valency) with complexes of transition metals ML^1 is carried out through ligand (complexes of nickel and copper with the Shiff bases, copper acetylacetonate: ML^1— M′L (bonds L^1— M′) (I type). In the other cases – anions of salts of metals of constant valency are coordinated with ions of transitional metal (acetylacetonates of nickel and cobalt: L^1M—L M′ (bonds M—L) (II type)). For example, in complexes [15-crown-5⊃Ca(II)–(μ-OH)–Mn^{III}MST]$^+$ calcium ions are bound with $[MST]^{3-}$ - ligand of Mn complex ($[MST]^{3-}$ - tripodal ligand N,N′,N′′-[2,2′,2′′-nitrilotris(ethane-2,1-diyl)]tris(2,4,6-trimethylbenzenesulfonamido) [14]:

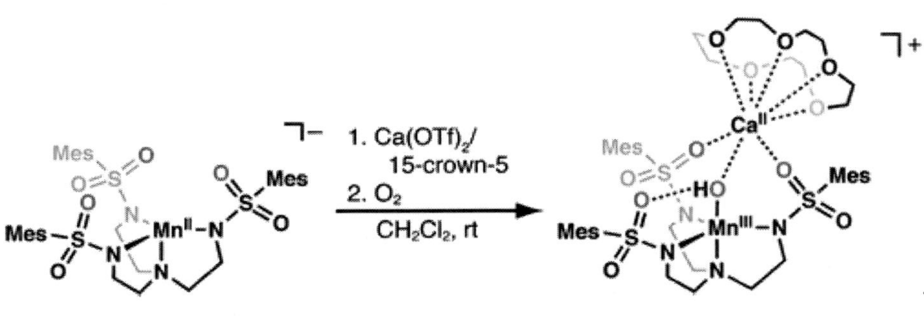

Earlier we have received kinetic and spectrophotometry (UV spectra) proofs in favor formation of complexes of bis(acetylacetonate)nickel, $Ni^{II}(acac)_2$, with NaSt of II type, M—L, namely, $(acac)_2$ Ni — St(Na) [2,3].

We discovered that the introduction of phenol together with the $\{Ni(acac)_2 + L^2\}$ catalyst in the reaction system in the initial stage of ethylbenzene oxidation is one of the most efficient methods of designing catalytic systems for the ethylbenzene selective oxidation to α-phenyl ethyl hydroperoxide. The high efficiency of three-component systems $\{Ni(acac)_2$+MSt+PhOH$\}$ (M = Na, Li) in the reaction of selective oxidation of ethylbenzene to α-phenyl ethyl hydroperoxide was associated with the formation of extremely stable heteroligand complexes $Ni(acac)_2 \cdot MSt \cdot PhOH$ [4]. The stability of these triple complexes throughout oxidation process allowed us to assume the formation of stable supramolecular structures due to H-bonding [5].

III. Experimental

AFM SOLVER P47/SMENA/ with Silicon Cantilevers NSG11S (NT MDT) with curvature radius 10 nm, tip height: 10 – 15 μm and cone angle ≤ 22° in taping mode on resonant frequency 150 KHz was used.

As substrate the polished Silicone surface special chemically modified was used.

Waterproof modified Silicone surface was exploit for the self-assembly-driven growth due to H-bonding of complexes $Ni^{II}(acac)_2$·NaSt(or LiSt)·PhOH with Silicone surface. The saturated chloroform ($CHCl_3$) solution of complex $Ni^{II}(acac)_2$·NaSt(or LiSt)·PhOH (1:1:1) was put on a surface, maintained some time, and then solvent was deleted from a surface by means of special method – spin-coating process.

In the course of scanning of investigated samples it has been found, that the structures are fixed on a surface strongly enough due to H-bonding. The self-assembly-driven growth of the supramolecular structures on the basis of complexes $Ni^{II}(acac)_2$·NaSt(or LiSt)·PhOH due to H-bonds and perhaps the other non covalent interactions was observed on Silicone surface. One can watch these structures on the basis of $Ni^{II}(acac)_2$·NaSt·PhOH with big height and volume. In check experiments it has been shown that for binary systems $\{Ni^{II}(acac)_2$+NaSt$\}$, and $\{Ni^{II}(acac)_2$+PhOH$\}$, the formation of the similar structures (exceeding on height of 2-10 nanometers) is not observed. In the case of $\{Ni^{II}(acac)_2$+LiSt$\}$ binary system the observed structures had interesting unusual form with height about 8-10 nm, unlike those which we received in the case of $Ni^{II}(acac)_2$·LiSt·PhOH complexes. The particles, observed in the case binary systems {LiSt(NaSt)+PhOH} had smaller height about 4-6 nm.

Ethylbenzene (RH) was oxidized with dioxygen at 120°C in glass bubbling-type reactor in the presence of three-component systems $\{Ni(II)(acac)_2+L^2$+PhOH$\}$ (L^2=NaSt, LiSt)) [4].

Analysis of oxidation products. α-Phenylethylhydroperoxide (PEH) was analyzed by iodometry. By-products, including methylphenylcarbinol (MPC), acetophenone (AP), and phenol (PhOH) as well as the RH content in the oxidation process were examined by GLC [4].

The order in which PEH, AP, and MPC formed was determined from the time dependence of product accumulation rate rations at $\mathbf{t \rightarrow 0}$. The variation of these rations with time was evaluated by graphic differentiation ([4], see Figure 2, 3). Experimental data processing was done using special computer programs Mathcad and Graph2Digit.

IV. Results and Discussion

Earlier we have established, that the increase in the initial rate of the ethylbenzene oxidation with dioxygen, catalyzed with $Ni^{II}(acac)_2$ in the presence of additives of metalloligand MSt (M=Li, Na, K), is due to higher activity of formed complexes $Ni^{II}(acac)_2$•MSt in the micro stages of chain initiation and/ or decomposition of PEH with free radical formation. As it was established the increase in rates of chain initiation and/ or radical decomposition of PEH (with free radical formation) seems to be due to the increase in DN of MSt as ligand-modifiers in row:

LiSt < NaSt < KSt.

At that the participation of catalyst $Ni^{II}(acac)_2$•MSt in micro steps of chain propagation and, probably, in chain termination must also be taken into account. [2]. The results found in [14] illustrate the possibility that redox-inactive metal ions can be used to facilitate the activation of dioxygen (see the abstract scheme of [14], above), which will be coordinated with our data.

At catalysis with triple complexes $Ni(acac)_2 \cdot L^2 \cdot PhOH$ (L^2 = NaSt, LiSt) the parallel formation of α-phenyl ethyl hydroperoxide, acetophenone and MPC was observed

($w_{AP(MPC)}/w_{PEH} \neq 0$ at $t \to 0$, $w_{AP}/w_{MPC} \neq 0$ at $t \to 0$) throughout the reaction of ethylbenzene oxidation) (see, for example, Figure 2, 3).

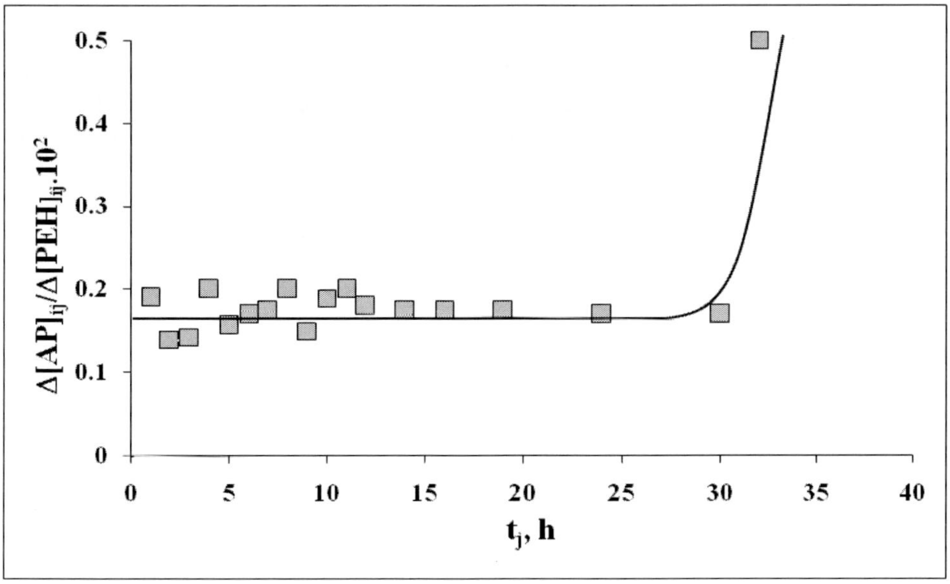

Figure 2. Dependence $\Delta[AP]_{ij}/\Delta[PEH]_{ij} \cdot 10^2$ on time t_j in the course of ethylbenzene oxidation, catalyzed with complexes $Ni(II)(acac)_2 \cdot NaSt \cdot PhOH$ (1:1:1), 120ºC.

A more considerable increase in the selectivity (S_{PEH}) at the catalysis by $Ni(acac)_2 \cdot L^2 \cdot PhOH$ (L^2 = NaSt, LiSt) complexes as compared with non-catalytic oxidation was associated with the change in the route of acetophenone and methylphenylcarbinol formation (AP and MPC form in parallel with PEH rather than as a result of its decomposition as observed in the non-catalytic ethylbenzene oxidation) and with the inhibition of the PEH heterolytic decomposition.

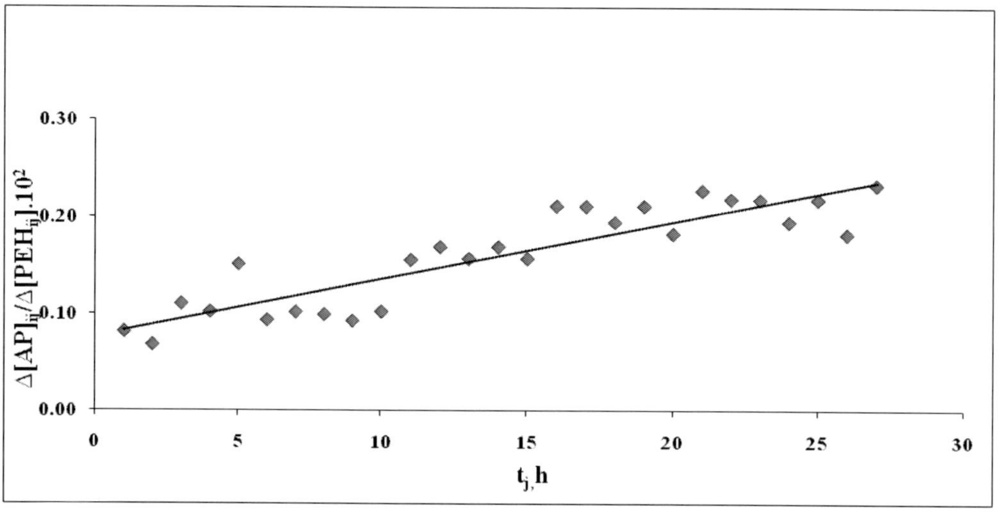

Figure 3. Dependence $\Delta[AP]_{ij}/ \Delta[PEH]_{ij} \cdot 10^2$ on time t_j in the course of ethylbenzene oxidation, catalyzed with complexes $Ni(II)(acac)_2 \cdot LiSt \cdot PhOH$ (1:1:1), 120ºC.

Thus we had shown that the triple complexes $Ni(II)(acac)_2 \cdot NaSt$ (or LiSt)·PhOH unlike binary complexes $Ni(II)(acac)_2 \cdot NaSt$ (or LiSt) seem to be not active in the reaction with hydroperoxide, but active in the reactions of chain initiation (O_2 activation) and the chain propagation ($Cat+RO_2^{\bullet}\rightarrow$) (and, probably, in chain termination). In these systems dioxygen activation may be promoted through the formation of intramolecular H-bonds [2]. The role of intramolecular H-bonds are established by us in mechanism of formation of triple catalytic complexes $\{Ni(II)(acac)_2 \cdot L^2 \cdot PhOH\}$ (L^2 = N-methylpirrolidon-2) in the process of ethylbenzene oxidation with molecular oxygen [4].

As mentioned above, the advantage of ternary systems $\{Ni(acac)_2+MSt+PhOH\}$ (M = Na, Li) is the long-term activity of the *in situ* formed complexes $Ni(acac)_2 \cdot L^2 \cdot PhOH$ (activity in chain initiation (O_2 activation) and the chain propagation ($Cat+RO_2^{\bullet}\rightarrow$)); the acac ligand does not undergo transformations in the course of ethylbenzene oxidation. So the ternary systems $\{Ni(acac)_2+MSt+PhOH\}$ (M = Na, Li) are the high efficient as catalysts of the selective ethylbenzene oxidation to α-phenyl ethyl hydroperoxide that impressed in the considerable increase in the degree of conversion of ethylbenzene to PEH at selectivity S_{PEH}=90% and in the yield of α-phenyl ethyl hydroperoxide. The high stability and long-term activity of heterobinuclear heteroligand complexes $Ni^{II}(acac)_2 \cdot MSt \cdot PhOH$ seems to be due to formation of active supramolecular structures $\{Ni^{II}(acac)_2 \cdot NaSt(\text{or } LiSt) \cdot PhOH\}_n$ as a result of intermolecular (phenol–carboxylate) H-bonds and, possibly, of the other non-covalent interactions [5].

The possibility of association of triple complexes $Ni^{II}(acac)_2$·NaSt(or LiSt)·PhOH to supramolecular structures due to H-bonding is followed from analysis of AFM data, received by us in this work. Results are presented on the next Figures (4-9) and Table 1.

Figure 4-5 demonstrated three-dimensional and two-dimensional AFM image (30x30 and 10×10(μm)) of the structures on the basis of triple complexes $Ni^{II}(acac)_2$·NaSt·PhOH formed at drawing of a uterine solution on a surface of modified silicone.

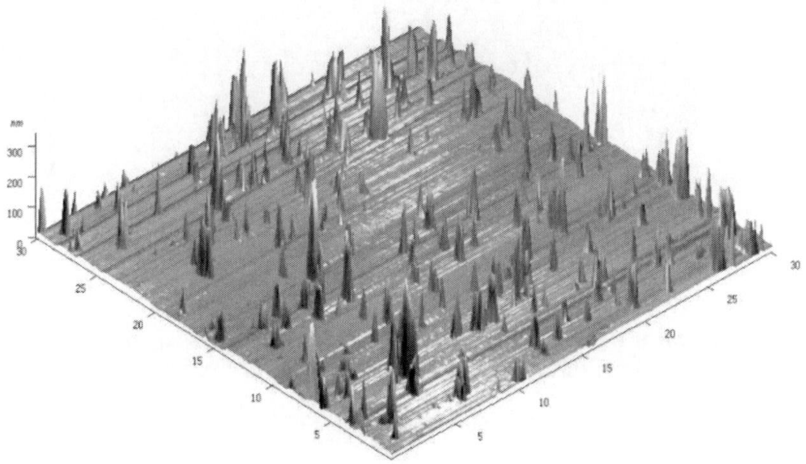

a

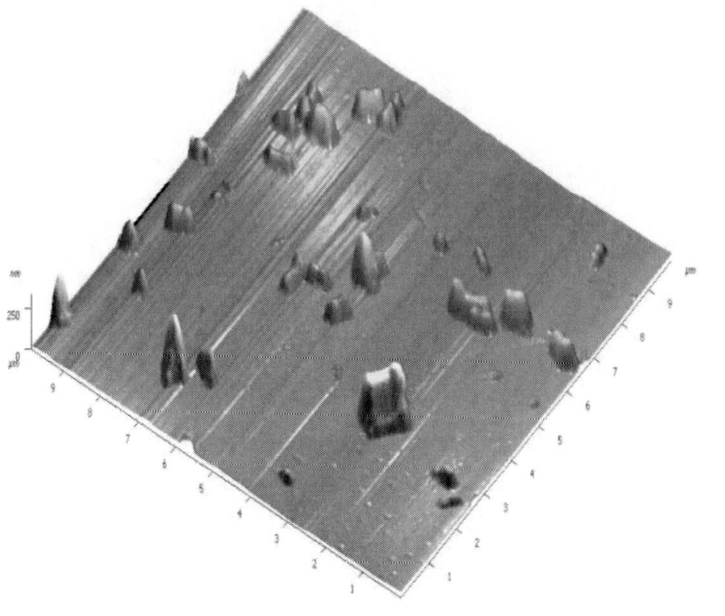

b

Figure 4. The AFM three-dimensional image (30×30 (a) and 10×10 (b) (μm)) of the structures formed on a surface of modified silicone on the basis of triple complexes $Ni^{II}(acac)_2$·NaSt·PhOH.

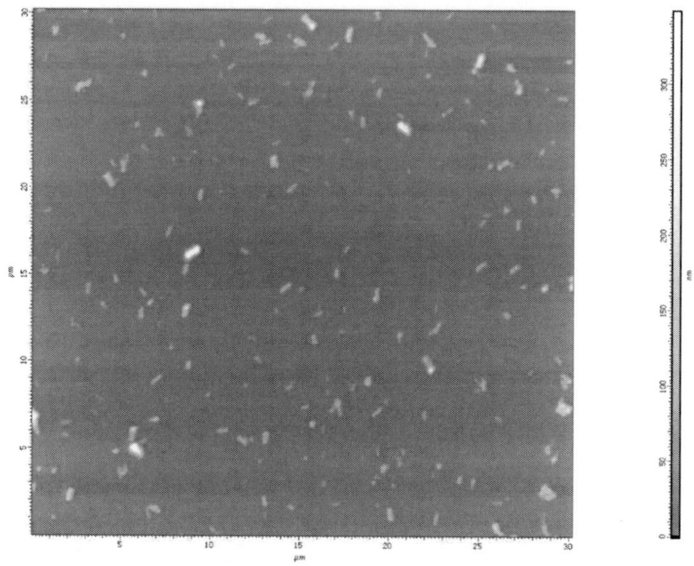

Figure 5. The AFM two-dimensional image (30×30 (μm)) of nanoparticles on the basis $Ni^{II}(acac)_2$·NaSt·PhOH formed on the surface of modified silicone.

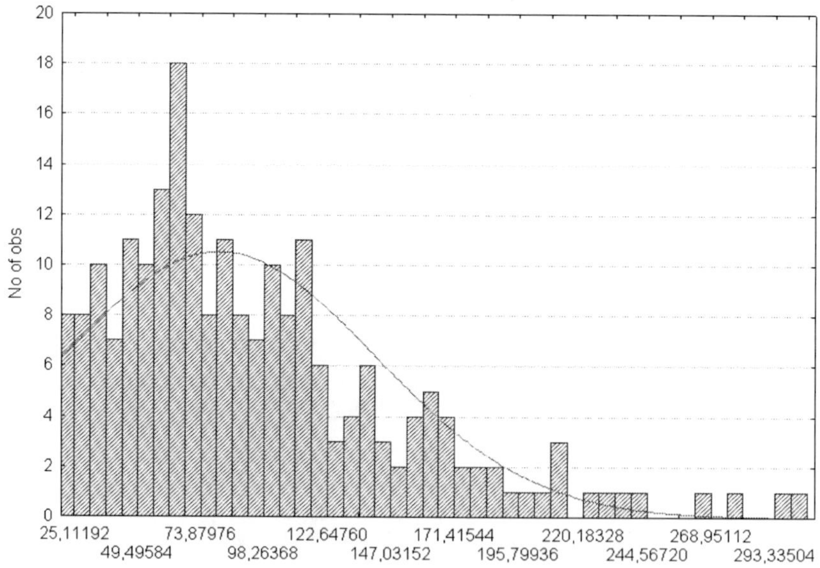

Figure 6. Histogram of mean values of height (nm) of the afm images of nano structures based on $Ni^{II}(acac)_2$·NaSt·PhOH, formed on the surface of modified silicone.

In the Figure 6 the histogram of mean height of nanoparticles on basis of $Ni^{II}(acac)_2$·NaSt·PhOH is presented. As can see, structures are various on heights from the 25 nm to ~ 250-300 nm for maximal values. The distribution histogram shows that the greatest number of particles - is particles of the mean size of 50-100 nm on height.

Table 1 shows the mean values of area, volume, height, width, length of nanoscale structures on basis of triple complexes $Ni^{II}(acac)_2 \cdot NaSt \cdot PhOH$ formed on the surface of modified silicone.

Table 1. The mean values of area, volume, height, length, width of the afm nanoparticle images on the basis of $Ni^{II}(acac)_2 \cdot NaSt \cdot PhOH$ formed on the surface of modified silicone

Variable	Mean values	Confidence -95.000%	Confidence +95,.000%
Area (μm^2)	0.13211	0.11489	0.14933
Volume (μm^3)	14.11354	11.60499	16.62210
Z (Height) (nm)	80.56714	73.23940	87.89489
Length (μm)	0.58154	0.53758	0.62549
Width (μm)	0.19047	0.17987	0.20107

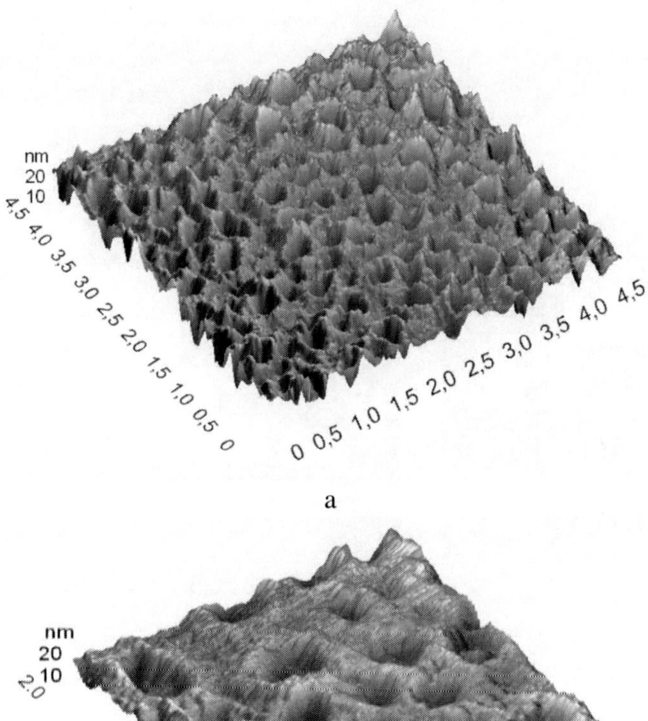

a

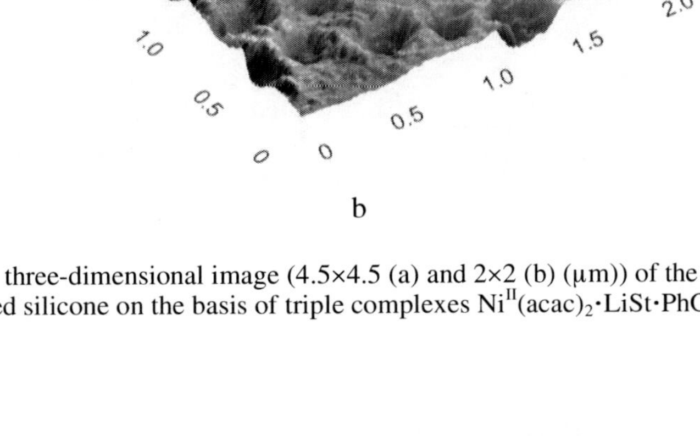

b

Figure 7. The AFM three-dimensional image (4.5×4.5 (a) and 2×2 (b) (μm)) of the structures formed on a surface of modified silicone on the basis of triple complexes $Ni^{II}(acac)_2 \cdot LiSt \cdot PhOH$.

a

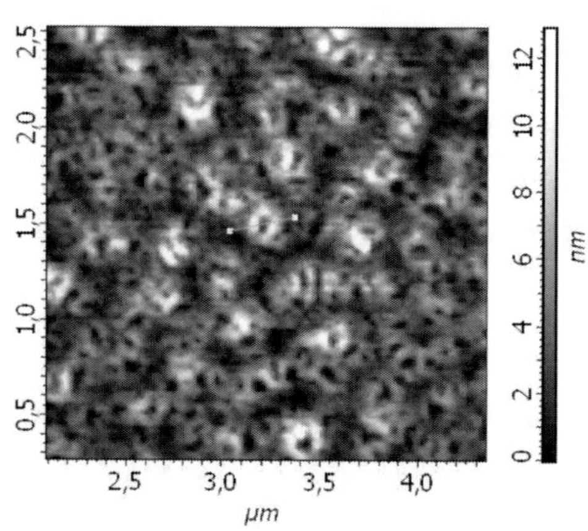

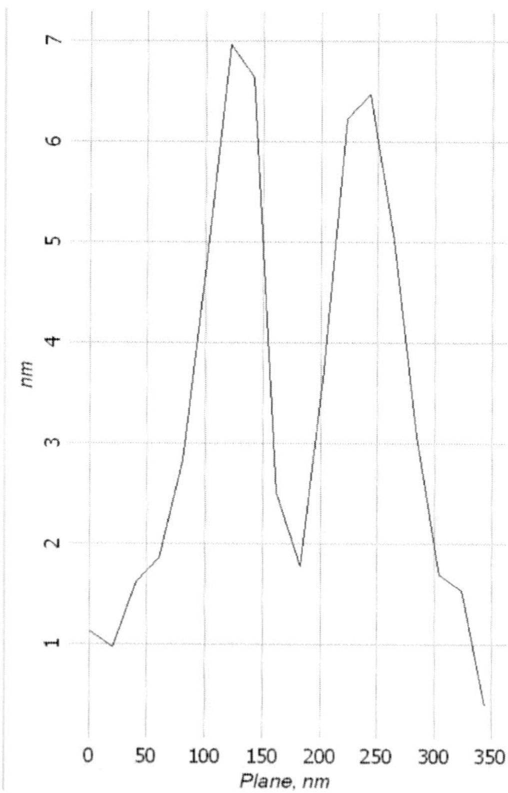

b c

Figure 8. The AFM three- and two-dimensional image (5×5 (a) and 2.5×2.5 (b) (μm)) of the structures with more simple form received on a surface of modified silicone on the basis of triple complexes $Ni^{II}(acac)_2$·LiSt·PhOH and profile of one of these structures (c).

a

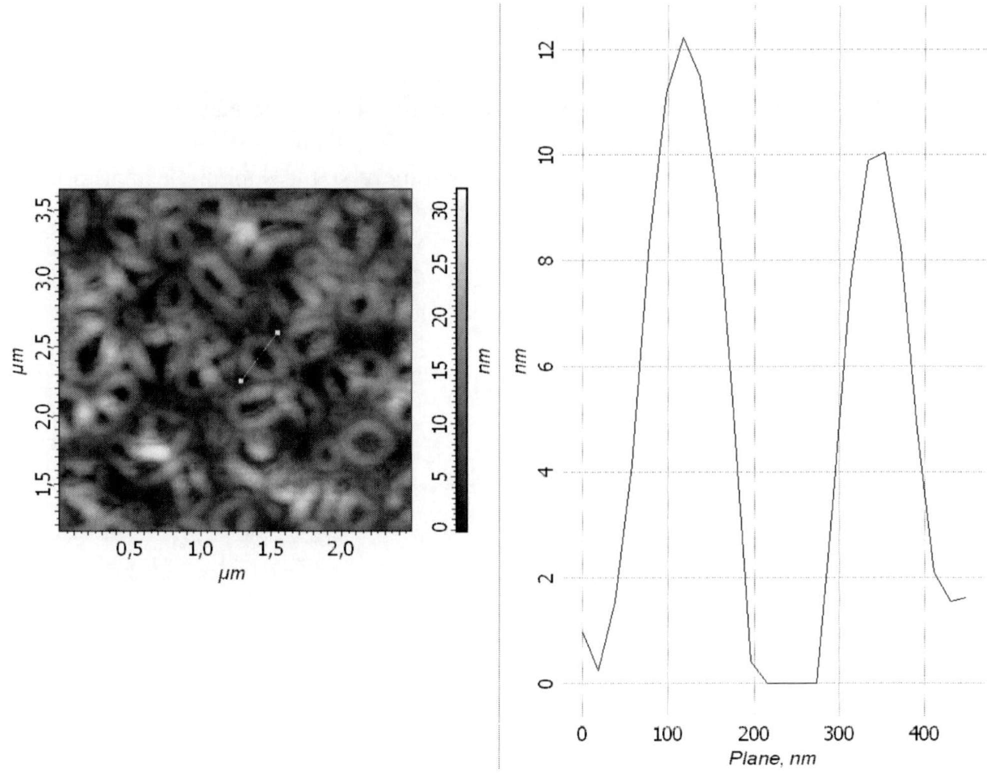

b

Figure 9. The AFM three-dimensional image (2.5×2.5 (μm)) (a), two-dimensional image (2.5×2.5 (μm)) and profile (b) of the nanostructures formed on a surface of modified silicone on the basis of $Ni^{II}(acac)_2{\cdot}LiSt$ complexes.

We have revealed an interesting fact that the length of the formed nanoparticles in the XY plane exceeds the width of the nanoparticles about three times (Table 1).

On the next pictures (Figure 7-9) the nanoparticle image on the basis of $Ni^{II}(acac)_2{\cdot}LiSt{\cdot}PhOH$ and $Ni^{II}(acac)_2{\cdot}LiSt$ complexes is demonstrated.

As one can see, nano structures on the basis of triple complexes $Ni^{II}(acac)_2$·LiSt·PhOH have the interesting cell form with cell height ~ 10 nm and cell width ~ 0.5 μm.

On the Figure 8 three- and two-dimensional image (5×5 and 2×2 (μm)) and profile of one of the nanostructures on the basis of triple complexes $Ni^{II}(acac)_2$·LiSt·PhOH with more simple form (cell height ~ 7 – 12 nm and cell width ~ 60 nm), which we observed also on a surface of modified silicone, are presented.

As can see, the nanostructures on the basis of $Ni^{II}(acac)_2$·LiSt·PhOH, presented with Figure 7 and 8, are different from nanostructures on the basis of triple complexes $Ni^{II}(acac)_2$·NaSt·PhOH by form and have lesser height (~ 10 – 12 nm unlike ~ 80 nm for nanostructures on the basis of triple complexes $Ni^{II}(acac)_2$·NaSt·PhOH) (Figure 6, Table 1).

In the case of binary complexes, {$Ni^{II}(acac)_2$·LiSt} and LiSt·PhOH, nanostructure were observed also, but had or lesser height (~4-6 nm in the case of LiSt·PhOH) or less neatly regular structure ($Ni^{II}(acac)_2$·LiSt, see, for example, Figure 8) in comparison with nanostructure on the basis of triple complexes $Ni^{II}(acac)_2$·LiSt·PhOH.

So, it can be concluded that the high the degrees of conversion of ethylbenzene to PEH and the yields of α-phenyl ethyl hydroperoxide in the case of catalysis of three-component systems {$Ni(acac)_2$+MSt+PhOH} (M = Na, Li) in the reaction of selective oxidation of ethylbenzene to α-phenyl ethyl hydroperoxide, as one of the reasons, may be associated with the formation of heterobinuclear heteroligand complexes $Ni^{II}(acac)_2$·MSt·PhOH, which are self organized during ethylbenzene oxidation to extremely stable supramolecular structures $\{Ni^{II}(acac)_2$·NaSt(or LiSt)·PhOH$\}_n$ as a result of intermolecular (phenol–carboxylate) H-bonds and, possible, the other non-covalent interactions [5]. The more effectivity of heterobinuclear heteroligand complexes $Ni^{II}(acac)_2$·NaSt·PhOH, including metalloligand NaSt, as selective catalysts as compared with $Ni^{II}(acac)_2$·LiSt·PhOH, seems to be due to formation of more stable supramolecular structures $\{Ni^{II}(acac)_2$·NaSt·PhOH$\}_n$. Data, presented on Figs.4-9, testify in favor of this supposition.

Conclusion

Thus, in the present work we applied AFM method in the analytical purposes to research the possibility of the formation of supramolecular structures on basis of heterobinuclear heteroligand triple complexes Ni(II)$(acac)_2$·NaSt(or LiSt)·PhOH.

We have shown what the self-assembly-driven growth seems to be due to H-bonding of triple complexes Ni(II)$(acac)_2$·NaSt(or LiSt)·PhOH with a surface of modified silicone, and further formation supramolecular nanostructures {Ni(II)$(acac)_2$·NaSt(or LiSt)·PhOH$\}_n$ due to directional intermolecular (phenol–carboxylate) H-bonds [5], and, possibly, other non-covalent interactions (van Der Waals-attractions and π-bonding).

These data support the very probable supramolecular structures appearance on the basis of heterobinuclear heteroligand triple complexes Ni(II)$(acac)_2$·NaSt(or LiSt)·PhOH in the course of the ethylbenzene oxidation with dioxygen, catalyzed by three-component catalytic system {Ni(II)$(acac)_2$+NaSt(or LiSt)+PhOH} and this can be one of the explanations of the high values of conversion of the ethylbenzene oxidation into α-phenyl ethyl hydroperoxide at selectivity S_{PEH} preservation at level not below S_{PEH}=90 % in this process. The more effectivity of heterobinuclear heteroligand complexes $Ni^{II}(acac)_2$·NaSt·PhOH, including metalloligand NaSt, as selective catalysts as compared with $Ni^{II}(acac)_2 \cdot LiSt \cdot PhOH$, seems to be due to formation of more stable supramolecular structures $\{Ni^{II}(acac)_2 \cdot NaSt \cdot PhOH\}_n$.

REFERENCES

[1] Weissermel, K., Arpe, H.-J. (1997). Industrial Organic Chemistry, 3nd ed., transl. by Lindley C.R. VCH: New York, USA.
[2] Matienko, L.I. (2007). In: Reactions and Properties of Monomers and Polymers, D'Amore, A. & Zaikov, G. (Eds.), Chapter 2, Nova Sience Publ., Inc.: New York, 21.
[3] Matienko, L.I., Mosolova, L.A., Zaikov, G.E. (2010). Selective Catalytic Hydrocarbons Oxidation. New Perspectives, Nova Science Publ., Inc.: New York, USA.
[4] Matienko L.I., Mosolova L.A. (2007). *Oxidation Commun.* 30, 282.
[5] Dubey M., Koner R.R., Ray M. (2009). *Inorg. Chem.*, 48, 9294.
[6] Leninger, St., Olenyuk, B., Stang, P.J. (2000). *Chem. Rev.*, 100, 853.
[7] Stang, P.J., Olenyuk, B. (1997). *Acc. Chem. Res.*, 30, 502.
[8] Beletskaya, I., Tyurin, V.S., Tsivadze, A.Yu., Guilard, R., Stem, Ch. (2009). *Chem. Rev.*, 109, 1659.
[9] Matienko, L.I., Binyukov, V.I., Mosolova, L.A., and Mil, E.M. (2011) *J. of Characterization and Development of Novel Materials*. 4, №1, 54 (Nova Publ., Canada).
[10] Fukuzumi, S., Ohkubo, K. (2010). *Coord. Chem.Rev.* 254, 372.
[11] Valentine, J.S., Doucette, P.A., Potter, S.Z. (2005). *Annu. Rev. Biochem.* 74, 63.
[12] Dani, C.M., Das, A.K. (1984). *Indian J. Chem. Sect.* A., 23, 1041.
[13] Aminabhovi, T.M., Biradar, N.S., Divaka, M.S. (1984). *Inorg. Chim. Acta*, 92, 99.
[14] Park, Y.J., Ziller, J.W., Borovik, A.S. (2011). *J. Am. Chem. Soc.*, 133, 9258.

In: Quantitative Chemistry, Biochemistry and Biology
Editors: G.E. Zaikov, O.V. Stoyanov et al.
ISBN: 978-1-62948-332-0

Chapter 33

EFFECT OF THE STEREOMETRY ON THE REACTIVITY OF TERPENEPHENOLS AND THEIR PHENOXYL RADICALS IN THE INITIATED OXIDATION REACTIONS

L. I. Mazaletskaya**, N. I. Sheludchenko, L. B. Dudnik and L. N. Shishkina

Emanuel Institute of Biochemical Physics of Russian Academy of Sciences, Moscow, Russia

ABSTRACT

The stoichiometric factors of inhibition and the rate constants of the terpenephenols (TP) with isobornyl and isocamphyl substituents were determined by the reaction with peroxy radicals of ethylbenzene. The reactivity was found to decrease for o-alkoxy compared with o-alkyl substituent caused by the intramolecular hydrogen bond formation that is conformed by FTIR-spectroscopy. The inhibitory activity for mixtures of terpenephenols with 2,6-di-tert-butyl phenols in the initiated oxidation of ethylbenzene was also studied. In spite of the similar antiradical activities of terpenephenols with isobornyl and isocamphyl sunstituents, the reactivity of phenoxyl radicals formed from them are substantially different that is resulted from the kinetic data for mixtures of terpenephenols with sterically hindered phenols.

Keywords: Terpenephenols, phenoxyl rasicals, antiradical activity, initiated oxidation, inhibition, hydrogen bond

* Emanuel Institute of Biochemical Physics of Russian Academy of Sciences, 4 Kosygin st., Moscow, Russia; e-mail: lim@sky.chph.ras.ru

AIMS AND BACKGROUND

Natural and synthetic antioxidants (AO) are known to be active regulators of the lipid peroxidation in biomembranes and of interest as the biologically active substances with a wide spectrum of the action [1]. Substituted phenols synthesized on the natural raw material base are considerable as the worthwhile biological active substances due to their comparatively low toxicity [2 - 4]. Earlier the high reactivity was found for terpenephenols (TP) which differ in the structure of terpene substituents and their arrangement in the aromatic ring with respect to the OH-group [5, 6]. However, values of the rate constant of TP with free radicals (k_7) were detected to depend substantially on the nature of the peroxy radicals, the rate initiation of the oxidative process and the oxidation substrate. Analysis of the literature data is allowed us to consume differences in the reactivity of both TP and their phenoxyl radicals in the dependence on the stereometric arrangement of their substituents.

The aim of this work was to study the effect of the stereometry on the kinetic parameters of terpenephenols and their phenoxyl radicals with the isobornyl (IBP) and isocamphyl (ICP) substituents in the initiated oxidation of ethylbenzene.

Table 1. Structure, name, abbreviation used terpenephenols, their stoichiometric factor (*f*) and rate constants of TP with ethylbenzene peroxy radical (k_7)

Structure	Name, abbreviation	$fk_7 \times 10^{-4}$, l/(mol s)	*f*	$k_7 \times 10^{-4}$, l/(mol s)
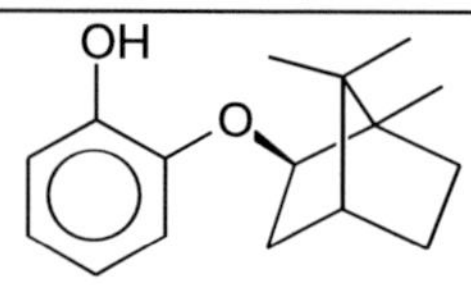	2-Isobornyloxyphenol, TP-1	0.8±0.05	–	0.4±0.03*
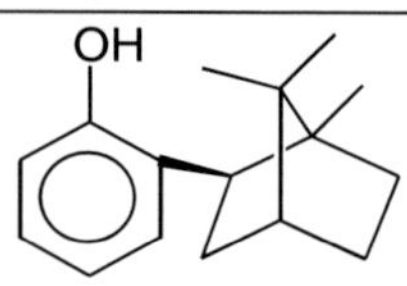	2-Isobornylphenol, TP-6	5.2±0.07	2.1	2.5±0.04
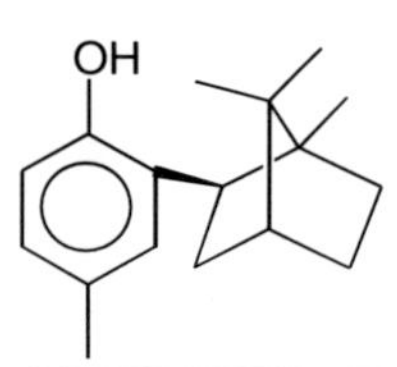	2-Isobornyl-4-methyl-phenol, TP-5	11.6±0.8	1.9	6.3±0.4
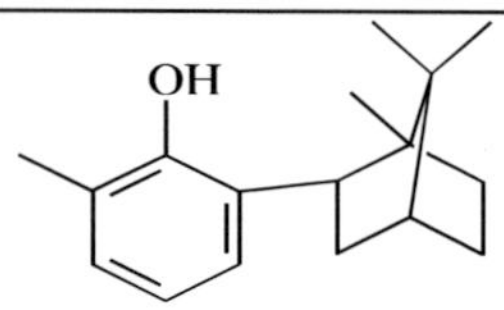	2-Isobornyl-6-methyl-phenol, TP-8	10.0±0.9	2.0	5.0±0.5

Structure	Name, abbreviation	fk7×10-4, l/(mol s)	f	k7×10-4, l/(mol s)
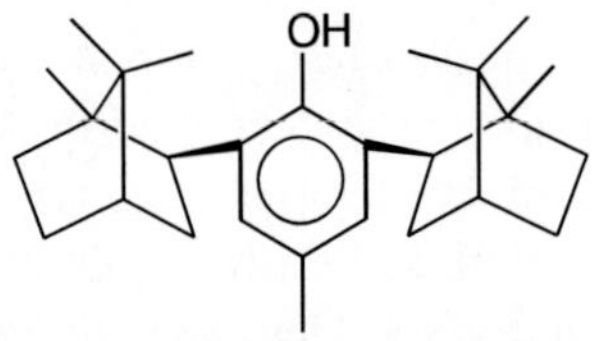	2,6-Diisobornyl-4-methylphenol, TP-7	20.7±0.3	1.8	11.5±0.2
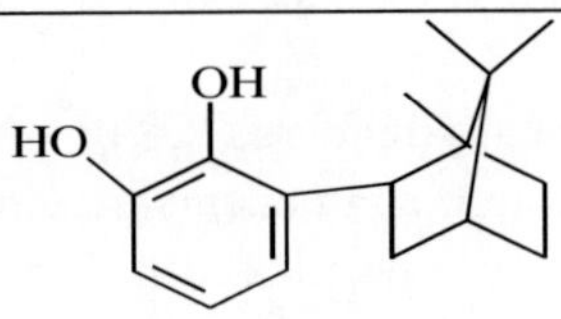	3-Isobornyl-1,2-hydroxybenzene, TP-2	54.9±2.6	2.0	27.5±1.3
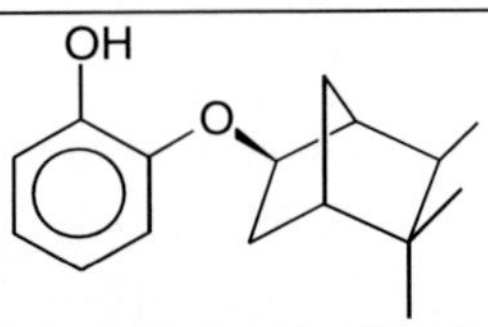	2-isocamphyloxyphenol, TP-28	1.06 ± 0.05	–	0.53±0.03*
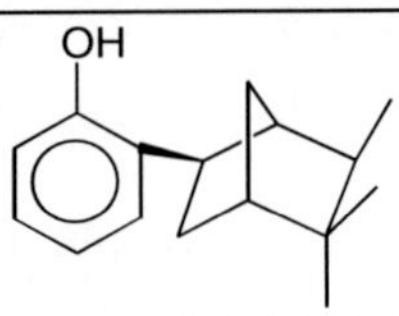	2-isocamphylphenol, TP-26	4.1 ± 0.04	1.9	2.05±0.03
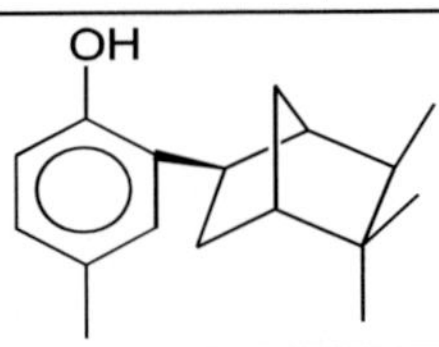	2-isocamphyl-4-methylphenol, TP-14	9.7 ± 0.8	1.1	8.8±0.8
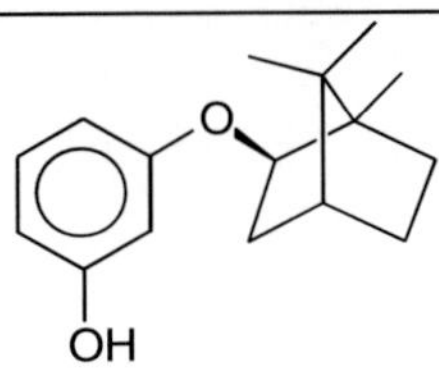	3-isocamphyloxyphenol, TP-29	2.6 ± 0.08	1.8	1.4±0.06
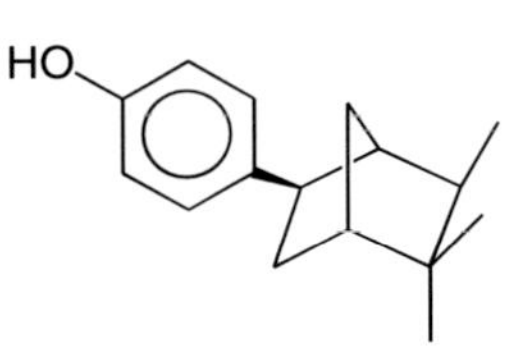	4-isocamphylphenol, TP-27	3.9 ± 0.1	2.0	1.95±0.8

*According to the assumption that $f = 2$.

EXPERIMENTAL

The AIBN-initiated oxidation of ethylbenzene, a model hydrocarbon, with air oxygen was studied by measuring the oxygen uptake on a sensitive volumetric Warburg setup at 333 K and the initiation rate of $W_i = 5\times10^{-8}$ mol/(l s). The induction period (τ) as described in [7] and the initial rate of oxygen uptake were determined from the kinetic curves of oxygen uptake in the presence of AO. The substrate of oxidation together with a dissolved initiator was preliminarily held at a constant temperature, and then an antioxidant was added. Ethylbenzene was purified according to the standard procedure.

The combined inhibitory action for mixtures of TP and the sterically hindered 2,6-di-*tert*-butylphenol was studied at a constant total concentration of the mixture components varying the ratio between the components. The effectiveness of the inhibitory action (a) of different components and their mixtures was estimated as the time when a certain amount of oxygen was absorbed.

The studied TP were synthesized by the research group led by A.V. Kutchin, a corresponding member of the Russian Academy of Sciences (Institute of Chemistry of the Ural Branch of the Russian Academy of Sciences) and were kindly provided for the present investigation. The synthesis of IBP was carried out according to the procedures described in [8, 9]. The synthesis of ICP was carried out according to the procedures described in [6]. The studied TP were used without the additional purification. The structures of the compounds under study are given in Table 1.

IR-spectra of the hexane solutions of the individual components at the concentration of 10^{-2} mol/l were registered on a Spectrum 100 IR-Fourier-spectrophotometer.

RESULTS AND DISCUSSION

The typical kinetic curves of the oxygen uptake in the initiated ethylbenzene oxidation in the absence and in the presence of TP-5 in dependence on its concentrations are shown in Figure 1. To calculate factor of inhibition (f), the dependence of τ on the initial AO concentration was plotted. It is seen that the f values calculated from the slopes of the straight lines (Figure 2) are close for all TP under study ($f \sim 2$) (Table 1), which agrees well with the mechanism of the action of phenolic AO, according to which two free radicals decay on a phenol molecule. It is necessary to note an unexpectedly low $f = 1.1$ value for TP-14 (Figure 2). So, $f = 1.9$ for a similar stereoisomer, TP-26, which does not contain the CH_3 group in the *o*-position and the f value is also 1.9 for TP-5 which has the isobornyl group in the *o*-position distinct from TP-14 (Table 1).

To calculate the rate constants of the interaction of AO with peroxy radicals (k_7), the values of the initial oxidation rates were presented in the coordinates of Eq. (I):

$$W_0/W - W/W_0 = f\,k_7[\mathrm{AO}]_0/\,k_6^{0.5}W_i^{0.5}, \qquad \text{(I)}$$

where W_0 and W are the oxidation rates in the absence and in the presence of AO, respectively, and k_6 = 1.9×10^7 1/(mol s) is the rate constant of the bimolecular recombination of the peroxy radicals [10].

The parameter fk_7 was calculated from the slope of the straight lines (Figure 3, 4). The calculated values are listed in Table 1. The phenols under study can be arranged in a row according to thc increase in their antiradical activity: TP-1 < TP-6 < TP-8 < TP-5 < TP-7 < TP-2 (for isobornylphenols) and TФ-28 < TФ-29 < TФ-27 <TФ-26 < TФ-14 for (for isocamylphenols).

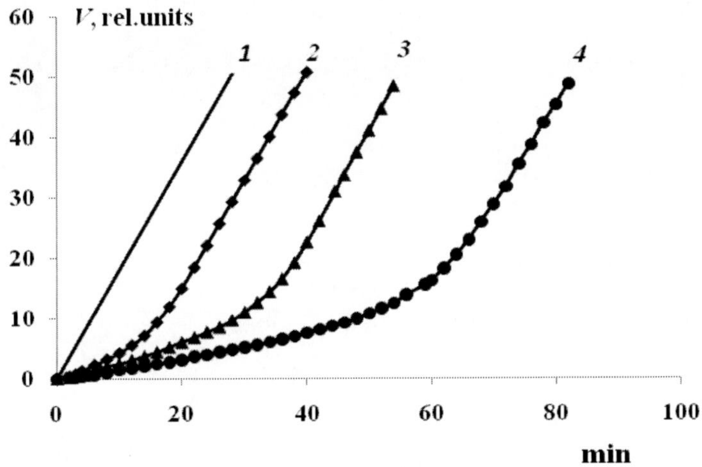

Figure 1. Kinetics of oxygen uptake in the oxidation of ethylbenzene without additives (*1*) and in the presence of $[TP\text{-}5]_0 \times 10^4$ mol/l 1: *2* – 0.3; *3* – 0.6; *4* – 1.0. Temperature 333 K; W_i = 5×10^{-8} mol/(l s).

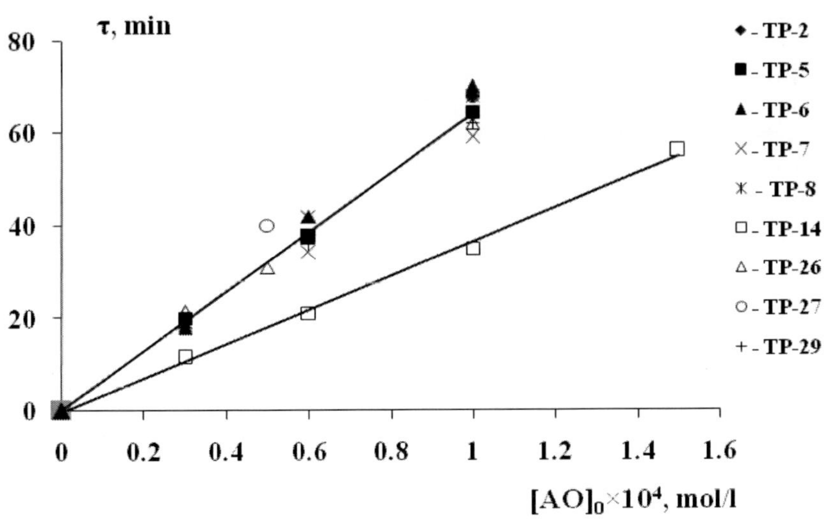

Figure 2. Induction time as a function of the initial AO concentration. Ethylbenzene, 333 K, W_i = 5×10^{-8} mol/(l s).

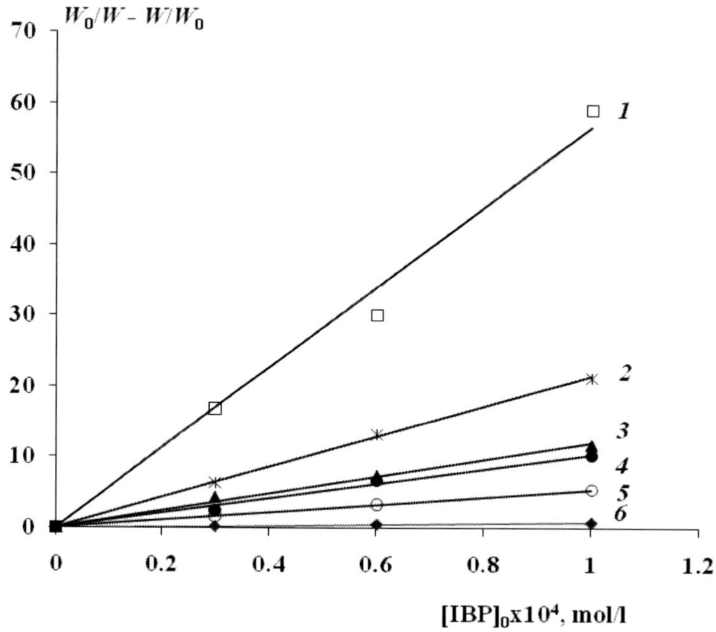

Figure 3. Initial rate of oxidation as a function of the initial IBP concentration in the coordinates of Eq. (I): 1 – TP-2, 2 – TP-7, 3 – TP-5, 4 – TP-8, 5 – TP-6, and 6 – TP-1. Temperature 333 K, ***Wi*** = 5×10-8 mol/(l s).

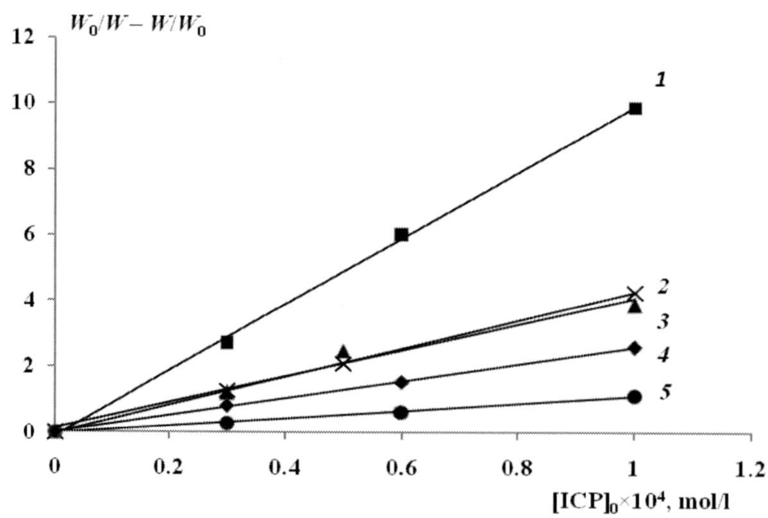

Figure 4. Initial rate of oxidation as a function of the initial ICP concentration in the coordinates of Eq. (I): 1 – TP-14, 2 – TP-26, 3 – TP-27, 4 – TP-29, and 5 – TP-28. Temperature 333 K, Wi = 5×10-8 mol/(l s).

TP-1 and TP-28 possess the lowest reactivity in respect to ethylbenzene peroxy radicals. A distinctive feature of these compounds in comparison with the other terpenephenols is the presence of 2-alkoxysubstituent. According to published data, alkoxy (–OR) substituents are characterized by the higher electron-donating properties in comparison with alkyl (–R) substituents: the Hammett constants for $-OCH_3$ and $-CH_3$ substituents are σ = -0.778 and -0.170 [11], respectively. The enhancement of the electron-donating properties of the substituent in the *o*-position to the hydroxyl group results in an increase in the rate constant k_7. For example, the values of k_7 (333 K) = 2×10^5 and 1.3×10^4 1/(mol s) for 4-methoxy- and 4-methylphenol, respectively [12].

The opposite effect of the substituents –OR and –R on the antiradical activity takes place when the substituents are introduced in the *o*-position. This follows from the data in Table 1. The replacement of the alkyl substituent (TP-6, TP-26) by the alkoxy substituent (TP-1, TP-28) results in a decrease in the value of fk_7 by a factor of 6.5 for IBP and about 3.9 for ICP. The decrease in the antiradical activity of TP with the alkoxy in comparison with TP with the alkyl substituent in the *o*-position is probably caused by the intramolecular hydrogen bond between the H atom of the hydroxyl group and the oxygen atom of the *o*-substituent. The shifts of the stretching mode from 3618.33 cm^{-1} (TP-6) and 3622.6 (TP-26) to 3561.51 cm^{-1} (TP-1) and 3560.3 (TP-28) respectively confirm the fact of hydrogen bond formation (Table 2). It is well documented that intramolecular hydrogen bonding with participation of the H atom of the phenolic OH group results in a drastic decrease in their reactivity [11]. This is confirmed by low values of k_7 for oxyanthraquinones [13] and ketophenol [14]. The decrease in the reactivity of 2-alkoxy-substituted phenols in comparison with their 4-substituted analogs in the reaction with the peroxy radicals agrees with the difference in the energy of the OH bond breaking for these AO.

Table 2. Stretching vibration band (ν) in substituted phenols

TP	TP-1	TP-5	TP-6	TP-7	TP-26	TP-28
ν, cm^{-1}	3561.6	3620.8	3618.8	3612.5	3622.6	3560.3

The values of bond breaking are 345.8 and 352.1 kJ/mol for 4-methoxy- and 2-methoxypenols, respectively [12]. As was found in [15], the linear correlation between the energy of the OH bond breaking and the log k_7 is observed. Thus, the formation of the intramolecular hydrogen bond between the hydrogen atom of the hydroxyl group and the oxygen atom of the *o*-substituent accounts for the low antiradical activity of TP-1 and TP-28. A decrease in antiradical activity caused by H-bond formation is very substantial; indeed, the fk_7 value for TP-29 with an alkoxy substituent in the *m*-position is higher by 2.45 times compared with TP-28.

The antiradical activity for monophenols with the alkyl substituents increases with an increase in the amount of donor substituents in the benzene ring. TP-6, TP-26 and TP-27 with an only substituent in the *o*- or *p*-position to the OH group is the less active of the alkyl-substituted phenols. According to the table 1, the antiradical activities of TP-26 and TP-27, which only contain one isocamphyl substituent in the *o*- or *p*-position, respectively, are similar. The fk_7 values for these TP are not for certain differ (table 1) and close to fk_7 for TP-6, the molecule of which also contains only one (isobornyl) substituent in the *o*-position

The addition of the CH_3 group at the *o*- (TP-8) or at the *p*-position (TP-5 and TP-14) results in an increase in fk_7, with the values of fk_7 being close to $(9.7 - 11.6) \times 10^4$ 1/(mol s).

The presence of two *o,o'*-isobornyl and *p*-CH_3 (TP-7) groups results in the following increase in fk_7 (Table 1).

Compound TP-2 containing two OH groups has the highest antiradical activity, which agrees with the literature data that polyphenols of the pyrocatechol series are characterized by the high values of the rate constant k_7 [16].

Thus, the reactivity of terpenephenols with respect to peroxy radicals of ethylbenzene increases in amount of alkyl substituents and OH-groups in aromatic ring, decreases when *o*-alkyl substituents are replaced by *o*-alkoxyl groups owing to the of intramolecular H-bond formation and almost do not change in the dependence on differences in the structure of the terpene substituents and their arrangement with respect to the OH group.

Special attention must be given to TP-7, the molecule of which contains two bulky substituents at the *o,o'*-positions. We assume that TP-7 belongs to sterically hindered phenol. 2,6-Di-*tert*-butylpenols are a classical example of these. The decrease in the rate constant k_7 in comparison to 2-*tert*-butyl-substituted analogs is characteristic. The antiradical activity of 2,6-di- *tert*-butylphenol is approximately fourfold lower than that for 2-*tert*-butylphenol (k_7 = 0.96×10^4 and 4×10^4 l/(mol s), respectively) [17]. If TP-7 with isobornyl substituents has the properties of the sterically hindered phenols, we might expect a decrease in the value of k_7 in comparison with TP-5. However, the experimentally obtained value of k_7 for TP-7 is not only almost twofold higher than that for TP-5, but significantly exceeds the antiradical activity of known sterically hindered phenols (2,6-di-*tert*-butyl-4-methylphenol $k_7 = 2.04 \times 10^4$ 1/(mol s) under similar conditions [15]) On the contrary, the antiradical activity of TP-7 is comparable with sterically unhindered (unshielded) phenols with alkyl substituents at 2,6-and 2,4,6-positions. For example, the values of k_7 are 8.6×10^4 and 1.9×10^5 l/(mol s) for 2,6-dicyclohexyl- and 2,4,6-trimethylphenol, respectively [17].

From the comparative data on the rate constants k_7, TP-7 can be classified as sterically unhindered phenol. According to [18], the synergetic action of the mixtures of TP-7 with sterically hindered phenols may be expected. It is known that unhindered and sterically hindered phenol form synergetic mixtures [18 – 20], with the efficiency changing in the dependence on the substrate and conditions of the oxidation and the antiradical activity of the mixture components. In particular, the efficiency of the inhibiting action of a mixture should not differ from the efficiency of the most active component (unhindered phenol), when the oxidation is carried out under condition of unbranched chain reaction and constant overall concentration of added antioxidants. This result is caused by regeneration of the unhindered phenol as a result of interaction of its phenoxyl radical with the sterically hindered phenol [18]. One can see in Figure 5 (curves *2, 3, 4*) that the inhibiting action of the equimolar mixture of TP-7 with sterically hindered 2,6-di-*tert*-butylphenol (DTBP) is additive, because the values a of the mixture are close to $a_{mix,\ add}$ For example, if we take the time during which the oxygen uptake is $V =$ 4 relative units as the efficiency of inhibition, then the experimental a_{mix} =11,5 min does not differ from $a_{mix,\ add} = (a_{DTBP} + a_{TP\text{-}7})/2 = (3.5 + 19.5)/2 = 11.5$ min. Hence, TP-7 being more active in the reaction with peroxy radicals does not regenerate in the presence of DTBP.

Probably, the two bulky isobornyl substituents at *o,o'*-positions shield the OH groups, and TP-7 plays the role of the sterically hindered phenol forming synergetic mixtures with unhindered phenols. Since TP-7 itself has a high value of k_7 in comparison with *o,o'*-di-*tert*-

butyl-substituted phenols, the synergetic mixtures on its basis should use AO with still higher reactivity of the OH group and capable of forming synergetic mixtures with sterically hindered phenols. α-Tocopherol (TPh) is one of these phenols with k_7 = 1×10[6] 1/(mol s) [21]. One can see in Figure 5 that TPh is more efficient (curve *7*) in comparison with TP-7 (curve *5*), and the inhibiting action of their equimolar mixture (*6*) is close to additive: at the start, TPh inhibits the reaction and oxygen absorption occurs at a rate typical of the process inhibited by TPh. When TPh is completely consumed in accordance to the stoichiometric factor, the oxygen uptake rate curve of the mixture (curve *6*) hardly differs from that in the presence of TP-7 (curve *5*), i.e., the inhibiting effects of the antioxidants are independent of one another.

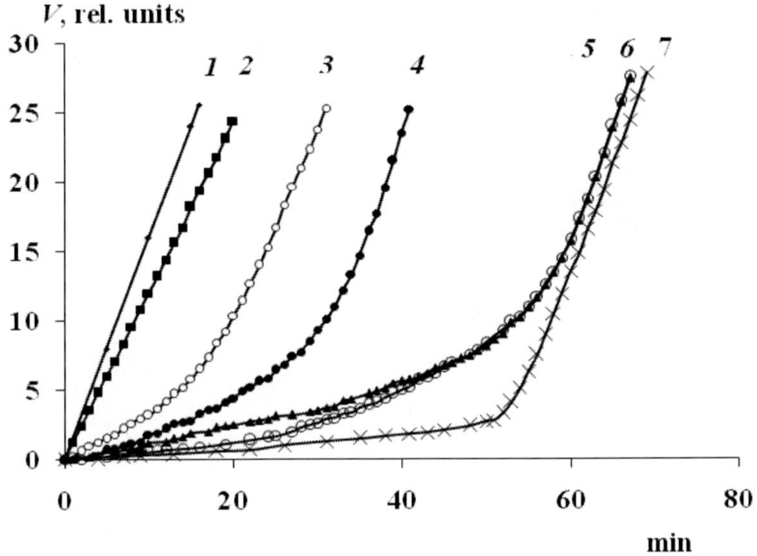

Figure 5. Kinetics of oxygen uptake in ethylbenzene oxidation without additives (1) and in the presence of AO, mol/1: 2 – 6x10-5 DTBP; 3 – 3x10-5 DTBP + 3x10-5 TP-7; 4 – 6x10-5 TP-7; 5 – 1x10-4 TP-7; 6 – 5x10-5 TPh + 5x10-5 TP-7; and 7 – 1x10-4 TPh; 333 K, Wi = 5x10-8 mol/(l s).

Thus, these data, the additivity of the inhibiting action of TP-7 mixtures with the unhindered phenol and the value of k_7 indicate that TP-7 can be classified as unhindered phenol.

Our results allow for the assumption that the absence of the expected synergistic effect of the mixtures of unhindered TP-7 with sterically hindered DTBP (Figure 5) is due to the specifics of the isobornyl substituents, which hinder hydrogen transfer from the OH group of DTBP to the phenoxyl radical formed from TP-7.

To verify this assumption, the joint action of DTBP with the other unhindered phenol TP-5 with an only *o*-isobornyl substituent was studied. The efficiency of the inhibition of the separate components and an equimolar mixture was estimated at a concentration of 1×10^{-4} mol/1. The time of oxygen uptake of V=4 relative units was used as a criterion for the inhibition period. The values of a are 5 and 25 min for DTBP and TP-5, respectively; a_{mix} = 15 min coincides with $a_{\text{mix,add}}$. Hence, the action of the mixture of TP-5 with the sterically hindered DTBP is additive similarly to TP-7.

Thus, the phenolic antioxidants containing one (TP-5) or two (TP-7) isobornyl substituents in the *o*-position, unlike TPh and unhindered phenols (with *tert*-butyl substituent at *o*-position, with methyl or cyclohexyl substituents at *o,o'*-positions), are not regenerated in the mixture with the sterically hindered phenols: the inhibiting effect of their mixtures is additive.

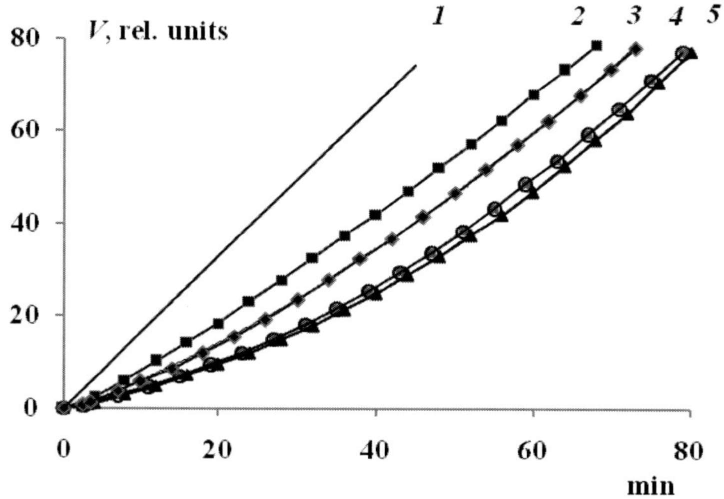

Figure 6. Kinetic curves of oxygen uptake in ethylbenzene oxidation without additives (1) and in the presence of (mol/1): 2 – 1×10-4 DTBP, 3 – a mixture of 0.8×10-4 DTBP and 0.2×10-4 TP-26, 4 – a mixture 0.5×10-4 DTBP and 0.5×10-4 TP-26 and 5 – 1×10-4 TP-26. Temperature 333 K, Wi = 5×10-8 mol/(l s).

Howerver, on the contrary, phenoxyl radicals of ICP can abstract a hydrogen atom from sterically hindered phenols. We studied the joint action of sterically hindered DTBP and *o*-isocamphylphenol (TP-26) and *p*-isocamphylphenol (TP-27) to reveal the influence of the terpene substituent structure and its spatial arrangement on the reactivity of the phenoxyl radials. As stated above, mixtures of two phenols, sterically hindered and unhindered, can exhibit a synergetic effect in the inhibition of the oxidation processes. The mechanism of synergism is caused by the regeneration of the more active mixture component (unhindered phenol) from its radical form because of the hydrogen abstraction from the sterically hindered phenol. Accordingly, we selected DTBP as a sterically hindered phenol, the antiradical activity of which (Figure 6, curve *2)* was lower than those of TP-26 (Figure 6, curve *5)* and TP-27. We see that the inhibiting action of a mixture is higher than additive and increases as the content of TP-26 (%) grows (Figure 6, curves *3, 4;* Figure 7, curve *1)*. The kinetics of oxygen absorption by an equimolecular mixture of *o*-isocamphylphenol and DTBP (Figure 6) and the effectiveness of mixture inhibition (Figure 7) are insignificantly different from those of the more active component, TP-26. This result is evidence of the interaction of the phenoxyl radical from TP-26 with DTBP resulting in the formation of initial TP-26 molecules. However, it is need to note that the regeneration of phenol TP-26 is incomplete and decreases as the fraction of DTBP in the mixture grows (Figure 7). This is explained by insignificantly different antiradical activities of mixture components. Indeed, the k_7 constant of DTBP is 0.95 × 10^4 l/(mol s) [17]. Therefore, at a DTBP : TP-26 = 4 : 1 ratio, we have fk_7[DTBP]$_0$ > fk_7[TP-26]$_0$, that is, DTBP successfully competes with TP-26 and is partially consumed in the reaction with the peroxy radicals. For an equimolecular mixture, fk_7[TP-26]$_0$ > fk_7[DTBP]$_0$. Therefore the peroxy radicals predominantly interact with TP-26 molecules. Phenoxyl radicals formed as a result of interaction react with DTBP with the regeneration of initial more active TP-26 molecules. The consumption of DTBP than mainly occurs in interaction with the phenoxyl radical from TP-26. As a result, the effectiveness of the equimolecular mixture action is close to the effectiveness of TP-26.

For TP-26 phenol with an *o*-isocamphyl substituent, as distinct from TP-5, we observe regeneration in a mixture with the sterically hindered phenol. The effectiveness of the action of a mixture of TP-26 and DTBP (Figure 7, curve *1*) is higher than additive (Figure 7, curve *2*), that is, mixtures exhibit synergetic effects.

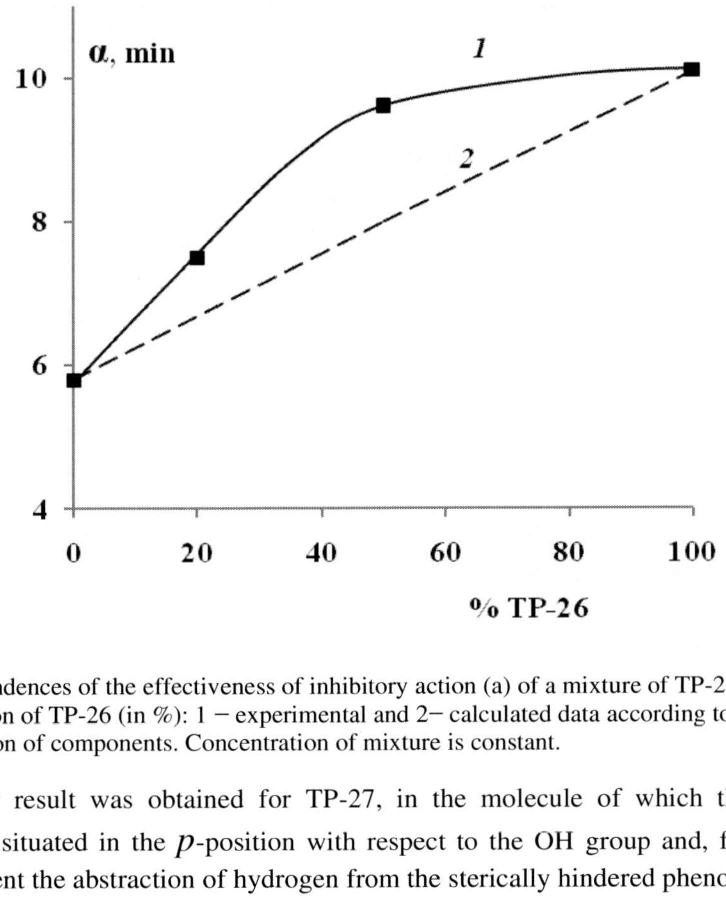

Figure 7. Dependences of the effectiveness of inhibitory action (a) of a mixture of TP-26 and DTBP on the concentration of TP-26 (in %): 1 – experimental and 2– calculated data according to the assumption of additive action of components. Concentration of mixture is constant.

A similar result was obtained for TP-27, in the molecule of which the isocamphyl substituent is situated in the *p*-position with respect to the OH group and, for this reason, does not prevent the abstraction of hydrogen from the sterically hindered phenol molecule.

Concluson

The rate constants of the reaction of the series of substituted isobornyl and isocamphyl phenols with the ethylbenzene peroxy radicals were measured. The phenols under study are arranged in a row according to the increase in their antiradical activity: TP-1 < TP-6 < TP-8 < TP-5 <TP-7 < TP-2 for isobomylphenols and TΦ-28 < TΦ-29 < TΦ-27 <TΦ-26 < TΦ-14 for isocamylphenols.

The presence of the oxygen atom in oposition with respect to the OH group of terpenephenols was found to decrease the antiradical activity of *o*-alkoxy compared *o*-alkyl substituted compounds by a factor of 6.5 for isobomylphenols and about 3.9 for isocamphylphenols owing to the intramolecular H-bond formation.

The differences in the structure of terpene substituents and their arrangement with respect to the OH group almost do not influence the antiradical activity of the investigated phenols, but change the reactivity of the phenoxyl radicals formed from them. When a phenoxyl radical interacts with a sterically hindered phenol molecule, its reactivity increases if the terpene substituent has the isocamphyl structure. Thanks to the interaction of the phenoxyl radicals formed from isocamphyl phenols with the sterically hindered phenols, mixtures of DTBP with TP-26 and TP-27 exhibit synergetic effects in the inhibition of the ethylbenzene oxidation.

Acknowledgments

The authors are very grateful to the scientific workers led by A.V. Kutchin, a corresponding member of the Russian Academy of Sciences, kindly provided terpenephenols for the investigation.

References

[1] Burlakova E. B. 2005. Bioantioxidants: yesterday, today, tomorrow/ In: *Chemical and Biological Kinetics. New Horizons.* Vol. 2.Biological Kinetics. Eds. Burlakova E.B., Varfolomeev S.D. Leden. Boston: VSP, p. 1-33.

[2] Plotnikov M.B., Smol'yakova V.I., Ivanov I.S., Kutchin A.V., Chukicheva I.Yu., Krasnov E. A. 2008. Antithrombogenic and antithrombocyte activity of the derivative of the orto-isobornylphenol. *Bull. Exp. Biol. Med.* 145 (3), 296-299 (in Russian).

[3] Plotnikov M.B., Chernysheva G.A., Smolyakova V.I., Ivanov I.S., Kutchin A.V., Chukicheva I.Yu., Krasnov E.A. 2009. Neuroprotective effects of dibornol and mechanism of action in the case of cerebral ischemia. *Herald of Russ. AMS.* No 11. 12-17 (in Russian).

[4] Chukicheva I. Yu., Fedorova I.V., Buravlev E.V., Lumpov A.E., Vikharev Yu.V., Anikina L.V., Grishko V.V., Kutchin A.V. 2010. Anti-immflamatory activity of isobornylphenol derivatives. 2010. *Chemistry of Natural Compounds.* No 3. 402-403 (in Russian).

[5] Mazaletskaya, L. I., Sheludchenko, N.I., Shishkina, L.N., Kutchin A.V., Fedorova I.V., Chukicheva I.Yu. 2011. Kinetic parameters of the reaction of isobornylphenols with peroxy radicals. *Petroleum Chemistry.* 51, (5), 348-353.

[6] Mazaletskaya, L.I., Sheludchenko, N.I., Shishkina, L.N., Kutchin A. V., Fedorova I.V., and Chukicheva I. Yu. 2012. Inhibiting activity of isocamphyl substituted phenols and

their mixtures with 2,6-di-*tert*-butylphenol in the initiated oxidation of ethylbenzene. *Russian Journal of Physical Chemistry A*. 86 (6), 929-926.

[7] Emanuel' N.M., Gladyshev G.P., Denisov E.T., Tsepalov V.F. , Kharitonov V.V., Piotrovskii K.B. 1976. Testing Procedure for Chemical Compounds as Stabilizers of Polymer Materials. Preprint. Chernogolovka, p. 11 (in Russian).

[8] Chukicheva I.Yu., Kutchin A.V. 2004. Natural and synthetic terpenophenols. *Ross. Khim. Zh.* 48 (3), 21-37 (in Russian).

[9] Chukicheva I.Yu., Timusheva I.V., Shirikhin L.V., and Kutchin A.V. 2007. Alkylahion of pyrocatechol and resorcinol by camphene. *Chem. Nat. Compounds* 43 (3), 245-249 (in Russian).

[10] Tsepalov V.F. and Shlyapintokh V.Ya. 1962. The rate constants of the elementary reactions of the ethylbenzene oxidation by molecular oxygen. *Kinet. Katal.* 3 (6), 870-876 (in Russian).

[11] Roginskii V.A. 1988, *Phenolic Antioxidants: Reactivity and Performance.* Moscow: Nauka 247 p. (in Russian).

[12] Denisov E.T. and Denisova T.G. 2000. *Handbook of Antioxidants; Bond Dissociation Energies, Rate Constants, Activation Energies and Enthalpies of Reaction,* 2nd Ed. CRS Press LLC. Boca Ration L. N.-York Washington, DC.

[13] Boguslavskaya L.V., Khrapova N.G., and Maksimov O.B. 1985. Polyhydroxyphenols - a new class of natural antioxidants. *Izv. Akad. Nauk SSSR, Ser. Khim.*, No. 7, 1471-1476 (in Russian).

[14] Tsepalov V. F., Kharitonova A.A., and Gladyshev G.P. 1977. Determination of the rate constants and the inhibition coefficients of phenols - antioxidants by means the model chain reaction. *Kinet. Katal.* 18 (5), 1261-1267 (in Russian).

[15] Belyakov V.A., Shanina E.L., Roginskii V.A., and Miller V. B. 1975. The binding energies of O-H and inhibitory ability of sterically hindered phenols *Izv. Akad. Nauk SSSR, Ser. Khim.* No. 12, 2685-2691 (in Russian).

[16] Azatyan N.A., Karpukhina G.V., Belostotskaya I.S., and Komissarova N.L. 1973. The mechanism of inhibition of the oxidation processes of diatomic phenols. *Neftekhimiya.* 13 (3), 435-440 (in Russian).

[17] Karpukhina G.V., Maizus Z.K., and Matienko L.I. 1966. Interaction of inhibitors – phenols and aromatic amines in the oxidation reactions of hydrocarbons. *Neftekhimiya.* 6 (4), 603-607 (in Russian).

[18] Karpukhina G. V., Emanuel N.M. 1984. Classification of synergistic mixtures of antioxidants and the mechanisms of synergy. *Dokl. Akad. Nauk SSSR*, 276 (5), 1163-1167 (in Russian).

[19] Makconey L.R. and Rooge M.A. 1967. Inhibitionof free-radical reactions. IV. The synergistic effect of 2,6-di-t-butylphenols on hydrocarcon oxidation retarded by 4-methoxyphenol. *J. Am. Chem. Soc.* 89 (22), 5619-5629.

[20] Azatyan N. A., Zolotova T.V., Karpukhina G.V., and Maizus Z. K. 1971. Mechanism of the inhibition of the oxidation reactions by phenol mixtures. *Neftekhimiya*, 11 (4), 568-573 (in Russian).

[21] Burlakova, E. B., Krashakov S.A., and Khrapova N.G. 1995. Kinetic features of tocopherols as antioxidants. *Khim. Fiz.*, 14 (10), 151-182 (in Russian).

INDEX

A

G

H

I

J

K

L

M

N

Q

R

T

U

V

W

X

Y

Z